Structure Identification and Functional Mechanism of Natural Active Components

Structure Identification and Functional Mechanism of Natural Active Components

Editor

Guowen Zhang

MDPI • Basel • Beijing • Wuhan • Barcelona • Belgrade • Manchester • Tokyo • Cluj • Tianjin

Editor
Guowen Zhang
Nanchang University
China

Editorial Office
MDPI
St. Alban-Anlage 66
4052 Basel, Switzerland

This is a reprint of articles from the Special Issue published online in the open access journal *Foods* (ISSN 2304-8158) (available at: https://www.mdpi.com/journal/foods/special_issues/Natural_Active_Components).

For citation purposes, cite each article independently as indicated on the article page online and as indicated below:

LastName, A.A.; LastName, B.B.; LastName, C.C. Article Title. *Journal Name* **Year**, *Volume Number*, Page Range.

ISBN 978-3-0365-5595-9 (Hbk)
ISBN 978-3-0365-5596-6 (PDF)

© 2022 by the authors. Articles in this book are Open Access and distributed under the Creative Commons Attribution (CC BY) license, which allows users to download, copy and build upon published articles, as long as the author and publisher are properly credited, which ensures maximum dissemination and a wider impact of our publications.

The book as a whole is distributed by MDPI under the terms and conditions of the Creative Commons license CC BY-NC-ND.

Contents

Zhaohua Huang, Miao Zhu and Guowen Zhang
Structure Identification and Functional Mechanism of Natural Active Components: A Special Issue
Reprinted from: *Foods* **2022**, *11*, 1285, doi:10.3390/foods11091285 1

Caili Tang, Zheng Wan, Yilu Chen, Yiyun Tang, Wei Fan, Yong Cao, Mingyue Song, Jingping Qin, Hang Xiao, Shiyin Guo and Zhonghai Tang
Structure and Properties of Organogels Prepared from Rapeseed Oil with Stigmasterol
Reprinted from: *Foods* **2022**, *11*, 939, doi:10.3390/foods11070939 5

Yilin Qian, Yuan Li, Tengteng Xu, Huijuan Zhao, Mingyong Zeng and Zunying Liu
Dissecting of the AI-2/LuxS Mediated Growth Characteristics and Bacteriostatic Ability of *Lactiplantibacillus plantarum* SS-128 by Integration of Transcriptomics and Metabolomics
Reprinted from: *Foods* **2022**, *11*, 638, doi:10.3390/foods11050638 19

Yijing Liao, Xing Hu, Junhui Pan and Guowen Zhang
Inhibitory Mechanism of Baicalein on Acetylcholinesterase: Inhibitory Interaction, Conformational Change, and Computational Simulation
Reprinted from: *Foods* **2022**, *11*, 168, doi:10.3390/foods11020168 37

Samo Lešnik and Urban Bren
Mechanistic Insights into Biological Activities of Polyphenolic Compounds from Rosemary Obtained by Inverse Molecular Docking
Reprinted from: *Foods* **2022**, *11*, 67, doi:10.3390/foods11010067 53

Qing-Hui Wen, Rui Wang, Si-Qi Zhao, Bo-Ru Chen and Xin-An Zeng
Inhibition of Biofilm Formation of Foodborne *Staphylococcus aureus* by the Citrus Flavonoid Naringenin
Reprinted from: *Foods* **2021**, *10*, 2614, doi:10.3390/foods10112614 81

Junpeng Gao, Yi Wang, Bo Lyu, Jian Chen and Guang Chen
Component Identification of Phenolic Acids in Cell Suspension Cultures of *Saussurea involucrata* and Its Mechanism of Anti-Hepatoma Revealed by TMT Quantitative Proteomics
Reprinted from: *Foods* **2021**, *10*, 2466, doi:10.3390/foods10102466 93

Yong Sun, Fanghua Guo, Xin Peng, Kejun Cheng, Lu Xiao, Hua Zhang, Hongyan Li, Li Jiang and Zeyuan Deng
Metabolism of Phenolics of *Tetrastigma hemsleyanum* Roots under In Vitro Digestion and Colonic Fermentation as Well as Their In Vivo Antioxidant Activity in Rats
Reprinted from: *Foods* **2021**, *10*, 2123, doi:10.3390/foods10092123 113

Xiang Gao, Yuhuan Jiang, Qi Xu, Feng Liu, Xuening Pang, Mingji Wang, Qun Li and Zichao Li
4-Hydroxyderricin Promotes Apoptosis and Cell Cycle Arrest through Regulating PI3K/AKT/mTOR Pathway in Hepatocellular Cells
Reprinted from: *Foods* **2021**, *10*, 2036, doi:10.3390/foods10092036 127

Huan Guo, Meng-Xi Fu, Yun-Xuan Zhao, Hang Li, Hua-Bin Li, Ding-Tao Wu and Ren-You Gan
The Chemical, Structural, and Biological Properties of Crude Polysaccharides from Sweet Tea (*Lithocarpus litseifolius* (Hance) Chun) Based on Different Extraction Technologies
Reprinted from: *Foods* **2021**, *10*, 1779, doi:10.3390/foods10081779 143

Editorial

Structure Identification and Functional Mechanism of Natural Active Components: A Special Issue

Zhaohua Huang, Miao Zhu and Guowen Zhang *

State Key Laboratory of Food Science and Technology, Nanchang University, Nanjing East Road 235, Nanchang 330047, China; zhaohuahuang@email.ncu.edu.cn (Z.H.); 13617912638@163.com (M.Z.)
* Correspondence: gwzhang@ncu.edu.cn; Tel.: +86-791-8830-5234; Fax: +86-791-8830-4347

Citation: Huang, Z.; Zhu, M.; Zhang, G. Structure Identification and Functional Mechanism of Natural Active Components: A Special Issue. *Foods* 2022, *11*, 1285. https://doi.org/10.3390/foods11091285

Received: 19 April 2022
Accepted: 26 April 2022
Published: 28 April 2022

Publisher's Note: MDPI stays neutral with regard to jurisdictional claims in published maps and institutional affiliations.

Copyright: © 2022 by the authors. Licensee MDPI, Basel, Switzerland. This article is an open access article distributed under the terms and conditions of the Creative Commons Attribution (CC BY) license (https://creativecommons.org/licenses/by/4.0/).

The natural active components derived from plants have attracted widespread attention due to their abundant species and source advantages. With the continuous deepening of research, studies have shown that many natural active components have broad-spectrum biological activities, such as antioxidant, antihypertensive, hypoglycemic, anti-inflammatory, antibacterial, anticancer, and enzyme-inhibiting activity properties, which are valuable sources of research and development in functional food factors and novel drugs. Systematical studies on the structure of components, physiological activities, the structure–activity relationship, and mechanisms of action for active components using modern scientific methods and experimental means are hot research topics. In addition, the exploration of the combined effect and mechanism of various natural bioactive substances will provide a theoretical basis for the further processing and comprehensive development of resources at multiple levels and from various points of view. This Special Issue of *Foods*, entitled "Structure Identification and Functional Mechanism of Natural Active Components", provides a forum for researchers to communicate some of their latest findings in this field. Subsequent to the peer review process, 9 original research articles were included in this Special Issue of *Foods*.

Tang et al. [1] used the natural ingredient stigmasterol as an oleogelator to explore the effect of concentration on the properties of organogels. Organogels based on rapeseed oil were investigated using various techniques (oil binding capacity, rheology, polarized light microscopy, X-ray diffraction, and Fourier transform infrared spectroscopy) to better understand their physical and microscopic properties. Results showed that stigmasterol was an efficient and thermoreversible oleogelator, which is capable of structuring rapeseed oil at a stigmasterol concentration as low as 2% with a gelation temperature of 5 °C. The oil binding capacity values of organogels increased to 99.74% as the concentration of stigmasterol was increased to 6%. The rheological properties revealed that organogels prepared with stigmasterol formed a pseudoplastic fluid with non-covalent physical crosslinking, and the G' of the organogels did not change as the frequency of scanning increased, showing its characteristic of a strong gel. The microscopic properties and Fourier transform infrared spectroscopy showed that stigmasterol formed rod-like crystals through the self-assembly of intermolecular hydrogen bonds, fixing rapeseed oil in its three-dimensional structure to form organogels. Therefore, stigmasterol can be considered as a good organogelator. It is expected to be widely used in food, medicine, and other biological-related fields.

Lactiplantibacillus plantarum could regulate certain physiological functions through the AI-2/LuxS-mediated quorum sensing (QS) system. Qian et al. [2] explored the regulation mechanism on the growth characteristics and bacteriostatic ability of *L. plantarum* SS-128. In their work, a *luxS* mutant was constructed using a two-step homologous recombination. Compared with ΔluxS/SS-128, the metabolites of SS-128 had stronger bacteriostatic ability. The combined analysis of transcriptomics and metabolomics data showed that SS-128 exhibited higher pyruvate metabolic efficiency and energy input, followed by a higher

LDH level and metabolite overflow compared to ΔluxS/SS-128, resulting in stronger bacteriostatic ability. The absence of *luxS* induces a regulatory pathway that burdens the cysteine cycle by quantitatively drawing off central metabolic intermediaries. To accommodate this mutations, ΔluxS/SS-128 exhibited lower metabolite overflow and abnormal proliferation. These results demonstrate that the growth characteristic and metabolism of *L. plantarum* SS-128 are mediated by the AI-2/LuxS QS system, which is a positive regulator involved in food safety. It would be helpful to further investigate the bio-preservation control potential of *L. plantarum*, especially when applied in food industrial biotechnology.

Alzheimer's disease (AD) is one of the most prevalent chronic neurodegenerative diseases in elderly individuals, which can cause dementia. Acetylcholinesterase (AChE) is regarded as one of the most popular drug targets for AD. Herbal secondary metabolites are frequently cited as a major source of AChE inhibitors. In the study of Liao et al. [3], baicalein, a typical bioactive flavonoid, was found to inhibit AChE competitively, with an associated IC_{50} value of 6.42 ± 0.07 µM through a monophasic kinetic process. AChE fluorescence quenching via baicalein was a static process. The binding constant between baicalein and AChE was an order of magnitude of 10^4 L mol^{-1}, and hydrogen bonding and hydrophobic interaction were the major forces in forming the baicalein−AChE complex. Circular dichroism analysis revealed that baicalein caused the AChE structure to shrink and increased its surface hydrophobicity by increasing the α-helix and β-turn contents and decreasing the β-sheet and random coil structure contents. Molecular docking revealed that baicalein predominated at the active site of AChE, likely tightening the gorge entrance and preventing the substrate from entering and binding with the enzyme, resulting in AChE inhibition. The preceding findings were confirmed by molecular dynamics simulation. The current study provides an insight into the molecular-level mechanism of baicalein interaction with AChE, which may offer new ideas for the research and development of anti-AD functional foods and drugs.

Rosemary (*Rosmarinus officinalis* L.) represents a medicinal plant known for its various health-promoting properties. Its extracts and essential oils exhibit antioxidative, anti-inflammatory, anticarcinogenic, and antimicrobial activities. The main compounds responsible for these effects are diterpenes carnosic acid, carnosol, and rosmanol, as well as the phenolic acid ester rosmarinic acid. However, surprisingly, little is known about the molecular mechanisms responsible for the pharmacological activities of rosemary and its compounds. To discern these mechanisms, Lešnik and Bren performed a large-scale inverse molecular docking study to identify their potential protein targets [4]. Listed compounds were separately docked into the predicted binding sites of all non-redundant holo proteins from the Protein Data Bank, and those with the top scores were further examined. Lešnik and Bren focused on proteins directly related to human health, including human and mammalian proteins, as well as proteins from pathogenic bacteria, viruses, and parasites. The observed interactions of rosemary compounds indeed confirm the aforementioned activities, whereas the authors also identified their potential for anticoagulant and antiparasitic actions. The obtained results were carefully checked against the existing experimental findings from the scientific literature, and as further validated using both redocking procedures and retrospective metrics.

Taking into consideration the importance of biofilms in food deterioration and the potential risks of antiseptic compounds, antimicrobial agents derived from natural products are a more acceptable choice for preventing biofilm formation and in attempts to improve antibacterial effects and efficacy. Citrus flavonoids possess a variety of biological activities, including antimicrobial properties. Therefore, in the study of Wen et al. [5], the anti-biofilm formation properties of the citrus flavonoid naringenin on the *Staphylococcus aureus* ATCC 6538 (*S. aureus*) were investigated using subminimum inhibitory concentrations (sub-MICs) of 5~60 mg/L. The results were confirmed using laser and scanning electron microscopy techniques, which revealed that the thick coating of *S. aureus* biofilms became thinner and finally separated into individual colonies when exposed to naringenin. The decreased biofilm formation of *S. aureus* cells may be due to a decrease in cell surface hydrophobicity and exopolysaccharide production, which is involved in the adherence or maturation of biofilms. Moreover, transcriptional results show that there was a downregulation in the expression of biofilm-related genes and alternative sigma factor *sigB* induced by naringenin. This work provides insight into the anti-biofilm mechanism of naringenin in *S. aureus* and suggests the possibility of naringenin use in the industrial food industry for the prevention of biofilm formation.

Saussurea involucrate (*S. involucrata*) was reported to have an anti-hepatoma function, but the mechanism is complex and unclear. To evaluate the anti-hepatoma mechanism of *S. involucrate* comprehensively and make a theoretical basis for the mechanical verification of later research, in the study of Gao et al. [6], the total phenolic acids from *S. involucrate* determined by a cell suspension culture (ESPI) was mainly composed of 4,5-dicaffeoylquinic acid, according to LC–MS analysis. BALB/c nude female mice were injected with HepG2 cells to establish an animal model of a liver tumor before being divided into a control group, a low-dose group, a middle-dose group, a high-dose group, and a DDP group. Subsequently, EPSI was used as the intervention drug for mice. Biochemical indicators and differences in protein expression determined by TMT quantitative proteomics were used to resolve the mechanism after the low- (100 mg/kg), middle- (200 mg/kg), and high-dose (400 mg/kg) interventions for 24 days. The results showed that EPSI can not only limit the growth of HepG2 cells in vitro, but can also inhibit liver tumors significantly, with no toxicity at high doses in vivo. Proteomics analysis revealed that the upregulated differentially expressed proteins (DE proteins) in the high-dose group were over three times that in the control group. ESPI affected the pathways significantly associated with the protein metabolic process, metabolic process, catalytic activity, hydrolase activity, proteolysis, endopeptidase activity, serine-type endopeptidase activity, etc. The treatment group showed significant differences in the pathways associated with the renin-angiotensin system, hematopoietic cell lineage, etc. In conclusion, ESPI has a significant anti-hepatoma effect, and the potential mechanism was revealed.

Tetrastigma hemsleyanum Diels et Gilg is a herbaceous perennial species distributed mainly in southern China. The *Tetrastigma hemsleyanum* root (THR) has been prevalently consumed as a functional tea or dietary supplement. In the study of Sun et al. [7], the digestion models in vitro including colonic fermentation were built to evaluate the release and stability of THR phenolics with the methods of HPLC–QqQ–MS/MS and UPLC–Qtof–MS/MS. From the oral cavity, the contents of total phenolic and flavonoid began to degrade. Quercetin-3-rutinoside, quercetin-3-glucoside, kaempferol-3-rutinoside, and kaempferol-3-glucoside were metabolized as major components, and they were absorbed in the form of glycosides for hepatic metabolism. On the other hand, the total antioxidant capacity (T-AOC), superoxide dismutase (SOD), glutathione peroxidase (GSH-Px) activity, and glutathione (GSH) content were significantly increased, while the malondialdehyde (MDA) content was decreased in the plasma and tissues of rats treated with THR extract in the oxidative stress model. These results indicated that the THR extract is a good antioxidant substance and has good bioavailability, which can effectively prevent some chronic diseases caused by oxidative stress. It also provides a basis for the effectiveness of THR as a traditional functional food.

4-hydroxyderricin (4-HD), as a natural flavonoid compound derived from *Angelica keiskei*, has largely unknown inhibition and mechanisms in liver cancer. Gao et al. [8] investigated the inhibitory effects of 4-HD on hepatocellular carcinoma (HCC) cells and clarified the potential mechanisms by exploring apoptosis and cell cycle arrest mediated via the PI3K/AKT/mTOR signaling pathway. The results showed that 4-HD treatment dramatically decreased the survival rate and activities of HepG2 and Huh7 cells. The protein expressions of apoptosis-related genes significantly increased, while those related to the cell cycle were decreased by 4-HD. 4-HD also downregulated PI3K, p-PI3K, p-AKT, and p-mTOR protein expression. Moreover, PI3K inhibitor (LY294002) enhanced the promoting effect of 4-HD on apoptosis and cell cycle arrest in HCC cells. Consequently, the authors demonstrated that 4-HD can suppress the proliferation of HCC cells by promoting PI3K/AKT/mTOR signaling pathway-mediated apoptosis and cell cycle arrest.

Guo et al. [9] used eight extraction technologies to extract sweet tea (*Lithocarpus litseifolius* (Hance) Chun) crude polysaccharides (STPs), and investigated and compared their chemical, structural, and biological properties. The results revealed that the compositions, structures, and biological properties of STPs varied based on different extraction technologies. Protein-bound polysaccharides and some hemicellulose could be extracted from sweet tea with diluted alkali solution. STPs extracted by deep-eutectic solvents and diluted alkali solution exhibited the most favorable biological properties. Moreover, according to the heat map, total phenolic content was the most strongly correlated with biological properties, indicating that the presence of phenolic compounds in STPs might be the main contributor to their biological properties. To the best of the authors' knowledge, this study reports the chemical, structural, and biological properties of STPs, and the results contribute to understanding the relationship between the chemical composition and biological properties of STPs.

In summary, the findings published in this Special Issue clearly indicate both the breadth and depth of the recent studies on the functional properties of natural active components. However, our understanding of natural active components is still far from adequate, and subsequent research must continue to build on previous studies. Finally, we thank the authors for their valuable contributions to this Special Issue.

Author Contributions: Conceptualization, Z.H., M.Z. and G.Z.; investigation, analysis and revision, Z.H. and M.Z.; Writing—original draft, Z.H. and G.Z.; resources, G.Z. All authors have read and agreed to the published version of the manuscript.

Funding: This research received no external funding.

Acknowledgments: The Guest Editors would like to thank the journal and its respective editors for the invitation and for providing the opportunity to develop this Special Issue, as well as the authors and reviewers who contributed.

Conflicts of Interest: The authors declare no conflict of interest.

References

1. Tang, C.; Wan, Z.; Chen, Y.; Tang, Y.; Fan, W.; Cao, Y.; Song, M.; Qin, J.; Xiao, H.; Guo, S.; et al. Structure and Properties of Organogels Prepared from Rapeseed Oil with Stigmasterol. *Foods* **2022**, *11*, 939. [CrossRef] [PubMed]
2. Qian, Y.; Li, Y.; Xu, T.; Zhao, H.; Zeng, M.; Liu, Z. Dissecting of the AI-2/LuxS Mediated Growth Characteristics and Bacteriostatic Ability of Lactiplantibacillus plantarum SS-128 by Integration of Transcriptomics and Metabolomics. *Foods* **2022**, *11*, 638. [CrossRef] [PubMed]
3. Liao, Y.; Hu, X.; Pan, J.; Zhang, G. Inhibitory Mechanism of Baicalein on Acetylcholinesterase: Inhibitory Interaction, Conformational Change, and Computational Simulation. *Foods* **2022**, *11*, 168. [CrossRef] [PubMed]
4. Lešnik, S.; Bren, U. Mechanistic Insights into Biological Activities of Polyphenolic Compounds from Rosemary Obtained by Inverse Molecular Docking. *Foods* **2022**, *11*, 67. [CrossRef] [PubMed]
5. Wen, Q.; Wang, R.; Zhao, S.; Chen, B.; Zeng, X. Inhibition of Biofilm Formation of Foodborne Staphylococcus aureus by the Citrus Flavonoid Naringenin. *Foods* **2021**, *10*, 2614. [CrossRef] [PubMed]
6. Gao, J.; Wang, Y.; Lyu, B.; Chen, J.; Chen, G. Component Identification of Phenolic Acids in Cell Suspension Cultures of Saussureainvolucrata and Its Mechanism of Anti-Hepatoma Revealed by TMT Quantitative Proteomics. *Foods* **2021**, *10*, 2466. [CrossRef] [PubMed]
7. Sun, Y.; Guo, F.; Peng, X.; Chen, K.; Xiao, L.; Zhang, H.; Li, H.; Jiang, L.; Deng, Z. Metabolism of Phenolics of Tetrastigma hemsleyanum Roots under In Vitro Digestion and Colonic Fermentation as Well as Their In Vivo Antioxidant Activity in Rats. *Foods* **2021**, *10*, 2123. [CrossRef] [PubMed]
8. Gao, X.; Jiang, Y.; Xu, Q.; Liu, F.; Pang, X.; Wang, M.; Li, Q.; Li, Z. 4-Hydroxyderricin Promotes Apoptosis and Cell Cycle Arrest through Regulating PI3K/AKT/mTOR Pathway in Hepatocellular Cells. *Foods* **2021**, *10*, 2036. [CrossRef] [PubMed]
9. Guo, H.; Fu, M.; Zhao, Y.; Li, H.; Li, H.; Wu, D.; Gan, R. The Chemical, Structural, and Biological Properties of Crude Polysaccharides from Sweet Tea (Lithocarpus litseifolius (Hance) Chun) Based on Different Extraction Technologies. *Foods* **2021**, *10*, 1779. [CrossRef] [PubMed]

Article

Structure and Properties of Organogels Prepared from Rapeseed Oil with Stigmasterol

Caili Tang [1,2], Zheng Wan [1,2], Yilu Chen [3], Yiyun Tang [1,2], Wei Fan [1,2], Yong Cao [4], Mingyue Song [4], Jingping Qin [1,2], Hang Xiao [3], Shiyin Guo [1,2] and Zhonghai Tang [1,2,*]

[1] College of Food Science and Technology, Hunan Agricultural University, Changsha 410128, China; cailitang@163.com (C.T.); jerrwzy@163.com (Z.W.); eviantyy420@163.com (Y.T.); weifan@hunau.edu.cn (W.F.); qinjingping@hunau.edu.cn (J.Q.); gsy@hunau.edu.cn (S.G.)
[2] Hunan Engineering Technology Research Center for Rapeseed Oil Nutrition Health and Deep Development, Changsha 410045, China
[3] Department of Food Science, University of Massachusetts, Amherst, MA 01003, USA; yiluchen@umass.edu (Y.C.); hangxiao@umass.edu (H.X.)
[4] College of Food Science, South China Agricultural University, Guangzhou 510642, China; caoyong2181@scau.edu.cn (Y.C.); songmy@scau.edu.cn (M.S.)
* Correspondence: tangzh@hunau.edu.cn

Abstract: This work used the natural ingredient stigmasterol as an oleogelator to explore the effect of concentration on the properties of organogels. Organogels based on rapeseed oil were investigated using various techniques (oil binding capacity, rheology, polarized light microscopy, X-ray diffraction, and Fourier transform infrared spectroscopy) to better understand their physical and microscopic properties. Results showed that stigmasterol was an efficient and thermoreversible oleogelator, capable of structuring rapeseed oil at a stigmasterol concentration as low as 2% with a gelation temperature of 5 °C. The oil binding capacity values of organogels increased to 99.74% as the concentration of stigmasterol was increased to 6%. The rheological properties revealed that organogels prepared with stigmasterol were a pseudoplastic fluid with non-covalent physical crosslinking, and the G' of the organogels did not change with the frequency of scanning increased, showing the characteristics of strong gel. The microscopic properties and Fourier transform infrared spectroscopy showed that stigmasterol formed rod-like crystals through the self-assembly of intermolecular hydrogen bonds, fixing rapeseed oil in its three-dimensional structure to form organogels. Therefore, stigmasterol can be considered as a good organogelator. It is expected to be widely used in food, medicine, and other biological-related fields.

Keywords: organogel; rapeseed oil; stigmasterol; network structure

1. Introduction

Traditional hydrogenated fats or saturated fats contain many saturated fatty acids (SFAs) and trans-fatty acids (TFAs) which have an impact on human health [1]. Their excessive intake increases the risk of cardiovascular and cerebrovascular diseases [2], obesity [3], diabetes [4], and other related diseases, making consumers aware of their serious threats to dietary health [5]. Numerous studies have focused on exploring ways to reduce the harmful content of SFAs and TFAs in foods [6–8]. Organogels have been considered as an appropriate strategy to reduce SFAs and eliminate TFAs in the diet while increasing the content of unsaturated fats [9].

Organogels are semi-solid systems; their liquid phase is fixed in a thermo-reversible three-dimensional network using various oleogelators which lead to the formation of lipid structures with obvious macroscopic properties (such as oil binding capacity, rheological properties, and thermostability) [10]. As a substitute for saturated fatty acids, and because of their properties, organogels have been widely applied in the food industry

and for shaping food products such as cakes, biscuits, meat products, chocolate, and ice cream [11]. Additionally, organogels can be utilized to stabilize and control the release of nutraceuticals and medicines [12]. They can be divided into low-molecular and polymeric organogels by the types of gelators used [13]. The former are called physical organogels formed by the low-molecules which can self-assemble to form supramolecular structures through weak non-covalent bond interactions such as hydrogen bonds, van der Waals forces, hydrophobicity, and π-π interactions [14]. The latter are referred to as chemical organogels, wherein the strong chemical bond between polymer chains form a swelling system with a cross-linked structure in organogels [15]. The physical organogels are more commonly used than the chemical organogels because they can provide a network structure to vegetable oils and are edible. However, as physical organogels have only recently been investigated, detailed information on gelation phenomena and intermolecular interactions is not yet available [16]. Moreover, the types of physical oleogelators are limited, mainly including natural waxes [17], fatty acids, fatty alcohols [18], and compounds of sterols and glutamine [19]. Therefore, a new oleogelator is required for further development.

Stigmasterol (ST), a natural 6-6-6-5 tetracyclic phytosterol [20], is a biosynthesized triterpene sterol. It is commonly found in various plants and deodorized distillates due to the refining of vegetable oils [21]. Recent studies have shown that stigmasterol exhibits a variety of biological activities as an antioxidant [22], anti-inflammatory [23], anti-tumor [24], and anti-diabetic [25]. Stigmasterol (Figure 1) has an amphiphilic structure with a large oleophilic surface and polar OH head group. It can be used as a gelator to immobilize liquid oil in the network structure by the self-assembly method [26].

Rapeseed oil is the second most abundantly produced edible oil in the world and is rich in unsaturated fatty acids such as oleic acid, linoleic acid, and linolenic acid [27]. The type and proportion of fatty acids are more in line with the dietary nutrition standards which can effectively reduce cholesterol and cardiovascular disease [28]. Therefore, we chose rapeseed oil as the base oil to prepare the edible organogels with stigmasterol as the gelator. The effects of different stigmasterol concentrations on the oil binding capacity (OBC), rheological properties, and microstructure of the organogels were researched. The mechanism of gel formation of organogels was studied by polarized light microscopy, X-ray diffraction (XRD), and Fourier transform infrared spectroscopy (FTIR). The results of this study can provide theoretical and technical support for the development and application of phytosterol organogels.

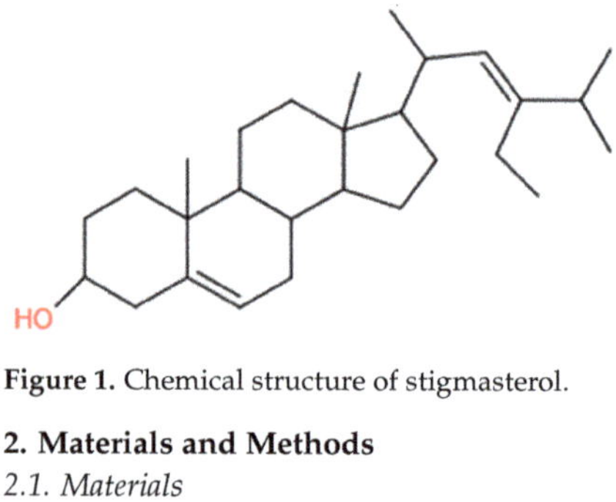

Figure 1. Chemical structure of stigmasterol.

2. Materials and Methods

2.1. Materials

Stigmasterol (90%) was obtained from Source Leaf Biotechnology (Shanghai, China); commercial grade rapeseed oil (approximately 6% saturated, 58% monounsaturated, and 36% polyunsaturated) was acquired from a local supermarket. The rest of the chemicals and reagents utilized in this experiment were of analytical grade.

2.2. Methods

2.2.1. Organogel Preparation

The organogels were prepared by mixing a certain concentration of stigmasterol (1%, 2%, 3%, 4%, 5%, 6%, and 7% (w/w)) to rapeseed oil. The mixture was heated and stirred at 100 °C for 40 min in oill bath at 200 rpm. After that, the hot mixtures were cooled at 5 °C for 24 h to form a gel. Physical properties of the samples were measured after this storage as described below.

2.2.2. Gelation Temperature Phase Diagram

The organogel samples prepared by mixing stigmasterol at different concentrations (1%, 2%, 3%, 4%, 5%, 6%, and 7% (w/w)) with rapeseed oil were poured into a serum bottle. After that, they were stored at 5, 10, 15, 20, 25, and 30 °C for 24 h. The self-sustaining ability of the samples was assessed visually by inverting the serum bottle. Samples were categorized as a gel, thickened liquid, or liquid, based on the appearance of behavior [29].

2.2.3. Oil Binding Capacity

By measuring the oil loss of organogels after centrifugation, the oil binding capacities (OBC) of the organogels were obtained. First, the weight of the Eppendorf tube (a) was measured and the Eppendorf tube containing 2 g of the melted organogel samples was weighed (b); after that, the tubes were stored at 5 °C for 24 h. Second, the tubes were centrifuged at 10,000 rpm for 15 min and then inverted to drain the separated rapeseed oil. The remaining organogel samples in the tube were then weighed (c). The oil binding capacity was calculated using the following formula:

$$\text{OBC}(\%) = \frac{(c-a)}{(b-a)} \times 100\% \qquad (1)$$

where a denotes the weight of the empty container, b represents the weight of the container containing the primary sample, and c denotes the weight of the container containing the sample after centrifugation. All the measurements were conducted in triplicate; the results were reported as mean $\pm$ standard deviation (SD).

2.2.4. Rheological Characterization

The rheological properties of the organogels were analyzed by a Kinexus pro advanced rheometer (Malvern Instruments Ltd., Malvern, UK) with a stainless steel cone-plate geometry (40 mm, 1° angle, 1 mm truncation). All the rheological tests were conducted within the linear viscoelastic range. Specifically, the frequency sweep experiments were carried out at 25 °C under a constant strain within the linear viscoelastic domain, ranging from 0.1 Hz to 100 Hz. The temperature sweeps tests were carried out at a constant frequency of 1 Hz and a heating rate of 2 °C/min in the 25–100 °C range. The apparent viscosity was measured with a constant shear strain with varying shear rates (from $0.01\ \text{s}^{-1}$ to $100\ \text{s}^{-1}$) at 25 °C.

2.2.5. Polarized Light Microscopy

Polarized light microscopy (#CX31., Olympus, Japan) was used to observe organogel crystal morphology. The organogel samples were lightly smeared on a microscope slide and a coverslip was carefully overlaid on the sample. After that, pictures were obtained with $100\times$ magnification using OLYCIA Series Imaging Analysis Software.

2.2.6. X-ray Diffraction Analysis

The XRD pattern was employed to analyze the crystallization patterns forms of the organogels by XRD spectroscopy (SHIMADZU., Kyoto, Japan) with reflection geometry and the Cu Kα radiation (λ = 1.542 Å) operating at 40 kV and 30 mA. The organogel samples were scanned at a scan rate of 2°/min with a 0.02° step size utilizing a 2θ range of 5° to 50°.

Each sample was tested in triplicate. The diffractograms were analyzed using MDI Jade 6.0 software (Materials Data Ltd., Livermore, CA, USA).

2.2.7. Fourier Transform Infrared Spectroscopy

The FTIR spectra of the samples were measured using an IRAffinity-1 model FTIR instrument (SHIMADZU, Kyoto, Japan) coupled with an attenuated total reflection (ATR) sampling accessory. The organogel samples, pure stigmasterol, and rapeseed oil were scanned within the 4000–400 cm^{-1} range to explore the interactions of the gel components [30].

2.2.8. Statistical Analysis

All experiments were performed in duplicate or triplicate. The data was analyzed using SPSS 20 (SPSS Inc., Chicago, IL, USA) software, calculating mean value and standard deviation (SD), the results were expressed as mean ±SD. The datasets were subjected to analysis of variance, and Duncan's multiple range test was used to assess the significant differences between the mean values (a difference of $p < 0.05$ was regarded as substantially different). Furthermore, the figures were drawn using Origin 2018 (OriginLab Corporation, Northampton, MA, USA) for basic data processing and mapping.

3. Results and Discussion

3.1. Gelation Phase Diagram

Visual observation of appearance was performed to ascertain the gelation of the organogels by simply inverting the serum bottle containing the samples. The systems that did not flow under the influence of gravity were named organogels [31]. Figure 2 shows the appearance and gelling behavior of organogels with different stigmasterol concentrations.

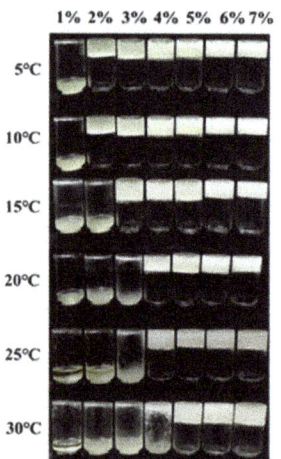
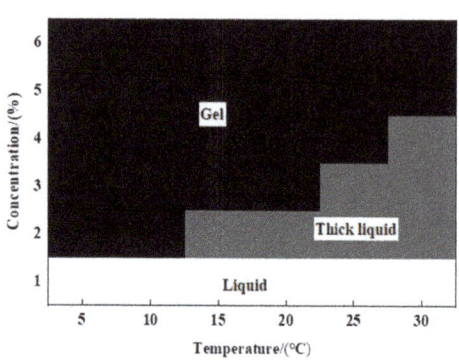

Figure 2. Temperature of Gelation Phase Diagram for different concentrations of stigmasterol organogels at different temperatures: gel (i.e., freestanding gel), thick liquid (liquid was clearly thickened, but freestanding gel was not observed), and liquid (i.e., no gelation observed).

It can be seen from Figure 2 that the formation of organogels was simultaneously affected by gelling temperature and stigmasterol concentration. The samples thickened at all temperatures under the stigmasterol concentration of 1%. When the stigmasterol concentration was ≥2%, the mixtures of stigmasterol and rapeseed oil could form organogels at low temperatures (5 °C). When the stigmasterol concentration was ≥4%, the organogels occurred at room temperature (25 °C). These results showed that the organogels only need a small amount of stigmasterol (2%) to fix rapeseed oil with a 5 °C gelling temperature.

With the increase in gelling temperature, the critical gelling concentration of stigmasterol required for the formation of organogels gradually increases. It was possible that the crystallization behavior and crystal structure of stigmasterol in rapeseed oil were extremely sensitive to gelling temperature. The internal structure of organogels by the intermolecular brownian motion decreased with the increasing gelation temperature [26]. According to this result, we could gather the critical concentration of stigmasterol to form organogels at different gelling temperatures. The organogels were prepared within the concentration range of 2–7% at a gelation temperature of 5 °C to further understand the physicochemical and microstructure properties of the organogels.

3.2. Physicochemical Properties of Stigmasterol Organogels

The physicochemical properties of stigmasterol organogels were studied by measuring the OBC and rheological properties.

3.2.1. Oil Binding Capacity

The oil binding capacity (OBC) is used to characterize the strength and ability of the organogels to decrease vegetable oil migration [32]. The OBC values of the organogels with different stigmasterol concentrations are shown in Figure 3. The OBC values were increased significantly from 50.74% to 99.74% when the concentration increased from 2% to 6%. It may be that with the increase in stigmasterol concentration, the internal system of the organogels could form more crystal structures through molecular interactions [33]. This further formed a three-dimensional network structure to fix up the rapeseed oil, resulting in significantly increased OBC values. It was worth noting that the OBC value (99.93%) of organogels did not change significantly with a 7% stigmasterol concentration.

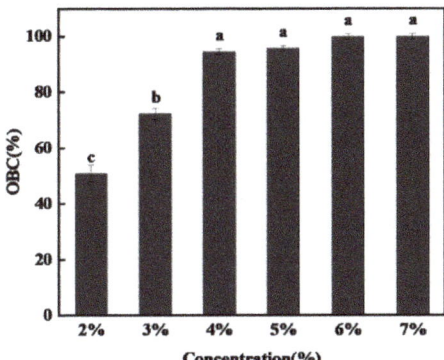

Figure 3. Effect of stigmasterol concentration on the oil binding capacity of the stigmasterol organogels. (Means with different letters in the same classification significantly differ at $p < 0.05$).

The self-assembled structure of stigmasterol may reach its saturation point [34] at 7% concentration with rapeseed oil at 100 °C. These results were similar to those obtained by Zefang Jiang, et al. It has been reported that the formation of organogels highly depends on the ability of the solubility to the gelator, it must be relatively dissolved in solution so that it can crystallize or self-assemble to form a microstructure in a solvent [7].

3.2.2. Rheological Properties

Rheological properties are also important physical and chemical characteristics of organogels. It is essential to understand these rheological properties for the application of organogels. In this experiment, the mechanical stability of the organogel was studied by the oscillatory rheological experiment and the variation law of the apparent viscosity of the organogels with the shear rate was studied by the static rheological experiment.

The viscoelasticity of the sample was reflected by frequency scanning. In rheological analysis, the G' is the elastic modulus of the sample and the G" is the viscous modulus of the sample. Frequency scanning is utilized to reflect the correlation between the viscoelastic modulus and frequency. If the G' > G" with the increase in frequency, the sample mainly exhibits elastic deformation, indicating that the sample presents solid state. If the G' = G", the sample presents a semi-solid state. If the G' < G", the viscosity modulus of the sample mainly has viscous deformation, indicating that the sample presents a liquid state. All organogel samples showed a solid-state behavior with the elastic modulus (G') higher than the viscosity modulus (G") within the frequency range of 0.1–100 Hz (see Figure 4). Additionally, the G' and G" values of organogels were independent of the increase in scanning frequency. These results showed that the organogels prepared from different concentrations of stigmasterol had a good tolerance in the test range of deformation frequency and were formed by a non-covalent physical cross-linked gel network structure [32]. Furthermore, the G' value was closely related to the stigmasterol concentration, the G' value increased notably when the concentration was increased from 2% to 6%, but the result was reversed at a concentration of 7%.

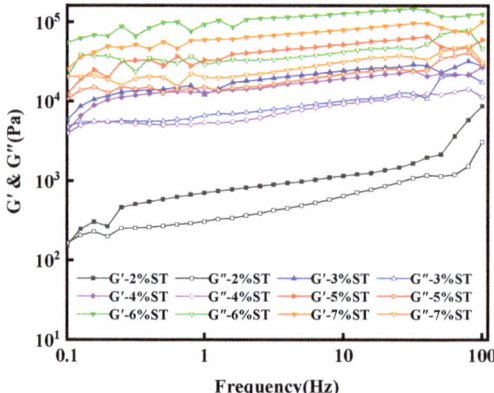

Figure 4. Viscoelastic properties of stigmasterol organogels from the different stigmasterol concentrations (2–7%) with the scanned frequency range from 0.1 Hz to 100 Hz at 25 °C.

Those results in rheology follow the same tendency observed in the oil binding capacity, that is, as the concentration of stigmasterol increased, G' values and the oil binding capacity increased. However, the G' value decreased at the stigmasterol concentration of 7%, which could mean that when the stigmasterol concentration reached 7%, the organogel system was under a supersaturated state. This supersaturation may affect the change of crystal structural units in the organogel [35], leading to the decrease in its structural integrity, decreasing the G' value. It was reported that the supersaturated state could increase the nucleation rate of crystals in the organogels [36], resulting in the formation of more individual networks in the organogels system; however, those crystal structures from different single networks were usually less entangled than the permanent junction of crystal structure in an organogels network. Therefore, the integrity of the structure and the overall mechanical properties decreased with the increasing nucleation rate and the number of structural elements. The results of frequency scanning showed the formation of a gel network and the physical interaction between organgeltor and vegetable oil.

A temperature ramp test of organogels is illustrated in Figure 5 to study the temperature-dependent flow behavior of the stigmasterol organogels. As the scanning temperature increased, the G' and G" values of organogels were significantly reduced and the critical phase transition temperature (G' = G") gradually emerged at the stigmasterol concentration of 2–6%. This result indicates that the organogels showed a viscous behavior at high temperatures in a completely molten state. The critical phase transition temperature increased prominently from 45 °C to 95 °C with the increasing concentration. However, the absence of the critical phase transition temperature was found in the organogels at a 7% stigmasterol concentration, which showed the organogels did not undergo a gel-sol transformation [37]. Therefore, the organogels have high thermal stability with a 7% stigmasterol concentration. This may be because the number of crystals was increased with the increase in the stigmasterol concentration; a higher temperature was needed to destroy the organogel structure [38].

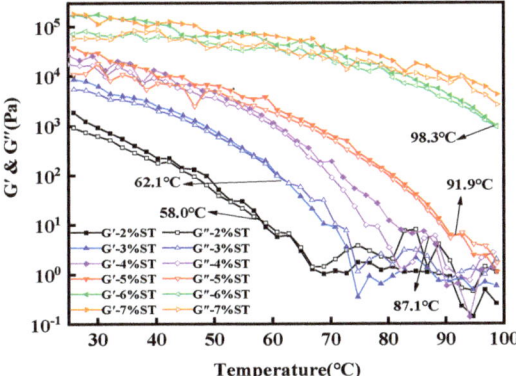

Figure 5. Different stigmasterol concentrations (2–7%) of stigmasterol organogels had viscoelastic properties in the temperature range of 25 to 100 °C.

Figure 6 shows that the initial apparent viscosity of organogels increases with the increasing stigmasterol concentration, forming stronger organogel structures. However, the complex viscosity of the organogel samples decreased exponentially as the shear rate was enhanced, reflecting its shear-thinning behavior [39]. This was likely due to the dynamic forces generated in the shearing process causing the fracture of the crystalline structure of stigmasterol organogels [40]. Similar results have been reported in many organogel structures with pseudoplastic properties [40–42]. The relationship between the apparent viscosity and shear rate of organogels prepared with different concentrations of stigmasterol is consistent with the power-law equation.

$$\eta = K\gamma^{(n-1)}, 0 < \gamma \leq 1 \quad (2)$$

where η denotes the apparent viscosity, γ represents the shear rate, K denotes the flow consistency index, and n is the degree of pseudo-plasticity index [43]. The fitting results (see Table 1) were shown that the organogels with a higher concentration of stigmasterol have higher consistency, the K value reached the maximum at 253.6 Pa·s at a stigmasterol concentration of 7% (see Table 1). More crystals were formed and cross-linked with increasing concentration; therefore, a stronger crystalline structure was provided to the entrapped oil molecules, resulting in higher resistance to shearing. The flow behavior index (n) < 1, between 0.03 to 0.47, indicated that organogel samples were a pseudoplastic fluid in this shear range. The crystal particles' gradual and orderly arrangement along the direction of shear depolymerization with the shear rate increased [44], which explained why the organogel system became more pseudoplastic and stronger with the increase in stigmasterol concentration. Additionally, the square value of the correlation coefficient (R^2) of the fitting function was between 0.991 and 0.999, indicating that the relationship between the apparent viscosity and shear rate test data conforms to the power-law equation.

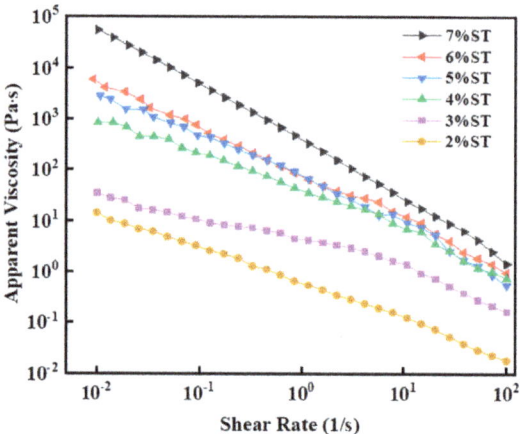

Figure 6. Changes in the apparent viscosity of stigmasterol organogels with different stigmasterol concentrations (2–7%) and shear rates ranging from 0.01 s^{-1} to 100 s^{-1}.

Table 1. Values of power-law parameters (K, n) of the stigmasterol organogels at different stigmasterol concentrations in the range of 0.01–100 s^{-1}.

ST Concentration	K/Pa·s	n	R^2
2%	0.42 ± 0.027	0.47 ± 0.087	0.997
3%	12.71 ± 0.088	0.42 ± 0.068	0.991
4%	39.49 ± 0.069	0.25 ± 0.025	0.991
5%	63.66 ± 0.032	0.14 ± 0.017	0.998
6%	81.59 ± 0.019	0.08 ± 0.022	0.997
7%	253.60 ± 0.048	0.03 ± 0.046	0.999

Note: Values are means ± standard of deviations.

3.3. Microstructure Properties of Stigmasterol Organogels

Morphology of the stigmasterol organogels was studied using a polarized light microscope (PLM), XRD, and FTIR.

3.3.1. Polarized Light Microscopy

The three-dimensional network structure is the basis of the mechanical properties of organogels [45]. The influence of gelator concentration on the crystal morphology and microstructure in the organogel system was observed using a polarizing microscope image. The results for the organogels are shown in Figure 7. The stigmasterol crystals were uniformly dispersed in the oil phase, appearing as birefringent patches against a black background [46]. The organogel prepared with 2–4% stigmasterol showed a randomly distributed rod-like crystal structure, while the organogels prepared from 5%–6% stigmasterol showed a rod-like crystal structure with close distribution, the crystal units of stigmasterol formed the three-dimensional structure of the organogels. The number of crystals increased significantly and the internal crystal size of the organogels gradually decreased from random crystal to tightly distributed rod structure with the increase of stigmasterol concentration. However, when the stigmasterol concentration was 7%, the crystal structure of organogels partially overlapped [47]. This is probably because a high oleogelator concentration leads to a higher degree of supersaturation which can accelerate nucleation and restricted the further growth of stigmasterol crystals.

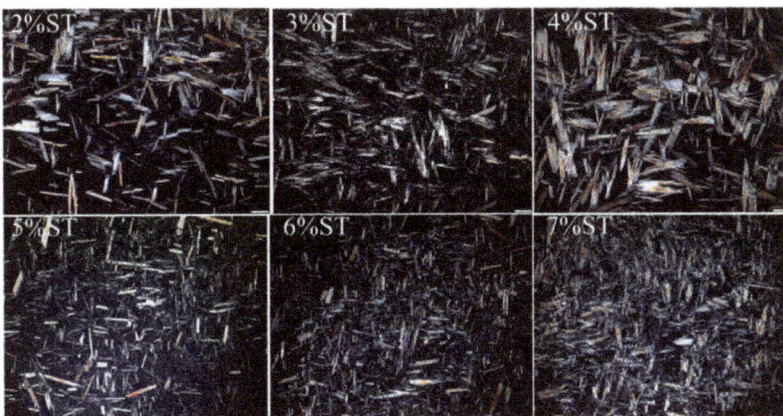

Figure 7. Microstructures of stigmasterol organogels observed at 100× after 24 h storage at 5 °C.

The polarized light results of stigmasterol organogels further confirm that the crystal structure of the organogels can self-assemble to form a compact rod-like fiber structure as the concentration of stigmasterol increases. This structure shows a stronger combination ability of the oil phase and can affect the mechanical resistance of the OBC and create a higher complex module (G′) in the rheological behavior, proving the correlation between the microstructure of the organogel and its mechanical resistance [48]. The crystal network formed by the independent assembly was rearranged when the concentration of stigmasterol reached the supersaturated state, leading to the instability of the organogel structure and the decline of the macroscopic properties. The concentration of the oleogelator plays a key role in controlling the non-covalent interaction-driven self-assembly of fibrillar networks in most cases.

3.3.2. X-ray Diffraction

The microstructure diagram of the gel system can only analyze the changes in crystal units in the gel system. However, the changes in cell parameters and crystal types can be obtained more accurately by XRD analysis. The d-spacing distance in the XRD analysis parameter represents the distance between two diffraction crystal planes of the sample and is used to reflect the crystal type of the sample and the homogeneous polycrystalline phenomenon of fat [49].

The diffraction patterns of rapeseed oil, neat stigmasterol, and organogels prepared with different concentrations of stigmasterol are shown in Figure 8a,b. Two major peaks at 4.52 Å and 4.26 Å were observed in pure stigmasterol and stigmasterol organogels, respectively. In the wide-angle region, the peak around 4.5 Å is usually considered as the characteristic peak of β-polymorph, and peak around 4.2 Å is the characteristic peak of α-polymorph [50]. In other words, the major peaks corresponding to pure stigmasterol and organogels reveal two distinct modes of parallel stacked arrangements, namely α and β. The organogel samples contained the positions and d-spacing of the main peaks corresponding to stigmasterol which indicated that the diffraction pattern of stigmasterol did not change during the formation of organogels. The diffraction patterns of both stigmasterol and organogel samples showed the existence of long and short spacing peaks. According to reports, the presence of long-spacing peaks provides information about the order of the molecular layers, while the presence of short-spacing peaks provides information about the lateral stacking of molecular layers [51]. Compared with the spectra of stigmasterol, the intensity of long-distance peaks in the organogel changed and the positions of some peaks shifted. This indicated that the addition of rapeseed oil in the stigmasterol caused a rearrangement of stigmasterol molecular packaging [52]. In addition, the long-distance peak of organogels enhanced with increasing concentrations of stigmasterol, indicating that the number of crystal structures of organogels increased. This corresponds to the results presented by polarizing microscopes. The variation of the spacing peak of organogels increases with the increasing stigmasterol concentration which further explains the effect of stigmasterol concentration on the structure of organogels.

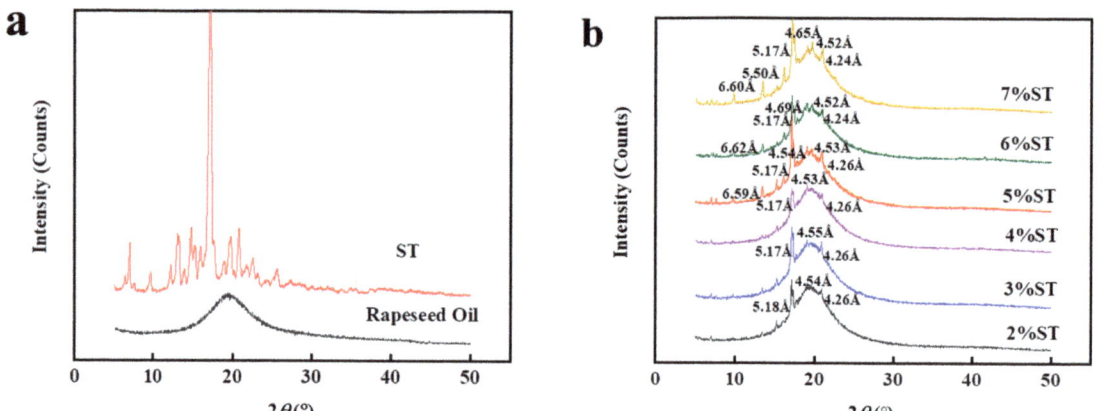

Figure 8. The XRD patterns of rapeseed oil and stigmasterol (**a**); organogels with diverse stigmasterol concentrations (**b**).

3.3.3. Fourier Transform Infrared Spectroscopy

FTIR spectroscopy is necessary to understand the interaction between the packing arrangements of organogelator molecules. The FTIR spectra of rapeseed oil, original stigmasterol, and the organogel samples prepared by different concentrations of stigmasterol are shown in Figure 9a,b. The spectra show the absorption bands of organogel owing to the functional groups in rapeseed oil and stigmasterol. The infrared absorption band of rapeseed oil ranged between 400–1800 cm^{-1} and 2800–3100 cm^{-1}. The peaks of C-H emerged at approximately 3000 cm^{-1}. Furthermore, the bands below 3000 cm^{-1} (2920 cm^{-1}) are attributed to the symmetric and anti-symmetric stretching of C-H in -CH$_3$ and -CH$_2$ functional groups [53], respectively. The characteristic absorption peak around 3346 cm^{-1} is the spectra of original stigmasterol, linked to the stretching of -OH groups [54]. The organogel samples only showed the characteristic peak around 3338 cm^{-1}, suggesting that the intermolecular hydrogen bonding observed in the oleogels comes from stigmasterol [55]. Furthermore, new covalent bonds did not form, which is consistent with the results of rheological frequency scanning. However, the characteristic peak of stigmasterol in the organogel samples shifted to a lower wavenumber with the stigmasterol increasing concentration. These results showed that the three-dimensional network structure of organogels was formed by stigmasterol aggregates through intermolecular hydrogen bonding and that the supramolecular aggregates are spontaneously formed through aggregation-nucleation-growth pathways of the stigmasterol crystals [56]. Stigmasterol was a kind of low molecular oleogelator that required the formation of a self-assembled network structure before supramolecular aggregation in an organogel structure could occur. Therefore, the self-assembly pathway of the stigmasterol determined the internal structure of the organogel which further affected its macroscopic properties. This is in line with previous research by Meng, Z et al. which showed that hydrogen bonds may have been responsible for the formation of the crystal structure that fixed the sunflower oil and provided the favorable physical characteristics of the PGE organogel [46]. Similar results were also observed in the SMS-PO organogels [57] reported by Suzuki, M. We thus conclude from the FTIR measurements that the hydrogen bonds play a significant role in the formation of the stigmasterol organogels.

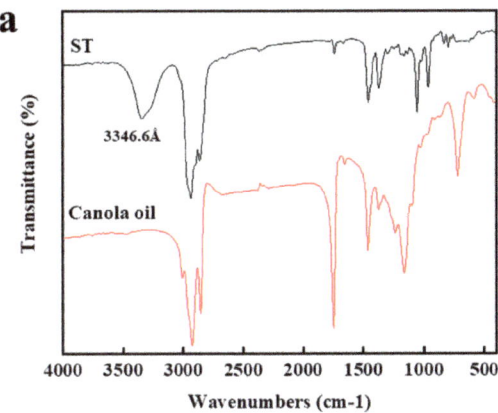

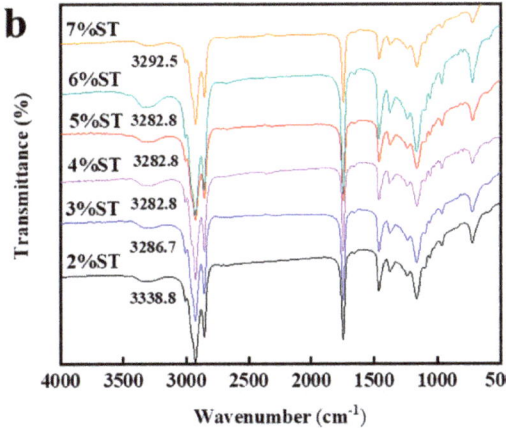

Figure 9. The FT-IR of rapeseed oil and stigmasterol (**a**); organogels with diverse stigmasterol concentrations (**b**).

4. Conclusions

We prepared the rapeseed oil-based organogels using stigmasterol as a self-assembly oleogelator. The formation of the organogel was related to the gelling temperature and the concentration of the oleogelator which had a critical gelation concentration of 2% at a gelling temperature of 5 °C. The results of macroscopic characteristics showed that the oil holding capacity increased to more than 99.74% when the stigmasterol concentration was 6%. The rheological properties revealed that the organogels prepared with stigmasterol were a pseudoplastic fluid with shear thinning. The results of microscopic characteristic tests showed that the stigmasterol concentration had a significant effect on the integrity and fineness of the crystal structure of the organogels, playing a key role in the densification of crystal network units and crystal size of the organogels. Stigmasterol impregnates rapeseed oil with intermolecular hydrogen bonding to form the crystal network structure, further proving that the concentration of stigmasterol had an important effect on the physical properties and microstructure of the rapeseed oil-based organogels.

This study demonstrates that stigmasterol is a preferable oleogelator to provide an effective approach for the preparation of highly unsaturated organogels. Stigmasterol traps liquid oil in its thermoreversible gel network through gelation, resulting in an organogel with plastic solid properties. Because these liquid oils have specific consistency and hardness without changing their chemical composition, they have great application potential for replacing saturated fats with plastic fats. More extensive applications of rapeseed oil-based organogels should be investigated in order to generate healthier and more spreadable food products in the future.

Author Contributions: Writing-original draft, C.T.; Investigation, Z.W., Y.T.; Data curation, Y.C. (Yilu Chen); Project administration, W.F.; Supervision, Y.C. (Yong Cao); Resources, M.S., J.Q.; Methodology, H.X.; Supervision, S.G.; Writing-review & editing, Funding acquistion, Z.T. All authors have read and agreed to the published version of the manuscript.

Funding: This research was funded by National Natural Science Foundation of China [No.31671858] and Changsha City's Major Research Plan [kq1801016]. And The APC was funded by Hunan Agricultural University.

Institutional Review Board Statement: Not applicable.

Informed Consent Statement: Not applicable.

Data Availability Statement: Data is contained within the article.

Conflicts of Interest: The authors declare no conflict of interest.

References

1. Qiu, B.; Wang, Q.; Liu, W.; Xu, T.C.; Liu, L.N.; Zong, A.Z.; Jia, M.; Li, J.; Du, F.L. Biological effects of trans fatty acids and their possible roles in the lipid rafts in apoptosis regulation. *Cell Biol. Int.* **2018**, *42*, 904–912. [CrossRef] [PubMed]
2. Te, M.L.; Montez, J.M. Health effects of saturated and trans-fatty acid intake in children and adolescents: Systematic review and meta-analysis. *PLoS ONE* **2017**, *12*, e0186672. [CrossRef]
3. Crupkin, M.; Zambelli, A. Detrimental impact of trans fats on human health: Stearic acid-rich fats as possible substitutes. *Compr. Rev. Food Sci. Food Saf.* **2018**, *7*, 271–279. [CrossRef]
4. Odegaard, A.O.; Pereira, M.A. Trans fatty acids, insulin resistance, and type 2 diabetes. *Nutr. Rev.* **2006**, *64*, 364–372. [CrossRef]
5. Esmaillzadeh, A.; Azadbakht, L. Consumption of hydrogenated versus nonhydrogenated vegetable oils and risk of insulin resistance and the metabolic syndrome among Iranian adult women. *Diabetes Care* **2008**, *31*, 223–226. [CrossRef] [PubMed]
6. Chaves, K.F.; Barrera-Arellano, D.; Ribeiro, A.P.B. Potential application of lipid organogels for food industry. *Food Res. Int.* **2018**, *105*, 863–872. [CrossRef] [PubMed]
7. Jiang, Z.; Gao, W.; Du, X.; Zhang, F.; Bai, X. Development of Low-calorie Organogel from sn-2 Position-modified Coconut Oil Rich in Polyunsaturated Fatty Acids. *J. Oleo Sci.* **2019**, *68*, 399–408.
8. Mert, B.; Demirkesen, I. Reducing saturated fat with oleogel/shortening blends in a baked product. *Food Chem.* **2016**, *199*, 809–816. [CrossRef]
9. Moghtadaei, M.; Soltanizadeh, N.; Goli, S.A.H. Production of sesame oil oleogels based on beeswax and application as partial substitutes of animal fat in beef burger. *Food Res. Int.* **2018**, *108*, 368–377. [CrossRef]
10. Patel, A.R.; Dewettinck, K. Edible oil structuring: An overview and recent updates. *Food Funct.* **2016**, *7*, 20–29. [CrossRef]
11. Pehlivanoğlu, H.; Demirci, M.; Toker, O.S.; Konar, N.; Karasu, S.; Sagdic, O. Oleogels, a promising structured oil for decreasing saturated fatty acid concentrations: Production and food-based applications. *Crit. Rev. Food Sci. Nutr.* **2018**, *58*, 1330–1341. [CrossRef]
12. Stortz, T.A.; Zetzl, A.K.; Barbut, S.; Cattaruzza, A.; Marangoni, A.G. Edible oleogels in food products to help maximize health benefits and improve nutritional profiles. *Lipid Technol.* **2012**, *24*, 151–154. [CrossRef]
13. Singh, V.K.; Pal, K.; Pradhan, D.K.; Pramanik, K. Castor oil and sorbitan monopalmitate based organogel as a probable matrix for controlled drug delivery. *J. Appl. Polym. Sci.* **2013**, *130*, 1503–1515. [CrossRef]
14. Vintiloiu, A.; Leroux, J.C. Organogels and their use in drug delivery—A review. *J. Control. Release* **2007**, *125*, 179–192. [CrossRef] [PubMed]
15. Bhattacharya, C.; Kumar, N.; Sagiri, S.S.; Pal, K.; Ray, S.S. Development of span 80-tween 80 based fluid-filled organogels as a matrix for drug delivery. *J. Pharm. Bioallied Sci.* **2012**, *4*, 155. [CrossRef] [PubMed]
16. Sagiri, S.S.; Singh, V.K.; Pal, K.; Banerjee, I.; Basak, P. Stearic acid based oleogels: A study on the molecular, thermal and mechanical properties. *Mater. Sci. Eng. C* **2015**, *48*, 688–699. [CrossRef]
17. Dassanayake, L.S.K.; Kodali, D.R.; Ueno, S.; Sato, K. Physical properties of rice bran wax in bulk and organogels. *J. Am. Oil Chem. Soc.* **2009**, *86*, 1163. [CrossRef]

18. Jaiyanth, D.; Ram, R. Organogelation of plant oils and hydrocarbons by long-chain saturated FA, fatty alcohols, wax esters, and dicarboxylic acid. *J. Am. Oil Chem. Soc.* **2003**, *80*, 417–421. [CrossRef]
19. Sawalha, H.; Venema, P.; Bot, A.; Flöter, E.; Van der Linden, E. The influence of concentration and temperature on the formation of γ-oryzanol+β-sitosterol tubules in edible oil organogels. *Food Biophys.* **2016**, *16*, 20–25.
20. Kaur, N.; Chaudhary, J.; Jain, A.; Kishore, L. Stigmasterol: A comprehensive review. *Int. J. Pharm. Sci. Res.* **2011**, *2*, 2259.
21. Panda, S.; Jafri, M.; Kar, A.; Meheta, B.K. Thyroid inhibitory, antiperoxidative and hypoglycemic effects of stigmasterol isolated from Butea monosperma. *Fitoterapia* **2008**, *80*, 123–126. [CrossRef] [PubMed]
22. Navarro, A.; Heras, B.; Villar, A. Anti-inflammatory and immunomodulating properties of a sterol fraction from Sideritis foetens Clem. *Biol. Pharm. Bull.* **2001**, *24*, 470–473. [CrossRef] [PubMed]
23. Gao, Z.; Maloney, D.J.; Dedkova, L.M.; Hecht, S.M. Inhibitors of DNA polymerase β: Activity and mechanism. *Bioorg. Med. Chem.* **2008**, *6*, 4331–4340. [CrossRef]
24. Wang, J.; Huang, M.; Yang, J.; Ma, X.; Zheng, S.; Deng, S.; Huang, Y.; Yang, X.K.; Zhao, P. Anti-diabetic activity of stigmasterol from soybean oil by targeting the GLUT4 glucose transporter. *Food Nutr. Res.* **2017**, *61*, 1364117. [CrossRef]
25. Batta, A.K.; Xu, G.; Honda, A.; Miyazaki, T.; Salen, G. Stigmasterol reduces plasma cholesterol levels and inhibits hepatic synthesis and intestinal absorption in the rat. *Metabolism* **2006**, *55*, 292–299. [CrossRef]
26. Bag, B.G.; Barai, A.C. Self-assembly of naturally occurring stigmasterol in liquids yielding a fibrillar network and gel. *RSC Adv.* **2020**, *10*, 4755–4762. [CrossRef]
27. Chew, S.C. Cold-pressed rapeseed (Brassica napus) oil: Chemistry and functionality. *Food Res. Int.* **2020**, *131*, 108997. [CrossRef]
28. Teh, S.S.; Birch, J. Physicochemical and quality characteristics of cold-pressed hemp, flax and canola seed oils. *J. Food Compos. Anal.* **2013**, *30*, 26–31. [CrossRef]
29. Fayaz, G.; Goli, S.A.H.; Kadivar, M.A. Novel propolis wax-based organogel: Effect of oil type on its formation, crystal structure and thermal properties. *J. Am. Oil Chem. Soc.* **2017**, *94*, 47–55. [CrossRef]
30. Patel, A.R.; Schatteman, D.; De Vos, W.H.; Lesaffer, A.; Dewettinck, K. Preparation and rheological characterization of shellac oleogels and oleogel-based emulsions. *J. Colloid Interface Sci.* **2013**, *411*, 114–121. [CrossRef]
31. Rocha, J.C.B.; Lopes, J.D.; Mascarenhas, M.C.N.; Arellano, D.B.; Guerreiro, L.M.R.; Da Cunha, R.L. Thermal and rheological properties of organogels formed by sugarcane or candelilla wax in soybean oil. *Food Res. Int.* **2013**, *50*, 318–323. [CrossRef]
32. Zhao, R.; Wu, S.; Liu, S.; Li, B.; Li, Y. Structure and Rheological Properties of Glycerol Monolaurate-Induced Organogels: Influence of Hydrocolloids with Different Surface Charge. *Molecules* **2020**, *25*, 5117. [CrossRef]
33. Wright, A.J.; Marangoni, A.G. Formation, structure, and rheological properties of ricinelaidic acid-vegetable oil organogels. *J. Am. Oil Chem. Soc.* **2016**, *83*, 497–503. [CrossRef]
34. Mustafa, O.; Nazan, A.; Emin, Y. Preparation and Characterization of Virgin Olive Oil-Beeswax Oleogel Emulsion Products. *J. Am. Oil Chem. Soc.* **2015**, *92*, 459–471. [CrossRef]
35. Huang, X.; Terech, P.; Raghavan, S.R.; Weiss, R.G. Kinetics of 5α-cholestan-3β-yl N-(2-naphthyl) carbamate/n-alkane organogel formation and its influence on the fibrillar networks. *J. Am. Chem. Soc.* **2005**, *127*, 4336–4344. [PubMed]
36. Li, J.L.; Liu, X.Y.; Wang, R.Y.; Xiong, J.Y. Architecture of a biocompatible supramolecular material by supersaturation-driven fabrication of its fiber network. *J. Phys. Chem. B* **2005**, *109*, 24231–24235. [CrossRef] [PubMed]
37. Marangoni, A.G. Organogels: An alternative edible oil-structuring method. *J. Am. Oil Chem. Soc.* **2012**, *89*, 749–780. [CrossRef]
38. Bascuas, S.; Hernando, I.; Moraga, G.; Quiles, A. Structure and stability of edible oleogels prepared with different unsaturated oils and hydrocolloids. *Int. J. Food Sci. Technol.* **2020**, *55*, 1458–1467. [CrossRef]
39. Dai, M.; Bai, L.; Zhang, H.; Ma, Q.; Luo, R.; Lei, F.; Fei, Q.; He, N. A novel flunarizine hydrochloride-loaded organogel for intraocular drug delivery in situ: Design, physicochemical characteristics and inspection. *Int. J. Pharm.* **2020**, *576*, 119027. [CrossRef]
40. Chauvelon, G.; Doublier, J.L.; Buléon, A.; Thibault, J.F.; Saulnier, L. Rheological properties of sulfoacetate derivatives of cellulose. *Carbohydr. Res.* **2003**, *338*, 751–759. [CrossRef]
41. Herazo, M.Á.; Ciro-Velásquez, H.J.; Márquez, C.J. Rheological and thermal study of structured oils: Avocado (*Persea americana*) and sacha inchi (*Plukenetia volubilis* L.) systems. *J. Food Sci. Technol.* **2019**, *56*, 321–329. [CrossRef] [PubMed]
42. Silva-Weiss, A.; Bifani, V.; Ihl, M.; Sobral, P.J.A.; Gómez-Guillén, M.C. Structural properties of films and rheology of film-forming solutions based on chitosan and chitosan-starch blend enriched with murta leaf extract. *Food Hydrocoll.* **2013**, *31*, 458–466. [CrossRef]
43. Rojas, J.; Cabrera, S.; Ciro, G.; Naranjo, A. Lipidic matrixes containing lemon essential oil increases storage stability: Rheological, thermal, and microstructural studies. *Appl. Sci.* **2020**, *10*, 3909. [CrossRef]
44. Zuo, F.; Li, X.; Yang, S.; Wang, P.; Yang, Q.; Wang, N.; Xiao, Z. Formation, rheological behavior and morphological structure of cinnamic acid based rice bran oil organogel. *Food Sci.* **2018**, *39*, 16–21. [CrossRef]
45. Wang, R.; Liu, X.Y.; Xiong, J.; Li, J. Real-time observation of fiber network formation in molecular organogel: Supersaturation-dependent microstructure and its related rheological property. *J. Phys. Chem. B* **2006**, *110*, 7275–7280. [CrossRef]
46. Meng, Z.; Guo, Y.; Wang, Y.; Liu, Y. Organogels based on the polyglyceryl fatty acid ester and sunflower oil: Macroscopic property, microstructure, interaction force, and application. *LWT* **2019**, *116*, 108590. [CrossRef]
47. Ojijo, N.K.; Neeman, I.; Eger, S.; Shimoni, E. Effects of monoglyceride content, cooling rate and shear on the rheological properties of olive oil/monoglyceride gel networks. *J. Sci. Food Agric.* **2004**, *84*, 1585–1593. [CrossRef]

48. Lan, Y.; Corradini, M.G.; Weiss, A.G.; Raghavan, S.R.; Rogers, M.A. To gel or not to gel: Correlating molecular gelation with solvent parameters. *Chem. Soc. Rev.* **2015**, *44*, 6035–6058. [CrossRef]
49. Mallia, V.A.; Terech, P.; Weiss, R.G. Correlations of properties and structures at different length scales of hydro-and organo-gels based on N-alkyl-(R)-12-hydroxyoctadecylammonium chlorides. *J. Phys. Chem. B* **2011**, *115*, 12401–12414. [CrossRef]
50. Teixeira, A.C.; Garcia, A.R.; Ilharco, L.M.; da Silva, A.M.G.; Fernandes, A.C. Phase behaviour of oleanolic acid, pure and mixed with stearic acid: Interactions and crystallinity. *Chem. Phys. Lipids* **2010**, *163*, 655–666. [CrossRef]
51. Gong, N.; Wang, Y.; Zhang, B.; Yang, D.; Du, G.; Lu, Y. Screening, preparation and characterization of diosgenin versatile solvates. *Steroids* **2019**, *143*, 18–24. [CrossRef] [PubMed]
52. Zeng, C.; Wan, Z.; Xia, H.; Zhao, H.; Guo, S. Structure and properties of organogels developed by diosgenin in canola oil. *Food Biophys.* **2020**, *15*, 452–462. [CrossRef]
53. Rogers, M.A.; Wright, A.J.; Marangoni, A.G. Nanostructuring fiber morphology and solvent inclusions in 12-hydroxystearic acid/canola oil organogels. *Curr. Opin. Colloid Interface Sci.* **2009**, *14*, 33–42. [CrossRef]
54. Xing, P.; Sun, T.; Li, S.; Hao, A.; Su, J.; Hou, Y. An instant-formative heat-set organogel induced by small organic molecules at a high temperature. *Colloids Surf. A Physicochem. Eng. Asp.* **2013**, *421*, 44–50. [CrossRef]
55. Baran, N.; Singh, V.K.; Pal, K.; Anis, A.; Pradhan, D.K.; Pramanik, K. Development and characterization of soy lecithin and palm oil-based organogels. *Polym.-Plast. Technol. Eng.* **2014**, *53*, 865–879. [CrossRef]
56. Kulkarni, C.; Balasubramanian, S.; George, S.J. What molecular features govern the mechanism of supramolecular polymerization? *ChemPhysChem* **2013**, *14*, 661–673. [CrossRef]
57. Suzuki, M.; Nakajima, Y.; Yumoto, M.; Kimura, M.; Shirai, H.; Hanabusa, K. Effects of hydrogen bonding and van der Waals interactions on organogelation using designed low-molecular-weight gelators and gel formation at room temperature. *Langmuir* **2003**, *19*, 8622–8624. [CrossRef]

Article

Dissecting of the AI-2/LuxS Mediated Growth Characteristics and Bacteriostatic Ability of *Lactiplantibacillus plantarum* SS-128 by Integration of Transcriptomics and Metabolomics

Yilin Qian, Yuan Li, Tengteng Xu, Huijuan Zhao, Mingyong Zeng and Zunying Liu *

College of Food Science and Engineering, Ocean University of China, Qingdao 266003, China; qianyl92@outlook.com (Y.Q.); liyuan@ouc.edu.cn (Y.L.); xtteng924@163.com (T.X.); zhj13583036006@163.com (H.Z.); mingyz@ouc.edu.cn (M.Z.)
* Correspondence: liuzunying@ouc.edu.cn; Tel./Fax: +86-532-8203-2400

Abstract: *Lactiplantibacillus plantarum* could regulate certain physiological functions through the AI-2/LuxS-mediated quorum sensing (QS) system. To explore the regulation mechanism on the growth characteristics and bacteriostatic ability of *L. plantarum* SS-128, a *luxS* mutant was constructed by a two-step homologous recombination. Compared with ΔluxS/SS-128, the metabolites of SS-128 had stronger bacteriostatic ability. The combined analysis of transcriptomics and metabolomics data showed that SS-128 exhibited higher pyruvate metabolic efficiency and energy input, followed by higher LDH level and metabolite overflow compared to ΔluxS/SS-128, resulting in stronger bacteriostatic ability. The absence of *luxS* induces a regulatory pathway that burdens the cysteine cycle by quantitatively drawing off central metabolic intermediaries. To accommodate this mutations, ΔluxS/SS-128 exhibited lower metabolite overflow and abnormal proliferation. These results demonstrate that the growth characteristic and metabolism of *L. plantarum* SS-128 are mediated by the AI-2/LuxS QS system, which is a positive regulator involved in food safety. It would be helpful to investigate more bio-preservation control potential of *L. plantarum*, especially when applied in food industrial biotechnology.

Keywords: quorum sensing; *Lactiplantibacillus plantarum*; bacteriostatic ability; transcriptomics; metabolomics

Citation: Qian, Y.; Li, Y.; Xu, T.; Zhao, H.; Zeng, M.; Liu, Z. Dissecting of the AI-2/LuxS Mediated Growth Characteristics and Bacteriostatic Ability of *Lactiplantibacillus plantarum* SS-128 by Integration of Transcriptomics and Metabolomics. *Foods* 2022, 11, 638. https://doi.org/10.3390/foods11050638

Academic Editor: Guowen Zhang

Received: 26 January 2022
Accepted: 19 February 2022
Published: 22 February 2022

Publisher's Note: MDPI stays neutral with regard to jurisdictional claims in published maps and institutional affiliations.

Copyright: © 2022 by the authors. Licensee MDPI, Basel, Switzerland. This article is an open access article distributed under the terms and conditions of the Creative Commons Attribution (CC BY) license (https://creativecommons.org/licenses/by/4.0/).

1. Introduction

Lactic acid bacteria (LAB) comprise a huge group of safe and widespread microorganisms in nature, and they are primarily applied as starter cultures and probiotics [1,2]. LAB are apprehended as ideal candidates for commercial exploitation in food industry with their status recognized as Generally Regarded As Safe (GRAS) and Qualified Presumption of Safety (QPS) in the European Union [3,4]. Aside from the health-promoting and probiotic properties, certain LAB producing organic acids, including lactic acid and phenyllactic acid (PLA), are also associated with food industrial biotechnology, providing food preservation for biocatalysis [5–9]. Today's food industry faces the tremendous problem in producing goods that are not only productive, but also wholesome for customers, as well as longer-lasting [4,5]. Consequently, the scientific research in the potential of LAB as bio-preservatives has attracted continuous interest. Strategies of enhancement on bio-preservation potential can be roughly classified into two major types: the strain development strategy and microbiological control strategy. The strain development strategy involves natural approaches or metabolic engineering, and it is used for strain transformation. The microbiological control strategy involves the regulation of LAB to microorganisms throughout the food system or environment. Both strategies are regulated by quorum sensing (QS), which provides new opportunities to enhance the safety and quality of foods by the "positive regulation" of QS.

QS, associated with some cellular activities, is a process in which cells socially coordinate gene expression, involving producing extracellular signaling molecules called autoinducers [10–12]. The QS mediated by autoinducer-2 (AI-2)/LuxS could be detected in various Gram-negative and -positive organisms [13,14]. The gene *luxS* converts S-ribosomal homocysteine (SRH) into homocysteine and 4,5-dihydroxy-2,3-pentanedione (DPD) that undergoes spontaneous rearrangements to form AI-2. The exogenous addition of AI-2 could show influences on the biofilm formation, bacteriostatic action, and stress tolerance by LAB have been reported [14–18]. However, to enhance the "positive regulation" of QS in LAB, focusing on the metabolic pathway regulated by *luxS* could be an effective way compared with the addition of expensive signaling molecules. Previous studies have shown that the inactivation of the *luxS* gene has shown greatly impacts on the *Lactobacillus reuteri* 100-23C behaviors, which are inhabited in mice intestinal tracts [19]. Lebeer et al. reported that the *luxS* mutation could show indirect impact on biofilm formation in *Lactobacillus rhamnosus* GG, while it was not impacted by exogenous AI-2 mediated QS [20]. The role of the *luxS* mediated molecular mechanisms from the proteomic analyses in bacteriocin production was reported by Jia et al. [21]. However, the growth characteristic of the wild strain and the *luxS* mutant in the above studies showed inconsistent trends. Meanwhile, the regulation of AI-2/LuxS on LAB growth characteristics and bacteriostatic ability has not been elucidated yet.

L. plantarum possesses the ability of rapid proliferation, due to the strong resistance against the environment of complex food matrix, indicating its high potential for industrial application [16,22]. Unlike other strains, *L. plantarum* can synthesize homocysteine from cysteine by direct sulfhydrylation involving the *cysK* enzyme, to ensure the regeneration of the methyl group of S-adenosyl methionine (SAM) [20]. Since the unique functions of the *cysK* enzyme carried out are pretty similar to the *luxS*, we hypothesized that it contributes to enhancing the growth ability and stress resistance of *L. plantarum*, and also helps to focus on the *luxS*.

Previously, we found that the cooperation of *L. plantarum* showed high biopreservative activity, which was related to the AI-2/LuxS system [23], but the role of the *luxS* system has not been clearly elucidated. In this study, we aimed to understand the regulation mechanism of the AI-2/LuxS-mediated QS system on the growth characteristics and bacteriostatic ability of LAB, to provide a basis for improving its bio-preservation function.

2. Materials and Methods

2.1. Construction of luxS-Mutant of SS-128

The *luxS*-mutant of SS-128, named Δ*luxS*/SS-128, was constructed using fusion PCR and a two-step homologous recombination. In brief, the *luxS* gene-flanked two fragments were amplified and fused by overlap extension PCR with a primer pair. The primers are shown in Table 1. The ligation products, transforming into competent *Escherichia coli* pFED760, generated the recombinant plasmids, which carried the homologous fragments for allelic exchange of flanking fragments of the *luxS* gene. The pFED760 was a temperature-sensitive plasmid and was a kind gift from Professor Yiyong Luo (Kunming University of Science and Technology, Kunming, China). Recombinant plasmids pFED760-*luxS* were extracted and transformed into SS-128 by electroporation. The Δ*luxS*/SS-128 was constructed through a two-step homologous recombination by changing the temperature to control the replication and suicide of pFED760.

2.2. AI-2 Assay

The AI-2 bioluminescence assay was operated according to the approach with some modifications [24]. The *Vibrio harveyi* strain BB170 was cultured in AB medium at 30 °C for 12 h and diluted with the AB medium at a ratio of 1:5000. The medium was added to the diluted *V. harveyi* culture at a final concentration of 10% (v/v). The white 96-well plates vibrated at 100 rpm at 30 °C (PBS was used as the negative control). Light production was recorded every 30 min using a Synergy H4 hybrid microplate reader (Bio-Tek, Winooski,

VT, USA) in the bioluminescence mode. AI-2 activity was calculated as the difference compared with the bioluminescence level in the control group and presented as relative luminescence units (RLUs).

Table 1. Primer sequences.

Gene Name		Oligonucleotide Sequence (5′-3′)
Up	Forward	TATCCC ACTACCTGAA ACTCG
	Reverse	GCACCACCATTACTTTTTATATTGTAGCACATTGCCCGTTA
Down	Forward	TAACGGGCAATGTGCATACAATATAAAAAGTAATGGTGGTGC
	Reverse	GCTGGTGCTTCGTAAACTTCC
16S rRNA	Forward	CGTAGGTGGCAAGCGTTGTCC
	Reverse	CGCCTTCGCCACTGGTGTTC
LuxS	Forward	CGGATGGATGGCGTGATTGACTG
	Reverse	CTTAGCAACTTCAACGGTGTCATGTTC
UD	Forward	GTTCTGCACGGACGCTATCT
	Reverse	ATTAACTTGCGTTGGTAGGC

The total RNA was directly extracted from *L. plantarum* according to the approach described by Zhang et al. at incubation times of 6, 8, 10, 12, 14, 16, 20, 24, 36, 48 h, respectively [25]. The bacteria were collected after being washed twice with PBS. According to the protocol, the obtained precipitate was used for total RNA extraction with TRIzol Reagent (Invitrogen, Carlsbad, CA, USA). The quality of the RNA was detected through the agarose gel electrophoresis, and the concentration was measured through the NanoDrop spectrophotometer (Nano-200, Hangzhou Austrian Sheng Instrument Co., Ltd., Hangzhou, China). RT-qPCR amplifications were measured in three biological replicates using a Bio-Rad CFX Connect System (Bio-Rad Laboratories, Inc., California, USA). The housekeeping gene, 16S rRNA, was chosen to normalize RNA amounts (internal control). The sequences coding for the 16S rRNA, *luxS* of the wild type (WT) SS-128 and the mutant strain Δ*luxS*/SS-128, were designed based on the *L. plantarum* WCFS1 gene sequences (GenBank, no. NC_004567.2, from https://www.ncbi.nlm.nih.gov/nuccore/NC_004567.2/, accessed on 10 April 2021). qPCR was performed with ChamQ SYBR Color qPCR Master Mix on a CFX Connect Real-Time PCR Detection System (Bio-Rad Laboratories, Inc., Hercules, CA, USA). The primer sequence is demonstrated in Table 1. Each experiment was repeated three times and analyzed using the $2^{-\Delta\Delta CT}$ method.

2.3. Bacterial Growth Assay

SS-128 (wild type) and Δ*luxS*/SS-128 (*luxS* mutant type) were first propagated twice in MRS broth at 37 °C overnight. The 1% overnight cultures (18 h) were inoculated into a 5 mL fresh medium under aerobic conditions at 37 °C for 48 h. During the incubation, the bacterial density of the samples was determined every 2 h at 600 nm. Plate counting was used to determine the live cell number of the samples. The WT and mutant strains cultured for 14 h were collected by centrifugation and then were immediately frozen and stored at −80 °C.

2.4. Flow Cytometry Assay

The live/dead cells were determined by flow cytometry employing fluorescence after being stained with SYTO 9 and propidium iodide (PI). The counting method was constructed by Leonard and Bensch et al. [25,26], with some modifications. Cells (10^2–10^3 cells/μL) were incubated with SYTO 9 (5 μg/mL) and PI (5 μg/mL) for 3 min in dark. Flow cytometric analyses were carried out using a BD FACSVerse™ flow cytometry (BD Biosciences, San Jose, CA, USA). Data were analyzed by using FlowJo software 10.6.1 (Tree Star Inc., Ashland, OR, USA).

2.5. Determination of Bacteriostatic Ability

The agar diffusion assay was used to identify the bacteriostatic ability of cell-free supernatant (CFS) by Zhang and Li et al. [27,28]. A 15 mL aliquot of 0.4% soft agar with 2% bacterial culture medium was poured into a sterile Petri dish, and sterile Oxford cups (ϕ = 7.64 mm) were put on the surface of the medium. The Oxford cup was filled with the CFS of *L. plantarum* (200 µL). The inhibition zone could be measured after incubation at 28 °C for 24 h.

2.6. Transcriptome Analysis

RNA isolation and high-throughput RNA sequencing (RNA-Seq) were conducted by Oebiotech Corp. (Shanghai, China). Total RNA was extracted using a Total RNA Purification Kit (Sangon Biotech, Shanghai, China), followed by evaluating the RNA integrity on Agilent 2100 bioanalyzer (Agilent Technologies, Santa Clara, CA, USA). The libraries were sequenced on an Illumina sequencing platform (HiSeq 2500), and 150-bp/125-bp paired-end reads were generated. The gene expression profiles were analyzed based on reads per kilobase of transcript per million mapped read (RPKM) normalization. DESeq was used to standardize the counts of genes in each sample, and NB was used to test the difference significance of reads. Differential genes were screened according to the difference multiple and different significance test results. Finally, the GO and KEGG enrichment of differentially expressed genes were analyzed under the cutoff of *p*-value < 0.05 using the BLAST program.

2.7. Non-Targeted Metabolomic Analysis

2.7.1. Sample Preparation

A total of 20 µL of 2-chloro-l-phenylalanine (0.3 mg/mL) dissolved in methanol as internal standard and methanol-water (4:1 = v/v) were added to each sample. Trichloromethane was added to each aliquot. The mixture was extracted in the ice water bath with ultrasonication for 20 min and then centrifugation (13,000 rpm, for 10 min at 4 °C). A total of 1 mL of supernatant was dried in a freeze concentration centrifugal dryer, sequentially redissolved in 25% methanol, and incubated at 4 °C for 2 min. After that, the mixtures were centrifuged again, using the same step as before. The supernatant of each tube was filtered to LC–MS and GC–MS for analysis. QC samples were prepared by mixing aliquots of all samples to be a pooled sample.

2.7.2. Gas Chromatography/Mass Spectrometry (GC/MS) Analysis

The derivatized samples were analyzed on an Agilent 7890B gas chromatography system coupled to an Agilent 5977A MSD system (Agilent Technologies Inc., CA, USA). A DB-5MS fused-silica capillary column (30 m × 0.25 mm × 0.25 µm, Agilent J&W Scientific, Folsom, CA, USA) was utilized to separate the derivatives. Helium (>99.999%) was used as the carrier gas at a 1 mL/min constant flow rate through the column. The QCs were injected at regular intervals throughout the analytical run to provide a set of data from which repeatability could be assessed.

2.7.3. Ultrahigh Performance Liquid Chromatography/Mass Spectrometry (UPLC/MS) Analysis

ACQUITY UPLC I-Class system (Waters Corporation Milford, Milford, MA, USA) coupled with VION IMS QTOF Mass spectrometer (Waters Corporation Milford, Milford, MA, USA) was used to analyze the metabolic profiling in both ESI positive and ESI negative ion modes. An ACQUITY UPLC BEH C18 column (1.7 µm, 2.1 × 100 mm) was employed in both positive and negative modes. Water and Acetonitrile/Methanol 2/3 (v/v), both containing 0.1% formic acid were used as mobile phases A and B, respectively. The injection volume was 1 µL. Data were collected in full scan mode (m/z ranges from 50 to 1000) combined with MSE mode, including two independent scans with different collision energies (CE) that were alternatively acquired during the running process.

2.7.4. Data Preprocessing and Analysis

The acquired LC–MS raw data were analyzed by the progenesis QI software (Waters Corporation, Milford, CT, USA) to identify the metabolites. The raw data were converted to .abf format, followed by processing on the MD-DIAL software through LUG database (Untargeted database of GC–MS rom Lumingbio).

Principle component analysis (PCA) and (orthogonal) partial least-squares-discriminant analysis (O) PLS-DA were carried out to visualize the metabolic alterations among experimental groups after mean centering (Ctr) and Pareto variance (Par) scaling, respectively. Variable importance in the projection (VIP) ranks the overall contribution of each variable to the OPLS-DA model, and those variables with VIP > 1 are considered relevant for group discrimination. In this study, the default seven-round cross-validation was applied with one/seventh of the samples being excluded from the mathematical model in each round.

The differential metabolites were selected based on the combination of a statistically significant threshold of VIP values obtained from the OPLS-DA model and p values from a two-tailed Student's t-test on the normalized peak areas, where metabolites with VIP values larger than 1.0 and p values less than 0.05 were considered as differential metabolites.

2.8. Statistical Analysis

All results were statistically analyzed with IBM SPSS Statistics version 19.0. Data were presented as means and standard deviations. One-way analysis of variance followed by Duncan's post hoc test was used to compare the mean differences. A p-value < 0.05 was considered significant.

3. Results

3.1. Confirmation of luxS Gene Knockout

In order to confirm the function of *luxS* as a S-ribosylhomocysteine lyase, a knockout mutant of *luxS* was constructed. Gene *luxS* was successfully deleted by a temperature-sensitive plasmid pFED760 (Figure 1a). The PCR in the *luxS*-mutant generated a short 3224-bp DNA fragment (lane 1) as expected, while a normal 3701-bp DNA fragment (lane 2) was amplified in WT strain when using UD-f and UD-r primer pair in Table 1 (Figure 1b). From lane 3, it can be seen that the primers of *luxS* did not amplify the gene. The sequencing of PCR products also confirmed that the *luxS*-mutant was successfully achieved.

To determine the effect of *luxS* knockout on the AI-2 production ability of the strains, we monitored the AI-2 levels in the supernatants of *L. plantarum* SS-128 and Δ*luxS*/SS-128 by bioluminescence assay using the *V. harveyi* BB170 strain. Lack of the *luxS* gene, which encodes an enzyme involved in methionine metabolism, ultimately leads to the loss of AI-2. As shown in Figure 1c, the accumulation concentration of AI-2 of *L. plantarum* SS-128 was at its highest in the late exponential phase and diminished slowly in the stationary phase. It was agreed with the results of Amandeep et al. [29], which reported that AI-2 activity increased till the late exponential phase. Moreover, the results demonstrated that AI-2 activities of SS-128 were significantly increased at 14 h compared to those of Δ*luxS*/SS-128 combined with transcript-level expression analysis (Figure 1d). These data suggested that the deletion of the *luxS* contributed to a disability in synthesizing AI-2.

3.2. Growth Characteristics of L. plantarum SS-128 and ΔluxS/SS-128

The behaviors of *L. plantarum* exposed to 37 °C were summarized in Figure 2. Results in Figure 2a showed that the tested SS-128 started to grow exponentially earlier compared with Δ*luxS*/SS-128. The pH of SS-128 reached its lowest value (3.54) after 14 h of cultivation, then began to increase, and the time was 2 h earlier than that of defective strain. These properties differ from the growth rate of *L. reuteri luxS* mutant [19]. This might be account for the *cysK* enzyme in *L. plantarum*. Gu et al. reported that the exogenous synthetic AI-2 led to different growth feedback on physiological behaviors of *Enterococcus faecium* and *Lactobacillus fermentum* in vitro [15,16]. However, the live cell number of the Δ*luxS*/SS-128 was significantly higher than that of the SS-128 at the late exponential phase (12–18 h)

(Figure 2b). It is speculated that the proportion of viable bacteria in Δ*luxS*/SS-128 was higher than that in the SS-128 within 12–18 h.

For determining the viable cell proportion (VCP) of *L. plantarum*, the SYTO 9 and PI staining was used to analyze the membrane integrity by flow cytometry. The results of double-staining on *L. plantarum* (Figure 2c) showed the live/dead cells, respectively. The VCP was 83.6% in SS-128 and 94.6% in Δ*luxS*/SS-128 at 14 h, and no apparent difference in the proportion of viable bacteria at 24 h was shown in Figure 2c ($p < 0.05$). The results demonstrated that the VCP in Δ*luxS*/SS-128 was higher than that in SS-128 during the late exponential phase. *L. plantarum* proliferates by binary fission, in which the cell must double its mass, replicate its DNA and divide equally to produce two daughter cells [30]. Based on this way of splitting and the adequate nutrition, the higher total number of living bacteria should exhibit an exponential increase with the normal proliferation rate of normal cells. However, the live cell number of the *luxS* mutant did not increase many times as much as that of the wild strain at the late exponential phase, contrary to the normal rate of the cell proliferation. Empirical evidence suggests that *E. coli* and *Schizosaccharomyces pombe*, all copied by binary division, avoid division by prolonging senescence when conditions are unfavorable [31,32]. Therefore, we hypothesized that the *luxS* gene might regulate cell proliferation, thereby affecting cell growth.

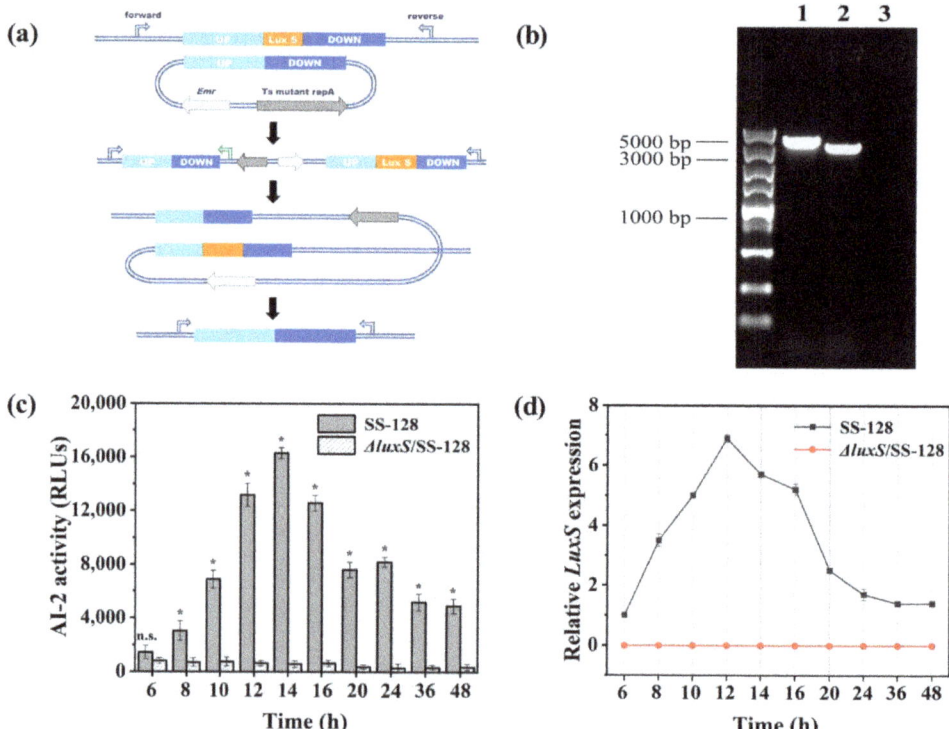

Figure 1. Two-step homologous recombination (**a**), PCR amplification (**b**), AI-2 activity (**c**), and transcription of the *luxS* gene (**d**) in SS-128 and Δ*luxS*/SS-128. All data points mean ± standard deviations ($n = 3$) with * denoting statistically significant differences ($p < 0.05$).

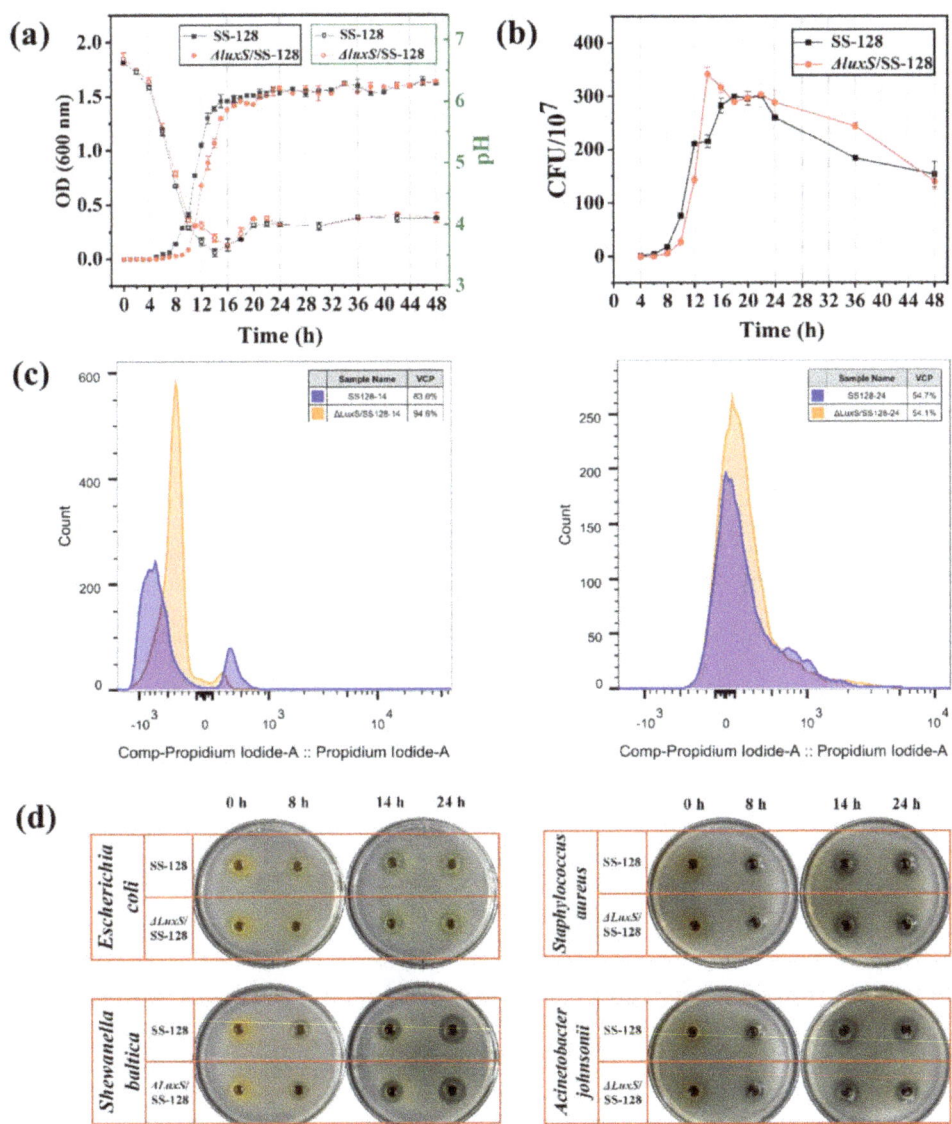

Figure 2. Cell density (**a**), live cell number (**b**), VCP of *L. plantarum* by flow cytometry (**c**), and images of inhibition zone of SS-128 and ΔluxS/SS-128 against *E. coli*, *S. baltica*, *S. aureus*, and *A. johnsonii* (**d**). All data points mean ± standard deviations (*n* = 3).

3.3. In Vitro Bacteriostatic Effect of L. plantarum SS-128 and ΔluxS/SS-128

LAB have great potential in hindering the activity of pathogenic and spoilage bacteria in food systems by producing bacteriostatic compounds mainly composed of organic acids [4,33]. Thus, we compared the bacteriostatic ability of cell-free supernatant against *E. coli*, *Shewanella baltica*, *Staphylococcus aureus* and *Acinetobacter johnsonii* of *L. plantarum* SS-128 and ΔluxS/SS-128 (Figure 2d). *E. coli* and *S. aureus* are common food spoilage and pathogenic microorganisms [34–36], while *A. johnsonii* and *P. fluorescens* are the main spoilage potential organism during the iced storage of aquatic products [37,38], which may pose considerable risks to the health of consumers. The activity was evaluated by Oxford cup tests with the sample from that of the CFS; the MRS broth without *L. plantarum* as the control. The inhibitory zone diameter of each sample was measured with calipers and summarized in Table S1. As shown in Figure 2d, the control group showed no bacteriostatic ring, indicating that MRS broth has no bacteriostatic ability against the food-borne spoilage and pathogenic bacteria. In contrast, SS-128 exhibited a clear antibacterial ring around the bacteria and was larger than that of $\Delta luxS$/SS-128 after 14 h. In particular, the CFS of SS-128 collected at 24 h showed significant antibacterial activity against *S. baltica* (16.41 mm) compared with $\Delta luxS$/SS-128 (15.18 mm) ($p < 0.05$). According to the above results, the metabolites of SS-128 had stronger bacteriostatic ability.

3.4. The Organic Acids Production of L. plantarum SS-128 and $\Delta luxS$/SS-128

To further explore the regulation of the *luxS* gene on the main bacteriostatic metabolites (organic acids) produced by *L. plantarum*, the organic acids production in *L. plantarum* SS-128 and $\Delta luxS$/SS-128 was measured. The data suggested that the organic acid production of *L. plantarum* SS-128 was significantly decreased upon knockout of the *luxS* gene (Figure 3a,b). The lactic acid production of *L. plantarum* SS-128 was 20.96% to 64.12% higher than that of the *luxS* mutant type in 12–16 h ($p < 0.05$) (Figure 3a). QS is widely believed to regulate the metabolism of bacteria (e.g., *Pseudomonas aeruginosa* and *S. baltica*) [39,40]. Similar results were observed for PLA production in 12–14 h. These results indicated that the production of lactic acid and PLA in *L. plantarum* might be related to the regulation of AI-2/LuxS system. In our study, a lower number of viable bacteria and stronger bacteriostatic ability persisted in SS-128 compared with $\Delta luxS$/SS-128 after 14 h, which could further confirm the regulation of *luxS* on the growth characteristics and metabolism of *L. plantarum*.

Lactate dehydrogenase (LDH), widely distributed in diverse sources including LAB, catalyzes the reduction of pyruvate to lactic acid. It can also catalyze the conversion of phenylpyruvic acid (PPA) to PLA due to its broad substrate specificity [41]. Therefore, the determination of LDH activity could reflect the effect of *luxS* gene on the acid production capacity of *L. plantarum*. Figure 3c shows the LDH activity in *L. plantarum* evaluated after 8, 10, 12, 14, 16, and 24 h of cultivation. Compared with $\Delta luxS$/SS-128 (2.32 mU/10^4 cell), LDH activity was significantly increased in SS-128 after 10 h (4.25 mU/10^4 cell) ($p < 0.05$). During the exponential growth period, the LDH activity of SS-128 remained stable and significantly higher than that of $\Delta luxS$/SS-128 ($p < 0.05$).

3.5. Transcriptomics Analyses between L. plantarum SS-128 and the $\Delta luxS$/SS-128

The *luxS* gene is involved in QS, which has been shown to regulate critical physiological traits and various adaptive processes in different bacteria [42,43]. Therefore, probing the relationship between AI-/LuxS-mediated QS and the cell transcriptional expression levels is the premise to influence bacterial metabolites accumulation. Based on this, the expression profile of *L. plantarum* without *luxS* gene was discerning, and transcriptome analysis was performed on wild type and mutant type. Overall, 22.92 and 22.83 million reads of SS-128 and $\Delta luxS$/SS-128 were used for further analysis, respectively, and 94.06–94.48% of the total reads mapped to the reference genome, among which 66.10–76.02% were coding sequence (CDS) mapped reads (Table S2). There were significant differences between $\Delta luxS$/SS-128 and SS-128 in PCA model (Figure 4a). The differences between the first and the second principal component were 77.93% and 11.78%, respectively. A total of 157 genes were identified in $\Delta luxS$/SS-128 from uniquely matched reads data compared to SS-128 expressing at different levels ($p < 0.05$) (Figure 4b).

The 157 differentially expressed genes (DEGs) were subsequently applied to functional analysis in Kyoto Encyclopedia of Genes and Genomes (KEGG) pathway functional enrichment. As shown in Figure 4c, KEGG enrichment analysis found that some pathways were remarkably changed after the *luxS* gene knocked out in the histogram. These pathways included "Membrane transport", "Carbohydrate metabolism", and "Amino acid metabolism". Our data indicated that, owing to the loss of the *luxS* gene, a series of transport-related pathways were adjusted, and the pathways involved in energy metabolism were also changed in the process of cell growth. The identified DEGs were combined with functional analysis results, such as KEGG annotations based on previous reports. These genes were considered the most critical genes related to cell growth and accumulation of metabolites, mainly involved in PTS, tricarboxylic acid (TCA), and methionine cycle (Table 2).

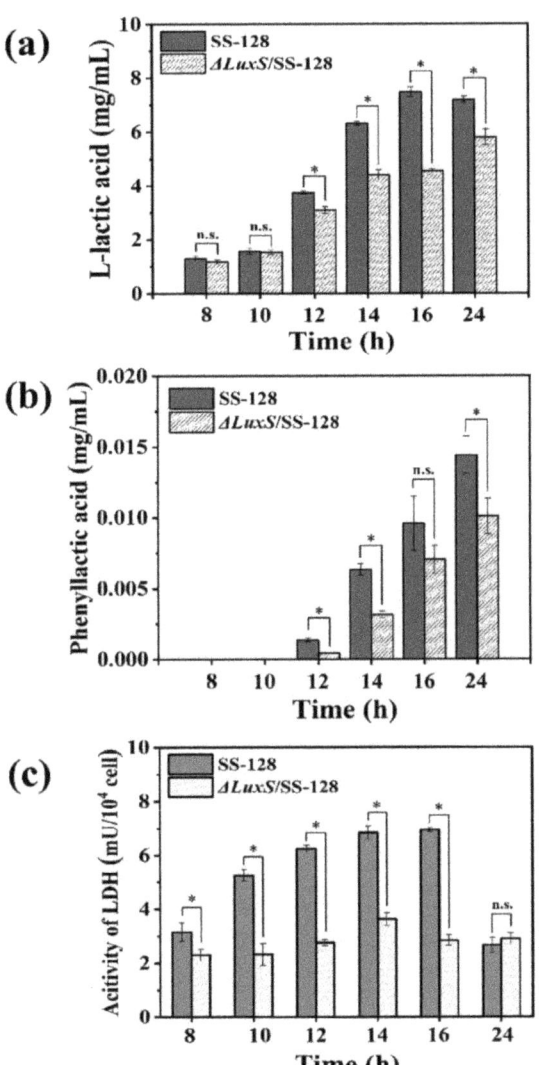

Figure 3. Extracellular concentrations of L-lactic acid (**a**) and PLA (**b**) detected by HPLC in SS-128 and ΔluxS/SS-128; LDH activity (**c**). All data points mean ± standard deviations (n = 3) with * denoting statistically significant differences ($p < 0.05$); n.s.: not significant.

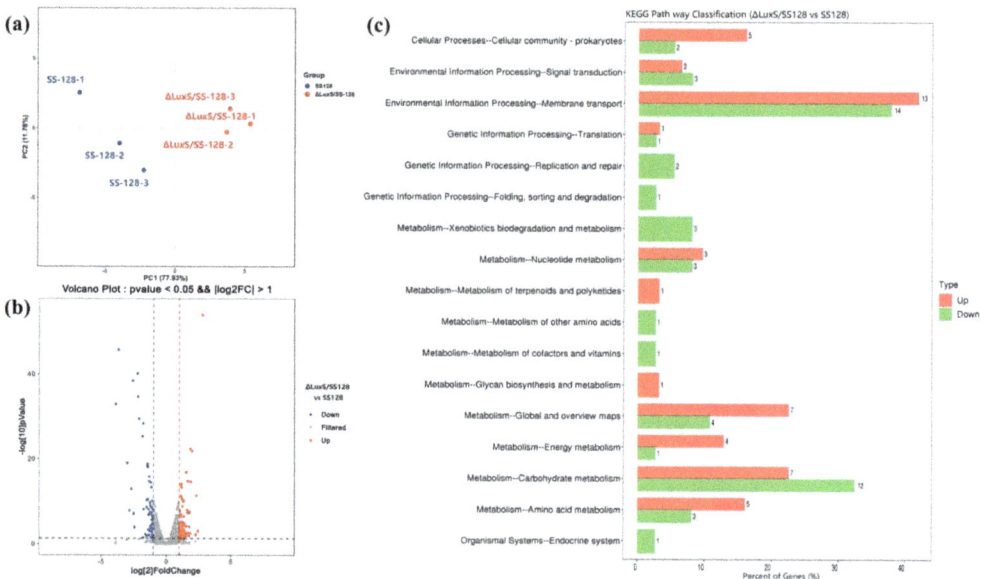

Figure 4. PCA analysis (**a**), volcano plot (**b**), and KEGG pathway of DEGs at level 2, (**c**) ($\Delta luxS$/SS-128 vs. SS-128).

Table 2. DEGs involved in PTS, pyruvate metabolism and quorum sensing.

Locus Tag	Entry	Gene Name (ID)	Definition	Fold Change	p-Value
PTS	K02793	manXa	mannose PTS system EIIA component	1.56 ↓	1.12×10^4
	K02768	fruB	fructose PTS system EIIA component	1.68 ↓	2.14×10^6
	K02810	scrA, ptsS	sucrose PTS system EIIBCA component	1.54 ↓	8.15×10^3
	K02761	celB, chbC	cellobiose PTS system EIIC component	3.36 ↓ *	2.21×10^4
	K02760	celA, chbB	cellobiose PTS system EIIB component	1.92 ↓	6.99×10^4
	K02759	celC, chbA	cellobiose PTS system EIIA component	3.19 ↓*	4.32×10^4
	K02773	gatA, sgcA	galactitol PTS system EIIA component	1.88 ↓	2.76×10^4
	K02774	gatB, sgcB	galactitol PTS system EIIB component	1.71 ↓	1.46×10^2
	K02755	bglF	beta-glucoside PTS system EIIA component	5.93 ↓ *	3.28×10^9
	K02798	cmtB	mannitol PTS system EIIA component	1.59 ↓	1.49×10^5
		LP_RS14755	PTS system EIIC component	1.74 ↓	1.79×10^2
		LP_RS12340	PTS system EIIA component	1.67 ↓	3.06×10^2
		LP_RS12650	PTS system EIIB component	1.52 ↓	4.89×10^2
		LP_RS12655	PTS system EIIC component	2.33 ↓*	9.25×10^{10}
		LP_RS13505	beta-glucoside PTS system EIIBCA component	1.54 ↓	8.15×10^3
		LP_RS14845	PTS system EIIB component	1.71 ↓	1.46×10^2
		LP_RS14850	PTS system EIIA component	1.88 ↓	2.76×10^4
Pyruvate metabolism	K00927	PGK, pgk	phosphoglycerate kinase	1.52 ↑	1.85×10^5
	K00016	ldh	L-lactate dehydrogenase	2.58 ↓ *	1.03×10^2
	K01610	pckA, PEPCK	phosphoenolpyruvate carboxykinase (ATP)	1.51 ↑	3.95×10^2
	K00027	ME, maeA	malate dehydrogenase	4.87 ↑ *	5.68×10^3
	K01676	fumA, fumB	fumarate hydratase, class I	3.39 ↑ *	1.10×10^4
	K00244	frdA	fumarate reductase flavoprotein subunit	4.08 ↑ *	3.14×10^2
	K01744	aspA	aspartate ammonia-lyase	1.70 ↑	1.99×10^7
	K01939	purA, ADSS	adenylosuccinate synthase	1.55 ↑	2.25×10^2
	K01512	acyP	acylphosphatase	1.70 ↑	2.95×10^5
Methionine cycle	K07173	luxS	S-ribosylhomocysteine lyase		2.53×10^4
	K01738	cysK	cysteine synthase	2.95 ↓	5.85×10^{10}
	K01999	livK	branched-chain amino acid transport system substrate-binding protein	2.01 ↓ *	2.75×10^2

Fold change was $\Delta luxS$/SS-128 vs. SS-128. ↑ and ↓ represented up and down regulation, respectively. * fold change was significantly higher than 2 ($p < 0.05$).

3.6. Metabolomics Analyses between L. plantarum SS-128 and the ΔluxS/SS-128

Except for transcriptome sequencing analysis, the untargeted GC–MS and LC–MS metabolomics analyses were used to further explore the adaptation of metabolic processes in *L. plantarum* SS-128 response to the *luxS* gene. The total number of metabolites detected in samples was 357, 2316, and 1330 for GC–MS, LC–MS positive ion mode, and negative ion mode, respectively. The tested samples were apparently separated from the wild type and mutant type in Figures 5a and S1a. Furthermore, the first principal components (PC) were clearly separated by the OPLS-DA model, corresponding a variation between groups of 38.7% (GC–MS) and 54.7% (LC–MS). These data indicated that the cell metabolism was changed significantly owing to the knocked out of the *luxS* gene (Figures 5b and S1b). The OPLS-DA model screened out differentially expressed metabolites in the two groups. The structure identities of 110 metabolites screened by GC–MS were assigned, among which 26 metabolites were significantly upregulated and 84 metabolites were significantly down-regulated ($p < 0.05$) (Figure 5c). LC–MS identified a total of 464 differentially expressed metabolites. Meanwhile, there were 17 metabolites substantial decreased and 33 metabolites increased in the first 50 metabolites that exhibited significant differences ($p < 0.05$) (Figure S1c).

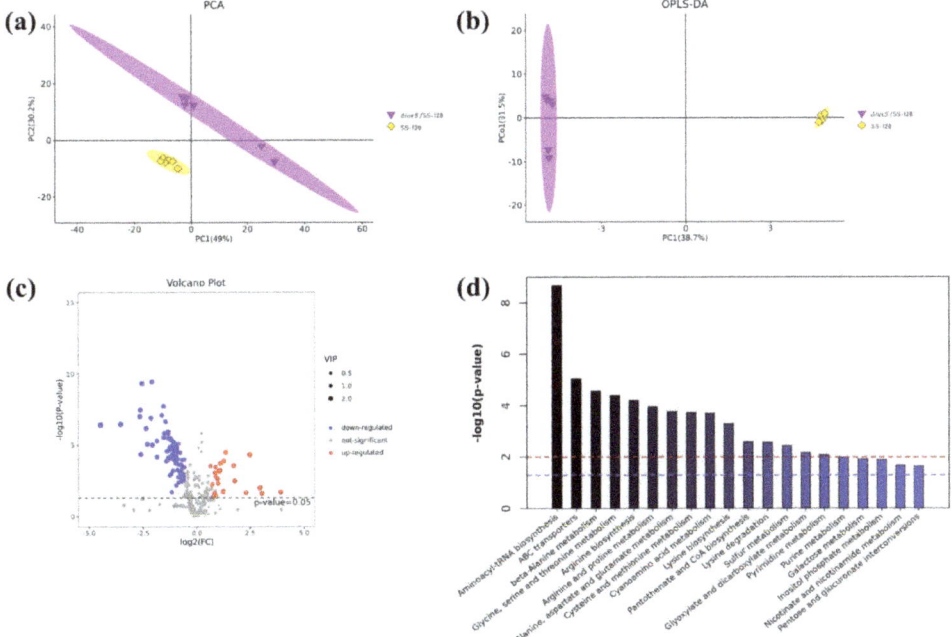

Figure 5. Changes in the metabolite contents compared the *luxS* mutant group with the control by GC–MS. The PCA (**a**), OPLS-DA (**b**), volcano plot (**c**), and metabolic pathway enrichment map-Top 20 (**d**) between ΔluxS/SS-128 and SS-128 under the cut off of $p < 0.05$ (ΔluxS/SS-128 vs. SS-128). The red line and blue line in (**d**) indicate $p = 0.01$ and 0.05, respectively.

To investigate the metabolic pathways involved, KEGG pathway analysis was performed. Among the top-20 metabolic pathway enrichment map, the differentially expressed metabolites (DEMs) in mutant type compared with WT strains were mainly enriched on aminoacyl-rRNA biosynthesis, ABC transporters, amino acid metabolism, pyrimidine metabolism and purine metabolism (Figure 5d). Based on the above, the key differential metabolites involved in the enriched pathways were listed in Table 3.

Table 3. Differential metabolites involving PTS, nucleotide synthesis, pyruvate metabolism, cysteine and methionine metabolism.

Metabolites	KEGG	KEGG Annotation	Dataclass	Formula	VIP	p-Value	FC
2′-Deoxyguanosine 5′-monophosphate	C00362	Purine metabolism	LC	C10H14N5O7P	1.12	3.46×10^8	1.53 ↑
Deoxyguanosine	C00330	Purine metabolism	LC	C10H13N5O4	1.26	2.38×10^6	1.84 ↑
Guanine	C00242	Purine metabolism	GC	C5H5N5O	1.38	6.32×10^4	2.05 ↑ *
Hypoxanthine	C00262	Purine metabolism	GC	C5H4N4O2	2.27	4.52×10^5	5.61 ↑ *
Xanthine	C00385	Purine metabolism	LC	C5H4N4O2	1.14	1.08×10^7	2.21 ↑ *
Cytosine	C00380	Pyrimidine metabolism	GC	C4H5N3O	1.43	5.73×10^4	2.15 ↑ *
dCMP	C00239	Pyrimidine metabolism	LC	C9H14N3O7P	1.62	1.58×10^4	1.62 ↑
dTMP	C00364	Pyrimidine metabolism	LC	C10H15N2O8P	5.82	1.20×10^7	1.73 ↑
Pseudouridine 5′-phosphate	C01168	Pyrimidine metabolism	LC	C9H13N2O9P	4.18	6.41×10^5	1.15 ↑
Uracil	C00106	Pyrimidine metabolism	GC	C4H4N2O2	1.67	3.17×10^5	2.59 ↑ *
Uridine	C00299	Pyrimidine metabolism	GC	C9H12N2O6	1.29	3.43×10^4	1.86 ↑
Uridine 5′-monophosphate	C00105	Pyrimidine metabolism	LC	C9H13N2O9P	2.20	5.38×10^4	1.42 ↑
Orotidine	C01103	Pyrimidine metabolism	LC	C10H12N2O8	1.42	7.33×10^7	4.61 ↓ *
L-glutamine	C00064	Purine metabolism pyrimidine metabolism	GC	C5H10N2O3	3.13	3.58×10^7	22.12 ↓ *
Flavin Mononucleotide	C00061	Oxidative phosphorylation Riboflavin metabolism	LC	C17H21N4O9P	1.04	4.29×10^3	1.66 ↑
Riboflavin	C00255	ABC transporters Riboflavin metabolism	LC	C17H20N4O6	1.23	3.40×10^3	2.22 ↑ *
Flavin adenine dinucleotide	C00016	Riboflavin metabolism	LC	C27H33N9O15P2	2.96	2.77×10^{10}	1.64 ↑
D-Glycerate 2-phosphate	C00631	Glycolysis/Gluconeogenesis	LC	C3H7O7P	1.81	2.85×10^7	1.84 ↑
D-Glycerate 3-phosphate	C00197	Glycolysis/Gluconeogenesis	LC	C3H7O7P	1.10	9.00×10^8	1.79 ↑
Phosphoenolpyruvic acid	C00074	Glycolysis/Gluconeogenesis Citrate cycle (TCA cycle)	LC	C3H5O6P	1.86	5.37×10^8	1.91 ↑
Isocitric acid	C00311	Citrate cycle (TCA cycle)	GC	C6H8O7	1.22	2.09×10^2	4.97 ↑ *
Succinic acid	C00042	Citrate cycle (TCA cycle)	GC	C4H6O4	1.47	1.91×10^4	2.17 ↑ *
L-aspartate	C00049	Cysteine and methionine metabolism	GC	C4H7NO4	1.29	4.77×10^3	2.02 ↑ *
L-glutamate	C00025	Alanine, aspartate and glutamate metabolism	LC	C5H9NO4	8.24	1.44×10^5	1.52 ↑
N-acetyl-glutamate	C01250	Arginine biosynthesis	GC	C7H11NO4	1.84	2.72×10^4	3.37 ↑ *
Glutathione (GSH)	C00051	Cysteine and methionine metabolism	ABC transporters	GC	C10H17N3O6S	1.25	8.57×10^4	1.82 ↑
L-gamma-glutamyl-L-valine	C03740	Glutathione metabolism	LC	C10H18N2O5	1.34	3.31×10^6	1.34 ↑
Malic acid	C00149	Citrate cycle (TCA cycle)	GC	C4H6O5	1.13	2.28×10^4	2.62 ↓
Fumaric acid	C00122	Citrate cycle (TCA cycle)	GC	C4H4O4	1.49	3.51×10^3	1.52 ↓
L-lactic acid	C00186	Pyruvate metabolism	GC	C3H6O3	2.05	5.33×10^4	4.56 ↓
L-Phenylalanine	C00079	Phenylalanine metabolism	LC	C9H11NO2	5.94	2.58×10^4	3.24 ↓
Phenyllactic acid	C05607	Phenylalanine metabolism	LC	C9H10O3	1.15	8.02×10^8	1.61 ↓
Aconitic acid	C00417	Citrate cycle (TCA cycle)	GC	C6H6O6	1.19	2.58×10^4	1.54 ↑
Citric acid	C00158	Citrate cycle (TCA cycle)	GC	C6H8O7	2.62	3.81×10^6	5.75 ↓ *
L-methionine	C00073	Cysteine and methionine metabolism	GC	C5H11NO2S	1.17	1.14×10^3	1.69 ↓
S-adenosyl-L-methionine (SAM)	C00019	Cysteine and methionine metabolism	LC	C15H22N6O5S	1.81	1.49×10^{10}	16.66 ↓ *
S-ribosyl-L-homocysteine (SRH)	C03539	Cysteine and methionine metabolism	LC	C9H17NO6S	5.76	1.00×10^8	4.71 ↑ *
L-cystathionine	C02291	Cysteine and methionine metabolism	GC	C7H14N2O4S	2.40	9.01×10^8	6.85 ↓ *
L-homocysteine	C00155	Cysteine and methionine metabolism	GC	C4H9NO2S	2.03	1.44×10^5	1.73 ↓
S-adenosyl-L-homocysteine (SAH)	C00021	Cysteine and methionine metabolism	GC	C14H20N6O5S	2.92	2.72×10^4	2.58 ↓

Table 3. Cont.

Metabolites	KEGG	KEGG Annotation	Dataclass	Formula	VIP	p-Value	FC
Serine	C00065	Cysteine and methionine metabolism	GC	C3H7NO3	1.58	6.93×10^5	0.42 ↓
Homoserine	C00263	Cysteine and methionine metabolism	GC	C4H9NO3	1.76	1.39×10^6	2.80 ↓ *
O-acetyl-L-serine	C00979	Cysteine and methionine metabolism	GC	C5H9NO4	1.51	1.02×10^5	2.89 ↓ *
Creatine	C00300	Glycine, serine and threonine metabolism	LC	C4H9N3O2	2.01	1.23×10^8	0.65 ↓
D-fructose 2,6-bisphosphate	C00665	Fructose and mannose metabolism	GC	C6H14O12P2	1.51	1.04×10^3	2.27 ↓ *
D-fructose	C02336	Phosphotransferase system (PTS) Fructose and mannose metabolism	GC	C6H12O6	1.74	2.06×10^7	2.70 ↓ *
Cellobiose	C00185	Phosphotransferase system (PTS)	GC	C12H22O11	1.73	1.40×10^6	1.59 ↓
Galactose	C00984	Galactose metabolism Phosphotransferase system (PTS)	GC	C6H12O6	1.48	5.98×10^6	2.08 ↓ *
Glucose-6-phosphate	C00092	Phosphotransferase system (PTS)	GC	C6H13O9P	1.35	1.41×10^4	1.93 ↓
D-ribulose 5-phosphate	C00199	Pentose phosphate pathway	GC	C5H11O8P	1.09	3.31×10^3	1.61 ↓
UDP-glucose	C00029	Pyrimidine metabolism Pentose and glucuronate	LC	C15H24N2O17P2	1.53	3.97×10^5	3.80 ↓ *
D-glucose	C00031	Glycolysis/Gluconeogenesis Phosphotransferase system (PTS)	LC	C6H12O6	2.04	2.79×10^5	2.33 ↓ *

Fold change was ΔluxS/SS-128 vs. SS-128. ↑ and ↓ represented up and down regulation, respectively. * fold change was significantly higher than 2 ($p < 0.05$).

4. Discussion

Since the above results showed that AI-2/LuxS positively regulated the bacteriostatic activity and growth of L. plantarum SS-128, to explore the correlation between two regulatory networks, the metabolome and transcriptome data were subjected to integrated analysis. The results suggest that the knockout of luxS gene resulted in the overall downregulation of methionine and cysteine cycling pathways. By comparing the ΔluxS/SS-128 with the SS-128, the metabolites (SAM, L-cystathionine, O-Acetyl-L-Serine, et al.) and genes (luxS and cysK) showed a lower expression in the mutant strain. Due to the disruption of the methionine cycle, L-homocysteine cannot be replenished-resulting in the consumption of a large SAM as methyl provider, and eventually, the dynamic cycle is disrupted. It is worth noting that it is converted to S-adenosine homocysteine (SAH) when methyl groups are transferred from SAM to nucleic acids, proteins, or other compounds via a SAM-dependent methyltransferase [44]. Therefore, SAM participated in DNA and protein methylation as a major methyl donor for methylation in living cells [45]. Thus, it can be inferred that luxS in the methionine cycle is involved in the regulation of methylation of DNA and other substances and cysteine metabolic pathways.

Bacteria, yeast, and mammalian cells are all dependent on an adequate supply of purines and pyrimidines to maintain proliferation. The proliferation of LAB requires a deoxyriboside bound to purine and pyrimidine bases [46]. As shown in Table 3, the DEMs, such as hypoxanthine, uracil, and riboflavin, were significantly increased by 5.61, 2.59, and 2.22 times, respectively. Therefore, the differential purine metabolites may be the response of the luxS mutant strain to changes in cellular metabolism, which is consistent with previous similar studies. It was reported that purine and purine precursors are related to cell growth under stress conditions [47–49]. Uracil dramatically affected L. plantarum growth incubated in ordinary air, and uracil sensitivity in aerobiosis was found in L. plantarum [50]. The accumulation of metabolites in the nucleotide synthesis pathway in cells reflects the delay of nucleotide metabolism in cell proliferation, further confirming the regulation of *luxS* on cell proliferation.

A significant increase in SRH was observed due to the interruption of activated methyl cycle (AMC) and store depletion of L-homocysteine. Intracellular L-cysteine levels have properly been regulated for proper cell growth [51]. As shown in Figure 6, L-cysteine was maintained only by the pyruvate, while the contents of metabolites in other related metabolic pathways were significantly decreased. The phenomenon is similar to the "anaplerotic sequences" described by Hans Kornberg in Biochemistry that some key intermediates are removed during the growth process, other related metabolites are used for the net production of the intermediate [52]. This is consistent with previous studies in which cysteine-deficient cells exhibited over 100-fold increased oxidative stress compared with the cysteine-added cells and was associated with protective mechanisms of cell growth [53]. Since pyruvate is a major metabolic junction linking carbohydrate or amino acid utilization to energy generation and biosynthetic pathways, the content of key intermediate metabolites requires timely replenishment through auxiliary reactions [54,55]. In response to the interruption of the methionine cycle, glycolysis and TCA cycle were upregulated to "maintain" the ATP for energy and NADPH for reducing power. DEGs related to pyruvate metabolism and TCA cycle, such as *pckA*, *ME2*, *fumA*, and *frdA*, were upregulated in Δ*luxS*/SS-128 compared to the control. Whereas operated by the "anaplerotic sequences", L-lactic, PLA, malate and fumarate synthesis were downregulated to maintain the overall rate of pyruvate, oxaloacetate, and phosphoenolpyruvate synthesis.

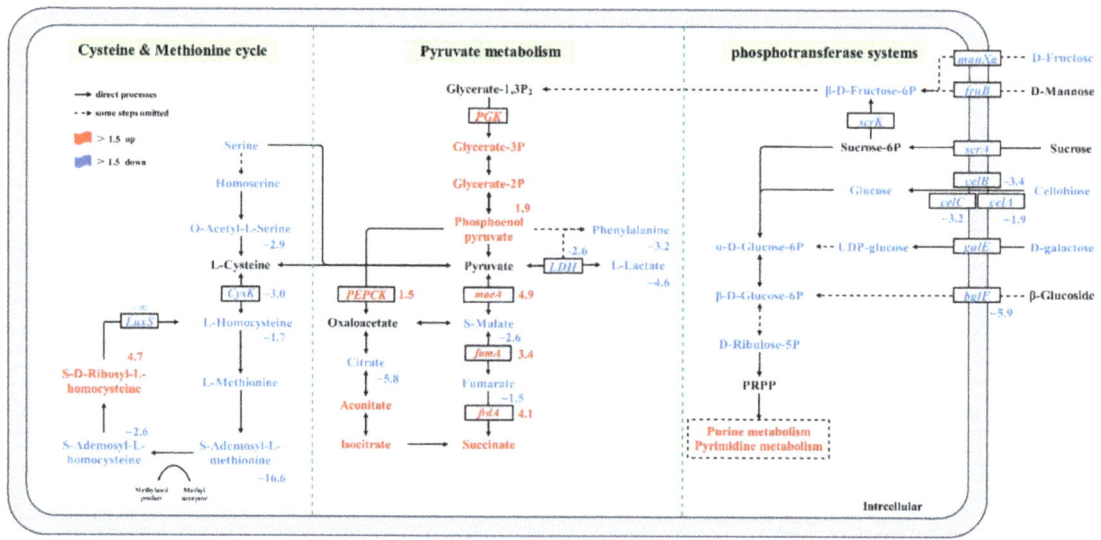

Figure 6. Changes of the expression levels of genes and metabolisms involved in L. plantarum (Δ*luxS*/SS-128 vs. SS-128). Pathways were constructed based on information provided in the KEGG database and previous studies. Microcompartments are depicted by dotted green lines.

The primary fermentation metabolite of LAB is lactic acid, which results in an environment of the lower pH which generally limits the growth of pathogenic and spoilage microbes [3]. The solubility of undissociated lactic acid within the cytoplasm membrane and the insolubility of dissociated lactic acid may affect the pH gradient across the membrane and reduce the available energy for cell growth [56]. As shown in Figure 6 and

Table 3, while succinic acid was accumulated in the TCA cycle, the lactic acid and PLA produced by fermentation of the ΔluxS/SS-128 was significantly lower than that of the SS-128. This conclusion was consistent with the result of lactic acid and PLA production by extracellular accumulation of SS-128 and ΔluxS/SS-128. Yang et al. reported that exogenous DPD could promote the growth and the production of PLA of L. Plantarum AB-1 cells [57]. Similarly, our data suggest that the lactic acid and PLA production of L. plantarum SS-128 was regulated by the AI-2/LuxS system. The crucial enzyme in lactic acid fermentation is LDH, which is mainly responsible for catalyzing the reversible reduction of pyruvate to lactic acid [58]. Recently, LDH has also been shown to be the main enzyme catalyzing the PPA to PLA, which is a broad-spectrum antibacterial compound [59,60]. The significant reduction of *ldh* of defective strains in transcriptomics results was also consistent with our previous experimental results ($p < 0.05$) (Figure 3c). The results indicated that the pyruvate production was maintained by increasing the directly related anabolism of the ΔluxS/SS-128, followed by a reduction in directly related catabolism. Thus, downregulation of the *ldh* reduces the production of the organic acid.

In addition, our results showed that the yield of malic acid, fumaric acid and citric acid in ΔluxS/SS-128 were also significantly downregulated ($p < 0.05$). The production of organic acids is undoubtedly the decisive factor to prolong the shelf life and safety of products by LAB, which was an important indicator to reflect the biocontrol ability. The significant decrease of organic acid production confirmed the regulation effect of *luxS* on the bacteriostatic ability of LAB.

Unlike pyruvate metabolism to output energy in the form of ATP, PTS is an ATP-dependent mechanism for absorbing sugars. It is dominant among the three mechanisms (PTS, secondary carriers and ABC transporters) of LAB, accounting for about 52%. It provides an essential carbon source to support lipid production and the biosynthesis of nucleotides and non-essential amino acids (NEAAs). Due to the various carbohydrate sources supplemented in MRS medium, the result of combined omics analysis could reflect the difference in carbohydrate transport capacity of SS-128 compared with ΔluxS/SS-128. More than 200 unigenes were involved in carbohydrate metabolism during the growth of *L. plantarum*, and 36 of these unigenes were involved in membrane transporter. Twenty-seven of these unigenes were DEGs in the PTS, and 17 of them in KEGG enrich top 20 (Figure 4) were summarized in Table 2. The results suggested that most of the DEGs associated with PTS were significantly downregulated ($p < 0.05$). Genes *celA*, *celB*, and *celC* encoding cellobiose transport system permease are responsible for the carbohydrates utilization in ΔluxS/SS-128, and significantly downregulated 1.92-, 3.36-, and 3.19-fold compared to SS-128. The *bglF* was downregulated 5.93-fold change in response to stimulating growth. Normal cells often increase the uptake of glucose to provide the significant carbon source for the production of lipids, nucleotides, and non-essential amino acids [61,62]. Therefore, lower transcript levels of genes encoding specific carbohydrate PTS components indicate that ΔluxS/SS-128 may only be able to uptake and utilize the limited range of energy substances, as part of a defense mechanism against the genetic mutations, and this could explain their less efficient for growth and lower biocontrol ability.

5. Conclusions

In summary, the comparative multiomic analyses demonstrated that the higher pyruvate metabolic efficiency and energy input, followed by higher LDH level and metabolite overflow in SS-128, resulted in stronger bacteriostatic ability compared to that of ΔluxS/SS-128. These results suggest that the AI-2/LuxS could positively regulate the bacteriostatic activity and growth of L. plantarum SS-128. Our data indicated that L. plantarum as biopreservatives was mediated by AI-2/luxS targeting to improve food safety, but further studies are needed to explore the strategies of enhancing positive regulation.

Supplementary Materials: The following supporting information can be downloaded at: https://www.mdpi.com/article/10.3390/foods11050638/s1, Figure S1: Changes in the metabolite contents compared the *luxS* mutant group with the control by LC-MS; Table S1: The mean inhibition zone diameter (mm) of CFS of *L. plantarum* against the *E. coli*, *S. baltica*, *S. aureus* and *A. johnsonii*; Table S2: Comparison of statistical results and reference genome.

Author Contributions: Conceptualization, Y.Q.; data curation, Y.Q. and H.Z.; formal analysis, Y.L. and H.Z.; funding acquisition, Z.L.; investigation, Y.Q. and Y.L.; methodology, Y.Q.; software, T.X.; supervision, M.Z. and Z.L.; visualization, T.X.; writing—original draft, Y.Q.; writing—review & editing, Y.L. All authors have read and agreed to the published version of the manuscript.

Funding: This research was funded by the National Natural Science Foundation of China (No. 31972141) and the National key research and development program (No. 2021YFD2100504).

Institutional Review Board Statement: Not applicable.

Informed Consent Statement: Not applicable.

Data Availability Statement: Not applicable.

Acknowledgments: This work was financially supported by the National Natural Science Foundation of China (no. 31972141) and the National key research and development program (No. 2021YFD2100504).

Conflicts of Interest: The authors declare no conflict of interest.

References

1. Hatti-Kaul, R.; Chen, L.; Dishisha, T.; Enshasy, H.E. Lactic acid bacteria: From starter cultures to producers of chemicals. *FEMS Microbiol. Lett.* **2018**, *365*, fny213. [CrossRef] [PubMed]
2. Barcenilla, C.; Ducic, M.; López, M.; Prieto, M.; Álvarez-Ordóñez, A. Application of lactic acid bacteria for the biopreservation of meat products: A systematic review. *Meat Sci.* **2022**, *183*, 108661. [CrossRef] [PubMed]
3. Crowley, S.; Mahony, J.; van Sinderen, D. Current perspectives on antifungal lactic acid bacteria as natural bio-preservatives. *Trends Food Sci. Technol.* **2013**, *33*, 93–109. [CrossRef]
4. Nasrollahzadeh, A.; Mokhtari, S.; Khomeiri, M.; Saris, P.E. Antifungal Preservation of Food by Lactic Acid Bacteria. *Foods* **2022**, *11*, 395. [CrossRef] [PubMed]
5. Michon, C.; Langella, P.; Eijsink, V.G.; Mathiesen, G.; Chatel, J.M. Display of recombinant proteins at the surface of lactic acid bacteria: Strategies and applications. *Microb. Cell Fact.* **2016**, *15*, 70. [CrossRef]
6. Gao, C.; Ma, C.; Xu, P. Biotechnological routes based on lactic acid production from biomass. *Biotechnol. Adv.* **2011**, *29*, 930–939. [CrossRef]
7. Gálvez, A.; Abriouel, H.; Benomar, N.; Lucas, R. Microbial antagonists to food-borne pathogens and biocontrol. *Curr. Opin. Biotechnol.* **2010**, *21*, 142–148. [CrossRef]
8. Kos, B.; Šušković, J.; Beganović, J.; Gjuračić, K.; Frece, J.; Iannaccone, C.; Canganella, F. Characterization of the three selected probiotic strains for the application in food industry. *World J. Microbiol. Biotechnol.* **2008**, *24*, 699–707. [CrossRef]
9. Lee, S.-J.; Jeon, H.-S.; Yoo, J.-Y.; Kim, J.-H. Some Important Metabolites Produced by Lactic Acid Bacteria Originated from Kimchi. *Foods* **2021**, *10*, 2148. [CrossRef]
10. Whiteley, M.; Diggle, S.P.; Greenberg, E.P. Progress in and promise of bacterial quorum sensing research. *Nature* **2017**, *551*, 313–320. [CrossRef]
11. Schauder, S.; Shokat, K.; Surette, M.G.; Bassler, B.L. The *LuxS* family of bacterial autoinducers: Biosynthesis of a novel quorum-sensing signal molecule. *Mol. Microbiol.* **2001**, *41*, 463–476. [CrossRef] [PubMed]
12. Zhang, Y.; Yu, H.; Xie, Y.; Guo, Y.; Cheng, Y.; Yao, W. Quorum sensing inhibitory effect of hexanal on Autoinducer-2 (AI-2) and corresponding impacts on biofilm formation and enzyme activity in *Erwinia carotovora* and *Pseudomonas fluorescens* isolated from vegetables. *J. Food Processing Preserv.* **2022**, *46*, e16293. [CrossRef]
13. Pereira, C.S.; Thompson, J.A.; Xavier, K.B. AI-2-mediated signalling in bacteria. *FEMS Microbiol. Rev.* **2013**, *37*, 156–181. [CrossRef] [PubMed]
14. Gu, Y.; Li, B.; Tian, J.; Wu, R.; He, Y. The response of LuxS/AI-2 quorum sensing in *Lactobacillus fermentum* 2-1 to changes in environmental growth conditions. *Ann. Microbiol.* **2018**, *68*, 287–294. [CrossRef]
15. Gu, Y.; Wu, J.; Tian, J.; Li, L.; Zhang, B.; Zhang, Y.; He, Y. Effects of Exogenous Synthetic Autoinducer-2 on Physiological Behaviors and Proteome of Lactic Acid Bacteria. *ACS Omega* **2020**, *5*, 1326–1335. [CrossRef] [PubMed]
16. Gu, Y.; Tian, J.; Zhang, Y.; Wu, R.; Li, L.; Zhang, B.; He, Y. Dissecting signal molecule AI-2 mediated biofilm formation and environmental tolerance in *Lactobacillus plantarum*. *J. Biosci. Bioeng.* **2021**, *131*, 153–160. [CrossRef]
17. Yeo, S.; Park, H.; Ji, Y.; Park, S.; Yang, J.; Lee, J.; Mathara, J.M.; Shin, H.; Holzapfel, W. Influence of gastrointestinal stress on autoinducer-2 activity of two *Lactobacillus* species. *FEMS Microbiol. Ecol.* **2015**, *91*, fiv065. [CrossRef]
18. Man, L.-L.; Meng, X.-C.; Zhao, R.-H.; Xiang, D.-J. The role of plNC8HK-plnD genes in bacteriocin production in Lactobacillus plantarum KLDS1.0391. *Int. Dairy J.* **2014**, *34*, 267–274. [CrossRef]

19. Tannock, G.W.; Ghazally, S.; Walter, J.; Loach, D.; Brooks, H.; Cook, G.; Surette, M.; Simmers, C.; Bremer, P.; Dal Bello, F.; et al. Ecological behavior of *Lactobacillus reuteri* 100-23 is affected by mutation of the *luxS* gene. *Appl. Environ. Microbiol.* **2005**, *71*, 8419–8425. [CrossRef]
20. Lebeer, S.; De Keersmaecker, S.C.; Verhoeven, T.L.; Fadda, A.A.; Marchal, K.; Vanderleyden, J. Functional analysis of *luxS* in the probiotic strain *Lactobacillus rhamnosus* GG reveals a central metabolic role important for growth and biofilm formation. *J. Bacteriol.* **2007**, *189*, 860–871. [CrossRef]
21. Jia, F.F.; Pang, X.H.; Zhu, D.Q.; Zhu, Z.T.; Sun, S.R.; Meng, X.C. Role of the *luxS* gene in bacteriocin biosynthesis by *Lactobacillus plantarum* KLDS1.0391: A proteomic analysis. *Sci. Rep.* **2017**, *7*, 13871. [CrossRef] [PubMed]
22. Sun, L.; Zhang, Y.; Guo, X.; Zhang, L.; Zhang, W.; Man, C.; Jiang, Y. Characterization and transcriptomic basis of biofilm formation by *Lactobacillus plantarum* J26 isolated from traditional fermented dairy products. *LWT* **2020**, *125*, 109333. [CrossRef]
23. Li, J.; Yang, X.; Shi, G.; Chang, J.; Liu, Z.; Zeng, M. Cooperation of lactic acid bacteria regulated by the AI-2/LuxS system involve in the biopreservation of refrigerated shrimp. *Food Res. Int.* **2019**, *120*, 679–687. [CrossRef]
24. Bassler, B.L.; Greenberg, E.P.; Stevens, A.M. Cross-species induction of luminescence in the quorum-sensing bacterium *Vibrio harveyi*. *J. Bacteriol.* **1997**, *179*, 4043–4045. [CrossRef]
25. Léonard, L.; Beji, O.; Arnould, C.; Noirot, E.; Bonnotte, A.; Gharsallaoui, A.; Degraeve, P.; Lherminier, J.; Saurel, R.; Oulahal, N. Preservation of viability and anti-Listeria activity of lactic acid bacteria, *Lactococcus lactis* and *Lactobacillus paracasei*, entrapped in gelling matrices of alginate or alginate/caseinate. *Food Control* **2015**, *47*, 7–19. [CrossRef]
26. Bensch, G.; Rüger, M.; Wassermann, M.; Weinholz, S.; Reichl, U.; Cordes, C. Flow cytometric viability assessment of lactic acid bacteria starter cultures produced by fluidized bed drying. *Appl. Microbiol. Biotechnol.* **2014**, *98*, 4897–4909. [CrossRef] [PubMed]
27. Zhang, X.; Zhang, S.; Shi, Y.; Shen, F.; Wang, H. A new high phenyl lactic acid-yielding *Lactobacillus plantarum* IMAU10124 and a comparative analysis of lactate dehydrogenase gene. *FEMS Microbiol. Lett.* **2014**, *356*, 89–96. [CrossRef]
28. Li, H.; Xie, X.; Li, Y.; Chen, M.; Xue, L.; Wang, J.; Zhang, J.; Wu, S.; Ye, Q.; Zhang, S.; et al. *Pediococcus pentosaceus* IM96 Exerts Protective Effects against Enterohemorrhagic *Escherichia coli* O157:H7 Infection. *In Vivo Foods* **2021**, *10*, 2945. [CrossRef]
29. Kaur, A.; Capalash, N.; Sharma, P. Expression of Meiothermus ruber *luxS* in *E. coli* alters the antibiotic susceptibility and biofilm formation. *Appl. Microbiol. Biotechnol.* **2020**, *104*, 4457–4469. [CrossRef]
30. Monahan, L.G.; Harry, E.J. You Are What You Eat: Metabolic Control of Bacterial Division. *Trends Microbiol.* **2016**, *24*, 181–189. [CrossRef]
31. Rang, U.; Peng, A.; Poon, A.; Chao, L. Ageing in *Escherichia coli* requires damage by an extrinsic agent. *Microbiology* **2012**, *158*, 1553–1559. [CrossRef] [PubMed]
32. Nakaoka, H.; Wakamoto, Y. Aging, mortality, and the fast growth trade-off of *Schizosaccharomyces pombe*. *PLoS Biol.* **2017**, *15*, e2001109. [CrossRef]
33. Rathod, N.B.; Phadke, G.G.; Tabanelli, G.; Mane, A.; Ranveer, R.C.; Pagarkar, A.; Ozogul, F. Recent advances in bio-preservatives impacts of lactic acid bacteria and their metabolites on aquatic food products. *Food Biosci.* **2021**, *44*, 101440. [CrossRef]
34. El-Saber Batiha, G.; Hussein, D.E.; Algammal, A.M.; George, T.T.; Jeandet, P.; Al-Snafi, A.E.; Tiwari, A.; Pagnossa, J.P.; Lima, C.M.; Thorat, N.D.; et al. Application of natural antimicrobials in food preservation: Recent views. *Food Control* **2021**, *126*, 108066. [CrossRef]
35. Shekarforoush, S.S.; Basiri, S.; Ebrahimnejad, H.; Hosseinzadeh, S. Effect of chitosan on spoilage bacteria, *Escherichia coli* and *Listeria monocytogenes* in cured chicken meat. *Int. J. Biol. Macromol.* **2015**, *76*, 303–309. [CrossRef]
36. Hong Tran, D.; Tran, T.; Pham, N.; Phung, H. Direct multiplex recombinase polymerase amplification for rapid detection of *Staphylococcus aureus* and *Pseudomonas aeruginosa* in food. *bioRxiv* **2021**. [CrossRef]
37. Ge, Y.; Zhu, J.; Ye, X.; Yang, Y. Spoilage potential characterization of *Shewanella* and *Pseudomonas* isolated from spoiled large yellow croaker (Pseudosciaena crocea). *Lett. Appl. Microbiol.* **2017**, *64*, 86–93. [CrossRef]
38. Wang, X.-Y.; Xie, J. Quorum Sensing System-Regulated Proteins Affect the Spoilage Potential of Co-cultured *Acinetobacter johnsonii* and *Pseudomonas fluorescens* From Spoiled Bigeye Tuna (*Thunnus obesus*) as Determined by Proteomic Analysis. *Front. Microbiol.* **2020**, *11*, 940. [CrossRef]
39. Flickinger, S.T.; Copeland, M.F.; Downes, E.M.; Braasch, A.T.; Tuson, H.H.; Eun, Y.-J.; Weibel, D.B. Quorum sensing between *Pseudomonas aeruginosa* biofilms accelerates cell growth. *J. Am. Chem. Soc.* **2011**, *133*, 5966–5975. [CrossRef]
40. Jie, J.; Yu, H.; Han, Y.; Liu, Z.; Zeng, M. Acyl-homoserine-lactones receptor LuxR of *Shewanella baltica* involved in the development of microbiota and spoilage of refrigerated shrimp. *J. Food Sci. Technol.* **2018**, *55*, 2795–2800. [CrossRef]
41. Li, X.; Jiang, B.; Pan, B.; Mu, W.; Zhang, T. Purification and partial characterization of *Lactobacillus* species SK007 lactate dehydrogenase (LDH) catalyzing phenylpyruvic acid (PPA) conversion into phenyllactic acid (PLA). *J. Agric. Food Chem.* **2008**, *56*, 2392–2399. [CrossRef]
42. Lin, M.; Zhou, G.-H.; Wang, Z.-G.; Yun, B. Functional analysis of AI-2/LuxS from bacteria in Chinese fermented meat after high nitrate concentration shock. *Eur. Food Res. Technol.* **2015**, *240*, 119–127. [CrossRef]
43. Moslehi-Jenabian, S.; Vogensen, F.K.; Jespersen, L. The quorum sensing *luxS* gene is induced in *Lactobacillus acidophilus* NCFM in response to *Listeria monocytogenes*. *Int. J. Food Microbiol.* **2011**, *149*, 269–273. [CrossRef] [PubMed]
44. Fontecave, M.; Atta, M.; Mulliez, E. S-adenosylmethionine: Nothing goes to waste. *Trends Biochem. Sci.* **2004**, *29*, 243–249. [CrossRef]

45. Chen, H.; Wang, Z.; Cai, H.; Zhou, C. Progress in the microbial production of S-adenosyl-L-methionine. *World J. Microbiol. Biotechnol.* **2016**, *32*, 153. [CrossRef] [PubMed]
46. Kilstrup, M.; Hammer, K.; Ruhdal Jensen, P.; Martinussen, J. Nucleotide metabolism and its control in lactic acid bacteria. *FEMS Microbiol. Rev.* **2005**, *29*, 555–590. [CrossRef] [PubMed]
47. Vilain, S.; Pretorius, J.M.; Theron, J.; Brözel, V.S. DNA as an adhesin: *Bacillus cereus* requires extracellular DNA to form biofilms. *Appl. Environ. Microb* **2009**, *75*, 2861–2868. [CrossRef] [PubMed]
48. Duwat, P.; Ehrlich, S.D.; Gruss, A. Effects of metabolic flux on stress response pathways in *Lactococcus lactis*. *Mol. Microbiol.* **1999**, *31*, 845–858. [CrossRef]
49. Rallu, F.; Gruss, A.; Ehrlich, S.D.; Maguin, E. Acid- and multistress-resistant mutants of *Lactococcus lactis*: Identification of intracellular stress signals. *Mol. Microbiol.* **2000**, *35*, 517–528. [CrossRef]
50. Nicoloff, H.; Hubert, J.-C.; Bringel, F. In *Lactobacillus plantarum*, Carbamoyl Phosphate Is Synthesized by Two Carbamoyl-Phosphate Synthetases (CPS): Carbon Dioxide Differentiates the Arginine-Repressed from the Pyrimidine-Regulated CPS. *J. Bacteriol.* **2000**, *182*, 3416–3422. [CrossRef]
51. Sperandio, B.; Polard, P.; Ehrlich, D.S.; Renault, P.; Guédon, E. Sulfur amino acid metabolism and its control in *Lactococcus lactis* IL1403. *J. Bacteriol.* **2005**, *187*, 3762–3778. [CrossRef] [PubMed]
52. Owen, O.E.; Kalhan, S.C.; Hanson, R.W. The key role of anaplerosis and cataplerosis for citric acid cycle function. *J. Biol. Chem.* **2002**, *277*, 30409–30412. [CrossRef] [PubMed]
53. Dijkstra, A.R.; Alkema, W.; Starrenburg, M.J.; Hugenholtz, J.; van Hijum, S.A.; Bron, P.A. Fermentation-induced variation in heat and oxidative stress phenotypes of *Lactococcus lactis* MG1363 reveals transcriptome signatures for robustness. *Microb Cell Fact.* **2014**, *13*, 148. [CrossRef] [PubMed]
54. Kornberg, H.L. The role and control of the glyoxylate cycle in *Escherichia coli*. *Biochem. J.* **1966**, *99*, 1–11. [CrossRef] [PubMed]
55. Petersen, S.; de Graaf, A.A.; Eggeling, L.; Möllney, M.; Wiechert, W.; Sahm, H. In Vivo Quantification of Parallel and Bidirectional Fluxes in the Anaplerosis of Corynebacterium glutamicum. *J. Biol. Chem.* **2000**, *275*, 35932–35941. [CrossRef] [PubMed]
56. Reis, J.A.; Paula, A.T.; Casarotti, S.N.; Penna, A.L.B. Lactic Acid Bacteria Antimicrobial Compounds: Characteristics and Applications. *Food Eng. Rev.* **2012**, *4*, 124–140. [CrossRef]
57. Yang, X.; Li, J.; Shi, G.; Zeng, M.; Liu, Z. Improving 3-phenyllactic acid production of *Lactobacillus plantarum* AB-1 by enhancing its quorum-sensing capacity. *J. Food Sci. Technol.* **2019**, *56*, 2605–2610. [CrossRef]
58. Feldman-Salit, A.; Hering, S.; Messiha, H.L.; Veith, N.; Cojocaru, V.; Sieg, A.; Westerhoff, H.V.; Kreikemeyer, B.; Wade, R.C.; Fiedler, T. Regulation of the activity of lactate dehydrogenases from four lactic acid bacteria. *J. Biol. Chem.* **2013**, *288*, 21295–21306. [CrossRef]
59. Rajanikar, R.V.; Nataraj, B.H.; Naithani, H.; Ali, S.A.; Panjagari, N.R.; Behare, P.V. Phenyllactic acid: A green compound for food biopreservation. *Food Control* **2021**, *128*, 108184. [CrossRef]
60. Jia, J.; Mu, W.; Zhang, T.; Jiang, B. Bioconversion of Phenylpyruvate to Phenyllactate: Gene Cloning, Expression, and Enzymatic Characterization of d-and l 1-Lactate Dehydrogenases from *Lactobacillus plantarum* SK002. *Appl. Biochem. Biotechnol.* **2010**, *162*, 242–251. [CrossRef]
61. Stoll, R.; Goebel, W. The major PEP-phosphotransferase systems (PTSs) for glucose, mannose and cellobiose of *Listeria monocytogenes*, and their significance for extra- and intracellular growth. *Microbiology* **2010**, *156*, 1069–1083. [CrossRef] [PubMed]
62. Zhu, J.; Thompson, C.B. Metabolic regulation of cell growth and proliferation. *Nat. Rev. Mol. Cell Biol.* **2019**, *20*, 436–450. [CrossRef] [PubMed]

Article

Inhibitory Mechanism of Baicalein on Acetylcholinesterase: Inhibitory Interaction, Conformational Change, and Computational Simulation

Yijing Liao [1,2], Xing Hu [1], Junhui Pan [1] and Guowen Zhang [1,*]

1 State Key Laboratory of Food Science and Technology, Nanchang University, Nanchang 330047, China; yijingliao@ncu.edu.cn (Y.L.); hx0726@ncu.edu.cn (X.H.); panjunhui@ncu.edu.cn (J.P.)
2 School of Pharmacy, Nanchang University, Nanchang 330006, China
* Correspondence: gwzhang@ncu.edu.cn

Abstract: Alzheimer's disease (AD) is the most prevalent chronic neurodegenerative disease in elderly individuals, causing dementia. Acetylcholinesterase (AChE) is regarded as one of the most popular drug targets for AD. Herbal secondary metabolites are frequently cited as a major source of AChE inhibitors. In the current study, baicalein, a typical bioactive flavonoid, was found to inhibit AChE competitively, with an associated IC_{50} value of 6.42 ± 0.07 µM, through a monophasic kinetic process. The AChE fluorescence quenching by baicalein was a static process. The binding constant between baicalein and AChE was an order of magnitude of 10^4 L mol^{-1}, and hydrogen bonding and hydrophobic interaction were the major forces for forming the baicalein−AChE complex. Circular dichroism analysis revealed that baicalein caused the AChE structure to shrink and increased its surface hydrophobicity by increasing the α-helix and β-turn contents and decreasing the β-sheet and random coil structure content. Molecular docking revealed that baicalein predominated at the active site of AChE, likely tightening the gorge entrance and preventing the substrate from entering and binding with the enzyme, resulting in AChE inhibition. The preceding findings were confirmed by molecular dynamics simulation. The current study provides an insight into the molecular-level mechanism of baicalein interaction with AChE, which may offer new ideas for the research and development of anti-AD functional foods and drugs.

Keywords: baicalein; acetylcholinesterase; inhibitory mechanism; conformational change; molecular docking; molecular dynamics simulation

Citation: Liao, Y.; Hu, X.; Pan, J.; Zhang, G. Inhibitory Mechanism of Baicalein on Acetylcholinesterase: Inhibitory Interaction, Conformational Change, and Computational Simulation. *Foods* **2022**, *11*, 168. https://doi.org/10.3390/foods11020168

Academic Editor: Gian Carlo Tenore

Received: 1 December 2021
Accepted: 4 January 2022
Published: 10 January 2022

Publisher's Note: MDPI stays neutral with regard to jurisdictional claims in published maps and institutional affiliations.

Copyright: © 2022 by the authors. Licensee MDPI, Basel, Switzerland. This article is an open access article distributed under the terms and conditions of the Creative Commons Attribution (CC BY) license (https://creativecommons.org/licenses/by/4.0/).

1. Introduction

Alzheimer's disease (AD), a progressive neurodegenerative disease featuring hidden onset, is clinically the most prevalent dementia form, which manifests as cognitive impairment, such as memory loss, out-of-control behaviour, and emotional disorder [1,2]. According to Alzheimer's Disease International's data in September 2021, the number of AD patients worldwide is 55 million. With the increase in the ageing population, the number of AD patients is estimated to reach 78 million by 2030 and 139 million by 2050, with a new AD case appearing every 3 s [3]. AD has become a major challenge impacting the health of older adults and has become a public health concern [4].

The aetiology of AD is currently unknown [5]. Scholars both at home and abroad have proposed multiple hypotheses about its pathogenesis, among which the "cholinergic hypothesis" is the most developed and well-known [6–8]. Degeneration of basal cholinergic neurons in the forebrain and loss of associated cholinergic neurotransmitters in the cerebral cortex and other regions lead to the degradation of cognitive function in AD patients [9]. Acetylcholinesterase (AChE, Acetylcholinesterase, E.C.3.1.1.7) is a serine protease that catalyses the hydrolysis of acetylcholine (ACh), an essential neurotransmitter in vivo, and decomposes it into acetic acid and choline, thus blocking nerve impulse transmission [10,11].

According to research, AD patients lack the cholinergic nerve cells in the basal forebrain and thus exhibit enhanced AChE activity and decreased acetylcholine content [12]. As a result, acetylcholinesterase inhibitors targeting AChE are mainly used to treat AD in the clinic, according to the "cholinergic hypothesis." Four of the five anti-AD drugs approved by the FDA include AChE inhibitors (namely tacrine, donepezil, galantamine, and rivastigmine, Figure 1) [13,14]. Tacrine, however, was withdrawn in 2013 owing to severe hepatotoxicity and other side effects [15]. The other three clinically utilised drugs have varying side effects and can only treat mild-to-moderate AD symptoms; therefore, in-depth research is required to achieve satisfactory therapeutic results [16]. The failure of clinical AD drug research teaches us that drug selection must be based on both safety and targeting, and screening AD inhibitors from natural products is an effective strategy [17].

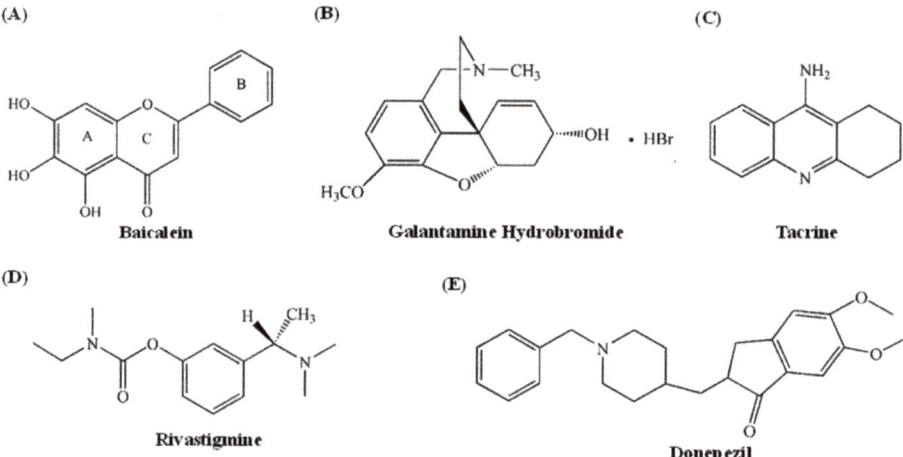

Figure 1. Structure of baicalein (A) and FDA-approved AChE inhibitors (B–E).

Foods high in flavonoids have been found to reduce the neurodegenerative pathological process of AD animal models and improve cognitive function [18]. Epidemiological studies have also demonstrated that eating a substantial amount of foods rich in flavonoids can improve normal cognitive function and reduce the incidence of dementia in the local population [19]. Flavonoids are highly effective in alleviating reversible neurodegenerative pathological processes and improving age-related cognitive decline [20] and thus have become a hot spot in the research and development of anti-AD drugs and functional food factors both at home and abroad.

Baicalein (5,6,7-trihydroxyflavone, as shown in Figure 1A) is an effective active component monomer of flavonoids extracted from the root of *Scutellaria baicalensis* Georgi [21], and it exerts numerous pharmacological effects, including antioxidant [22], scavenging oxygen free radicals [23], anti-inflammatory [24], antimicrobial [25], anti-carcinogenic [26] and neuroprotective effects [27]. The effects of baicalein on alleviating AD and improving cognition and brain protection have been reported [28–30]. Baicalein can activate the PI3K pathway, inhibit the levels of GSK3β and BACE1, reduce the content of total Aβ, and play a neuroprotective role in mice, thereby enhancing memory [31]. Wei [32] confirmed that baicalein can effectively alleviate cognitive impairment induced by Aβ $_{1-40}$ in rats and change the protein expression level in the cerebral cortex and hippocampus. Studies by Han [33] and other scholars have shown that baicalein is a selective and specific inhibitor of BACE and AChE that can be utilised to prevent and treat AD. Xie [34] investigated the inhibitory effect of 20 flavonoids on AChE and explored their structure−activity relationship in the preliminary stages. However, the detailed molecular mechanism of baicalein against AChE is yet to be explored.

In the current study, UV–Vis absorption, fluorescence, and circular dichroism (CD) spectroscopy were used to investigate the inhibitory activity, inhibitory kinetics, binding properties, number of binding sites, type of acting force, and effect on the conformation of AChE by baicalein. A PC12 cytotoxicity test was performed to assess the safety of baicalein. For the interaction of baicalein with AChE, the molecular simulation was utilised to predict binding sites, binding conditions, and major amino acid residues. Molecular dynamics (M.D.) simulation was used to investigate the stability of protein skeleton, peptide flexibility, and free energy of amino acid residues before and after the binding of baicalein with AChE. The findings may provide novel insights into the mechanism of baicalein in AChE inhibition and serve as a reference for the research and development of flavonoids, such as baicalein, as anti-AD food functional factors and drugs.

2. Materials and Methods

2.1. Materials

AChE (137 U/mg, Electrophorus electricus) was provided by Sigma-Aldrich Co. (St. Louis, MO, USA), and their stock solutions with 10 U/mL concentrations were prepared with 0.1 M phosphate-buffered saline (pH 7.6) and stored at −20 °C. Baicalein (purity ≥ 98%) was obtained from the National Institute for the Control of Pharmaceutical and Biological Products (Beijing, China) and dissolved in ethanol to prepare the stock solution (5 mM). 5,5-Dithiobis-(2-nitrobenzoic acid) (DTNB), acetylthiocholine iodide (ATCI), galathamine hydrobromide, and 3-(4, 5-dimethylthiazol-2-yl)-2,5-diphenyltetrazolium bromide (MTT) were provided by Aladdin Chemistry Co., Ltd. (Shanghai, China). Foetal bovine serum (FBS) was purchased from Gibco (Carlsbad, CA, USA). Dulbecco's modified Eagle's medium (DMEM), trypsin, and other cell culture reagents were purchased from Solarbio (Beijing, China). All remaining reagents were of analytical grade.

2.2. Enzyme Activity Assay

AChE activity was evaluated per Ellman's method, albeit with a slight modification [35]. Quartz cuvettes (3 mL) were utilised for measuring the assay mixture activity. Solutions containing varying amounts of baicalein, phosphate-buffered saline (0.1 M, pH 7.6), and 50 µL of AChE (2 U/mL, 3.4×10^{-8} M) were cultivated in 30 min at 25 °C. After the pre-incubation period, 50 µL of 5, 5-dithiobis-(2-nitrobenzoic) acid (DNTB, 5 mM) was added. The reaction was then initiated with 50 µL of acetylthiocholine iodide (ATCI, 15 mM). The absorbance at 412 nm was monitored every 5 s with the TU−1901 dual-beam UV–Vis Spectrophotometer (Persee; Beijing, China). The samples were measured thrice, and the average value was considered. Enzyme activity without baicalein was identified to be 100%. Herein, the measured half-inhibitory concentration (IC_{50}) for baicalein and the relative enzymatic activity agreed with those reported previously [36]. Galantamine hydrobromide served as the positive control.

2.3. Kinetic Analysis of the Inhibitory Type

To assess the kinetic mode of AChE inhibition displayed by baicalein, the inhibitory effect was determined with four different concentrations of baicalein (0, 4, 8, and 10 µM) when different substrate concentrations of ATCI coexisted in the same way of as the AChE activity assay. The Lineweaver–Burk plots and Michaelis–Menten enzyme kinetics were applied to analyse the inhibitory type. For competitive type inhibition, the dynamic double reciprocal equation was used as follows [37]:

$$\frac{1}{v} = \frac{K_m}{v_{\max}}\left(1 + \frac{(I)}{K_i}\right)\frac{1}{(S)} + \frac{1}{v_{\max}} \qquad (1)$$

$$K_m^{\text{app}} = \frac{K_m(I)}{K_i} + K_m \qquad (2)$$

where (I) and (S) denote the inhibitor and substrate concentrations, respectively; v represents the enzyme reaction rate irrespective of the presence of baicalein; v_{max} is the maximum catalytic reaction rate; K_i, K_m, and K_m^{app} stand for the inhibition constant, Michaelis−Menten constant, and apparent Michaelis constant, respectively.

The kinetic process and rate constants for enzyme inactivation were determined through the time-course test. Baicalein with concentrations of 2, 4, 8, and 16 µM was used for analysing the time-course. Enzyme activity at each concentration was measured every 3 min for 0–30 min, every 6 min for 30–60 min, and then at 70 and 80 min.

2.4. Fluorescence Spectrum Measurement

A spectrofluorometer (model F-7000; Hitachi, Tokyo, Japan) containing a thermostat bath was used to measure the fluorescence spectra. The AChE solution (2.5 mL; 0.17 µM) was introduced into a 1.0 cm path length quartz cell, followed by titration through the successive addition of 2.0 mM of diluted baicalein solution (to obtain the concentration range of 0–16 µM). The well-mixed solutions were set aside for three min for equilibration purposes. The fluorescence spectra were measured in the 290–450 nm range, and the relevant excitation wavelength set was 280 nm at 25 °C, 31°C, and 37 °C. Excitation and emission bandwidths were 2.5 nm. In addition, synchronous fluorescence spectra for AChE with baicalein were conducted by fixing the interval of excitation and emission wavelength constants at 15 nm and 60 nm. Fluorescence data were amended according to our previous reports to erase the re-absorption possibility and the inner filter effects of UV absorption [38].

2.5. CD Spectra Measurements

Far-UV CD spectra were measured using a spectrometer (MOS 450 CD; Bio-Logic, Claix, France) with a 1.0 mm path length quartz cuvette. All the spectra were measured in the 200–240 nm range at 25 °C, and the associated scan speed was 60 nm min^{-1}. The AChE in the PBS was completely blended with baicalein at varying concentrations and then equilibrated for three min before CD measurements. The spectra were amended by subtracting the spectrum of the buffer with that of baicalein at the same concentration as the sample solution. Three scans were conducted, and the ellipticity in millidegrees indicated the results. The contents of the discrepant secondary structures (α-helix, β-sheet, β-turn, and random coil) from AChE were measured based on CD spectroscopic data following the online SELCON3 program (http://dichroweb.cryst.bbk.ac.uk/html/home.shtml. Accessed on: 18 August 2021) [39].

2.6. Cytotoxicity of Baicalein in PC12 Cells

PC12 cells were purchased from the NanJing KeyGen Biotech Co. Ltd. (Nanjing, China). These cells were cultivated in high-glucose DMEM supplemented with 10% FBS, 100 U/mL of penicillin and 100 µg/mL of streptomycin at 37 °C under a humidified 5% CO_2 atmosphere. Baicalein was dissolved in DMSO to prepare a 5 mM stock solution and diluted with serum-free media before use. PC12 cells were seeded into a 96-well plate at the density of 2×10^4 cells/well (100 µL) and cultured for 24 h. Next, the medium was changed, and the cells were treated with baicalein at 0, 1, 10, 20, 40, and 80 µM. Each concentration had six replicate wells, and the culture was continued for 24 h. To determine the effect of baicalein on cell viability, 20 µL of MTT (5 mg/mL) was added to each well. After 4 h, the resulting formazan crystals were dissolved by adding 150 µL of DMSO for 10–15 min. The absorption (O.D.) of each hole at 570 nm wavelength was measured using an enzyme-labelled instrument (Varioskan LUX; Thermo Fisher Scientific, Waltham, MA, USA). Cell viability was expressed as the percentage of the control.

2.7. Molecular Docking

Molecular docking was adopted for visualising the interaction of baicalein with AChE through the Discovery Studio 4.5 program (neotrident, Beijing, China) [40]. The crystallographic structure of the donepezil–AChE complex (PDB ID: 4EY7) was acquired through the RCSB Protein Data Bank (http://www.rcsb.org/. Accessed on: 28 September 2021) [41]. According to the special tetramer structure in AChE, Chain A was selected for simulation optimisation, and hydrogenation and polarity addition were conducted. The 3D structure of baicalein was plotted using the Chem3D Ultra 8.0 (Cambridge Soft, Waltham, MA, USA), and molecular optimisation was performed to obtain the conformation with minimum energy. Then, the CHARMm force field algorithm was used, wherein AChE and baicalein served as the receptor and ligand, respectively. Therefore, to validate the accuracy of docking, the co-crystallised ligand donepezil was re-docked into the active site of AChE. The molecular docking was performed using the CDOCKER algorithm, with 100 running times, and the docking tolerance was set to 0.25 Å. The best binding pose with the lowest energy was selected for examining the interplay of baicalein with AChE.

2.8. M.D. Analysis

M.D. simulation was conducted using the GROMACS 4.5.6 software, with the AMBER 99SB-ILDN force field [42,43]. The PDB file for AChE was complied with the PDB file for molecular docking. The AChE– and AChE–baicalein complex systems were solvated using the explicit SPC solvent model [44] and then placed within a dodecahedron box full of water molecules, ensuring a minimum of the 1 nm distance between all protein atoms and the box wall. The system surface charges were neutralised by adding 18 Na^+ and nine Cl^-, followed by energy minimisation. The minimised system was equilibrated through NVT at 300 K and one bar for 100 ps, whereas M.D. simulation was conducted by NPT ensemble. Finally, the running time of M.D. simulations was set to 60 ns [45].

2.9. Statistical Analysis

All samples were tested in triplicate, and the results are indicated as mean ± standard deviation. Data analysis was performed with a one-way analysis of variance in Origin 9.0 (Origin Lab, Northampton, MA, USA), and a *p*-value of <0.05 was considered statistically significant.

3. Results and Discussion

3.1. Inhibitory Effect of Baicalein on AChE

Baicalein and galanthamine hydrobromide (Figure 1) significantly prohibited AChE activity in a dose-dependent manner (Figure 2A), with IC_{50} values of 6.42 ± 0.07 µM and 0.29 ± 0.02 µM, respectively. The AChE activity was almost completely inhibited by galanthamine. However, AChE activity inhibition by baicalein displayed a decreasing trend. When baicalein reached a concentration of 26.7 µM, the final inhibition rate was 90%. Although baicalein inhibited AChE less effectively than galanthamine, it was recognised as a compound with potential bioactivity that might prevent both Aβ aggregation and amyloid fibril plaque formation [33]. Moreover, baicalein demonstrated a lower IC_{50} value than baicalin (204.1 ± 16.5 µM), a flavone glycoside of baicalein. In vitro glycosylation of dietary flavonoids decreased plasma protein affinity [46]. Flavone glycosylation also altered the distribution and density of the electronic cloud among the rings, causing significant steric hindrance [47]. Thus, baicalein's inhibitory activity was substantially greater than its glycosylation activity. It may be deduced that glycosylation reduces AChE inhibitory activity, and the conclusion was reached in our prior study [40]. Furthermore, the hydroxyl groups in the A ring may play a key role in AChE inhibition by baicalein, thereby supporting the docking results, suggesting that the hydroxyl groups of flavonoids form hydrogen bonds containing AChE active site residues [48].

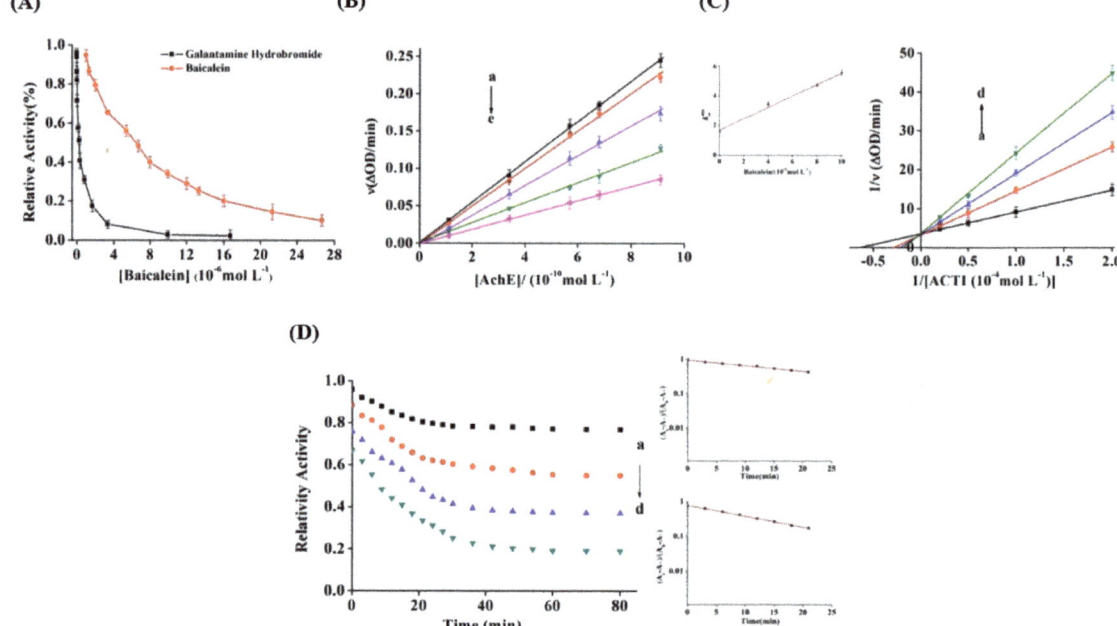

Figure 2. (**A**) Effect of baicalein and galantamine hydrobromide on the activity of AChE at 25 °C (pH 7.6); c(AChE) = 0.57 nM, c(ACTI) = 0.25 mM. (**B**) Plots of v versus (AChE). c(ACTI) = 0.25 mM and c (baicalein) = 0, 2, 4, 8, and 16 μM for curves a→e, respectively. (**C**) Lineweaver–Burk curve of baicalein, c(AChE) = 0.57 nM; c (baicalein) = 0, 4, 8, and 10 μM for curves a→d. (**D**) Time course for the relative activity of AChE in the presence of baicalein at the concentrations of 2, 4, 8, and 16 μM for curves a→d, respectively. c(AChE) = 0.57 nM and c(ACTI) = 0.25 mM. Semilogarithmic plot analysis for baicalein at 2 μM (the upper-right panel) and 16 μM (the lower-right panel), and the slope of the curves suggests the inactivation rate constants.

The solvent and temperature influencing factor tests were performed to optimise the experimental conditions, considering the impact of solvent and temperature on AChE activity. No evidence of AChE inactivation was available in the ethanol concentration range of 3%, as shown in Figure S1 (supplementary data). The enzyme activity revealed a prolonged inactivation trend with an increase in the solvent concentration. Thus, the final ethanol concentration in the reaction system was regulated by 3%. The activity of AChE slowly increased between 19–37 °C and was relatively stable between 37–43 °C, whereas it decreased rapidly at >43 °C temperature (Figure S2). Baicalein is a yellow, reductive component that darkens during incubation. With an increase in the incubation temperature, the colour darkens more noticeably, interfering with the UV measurement. Therefore, the incubation temperature was adjusted to room temperature (25 °C).

3.2. Inhibition Kinetics of Baicalein

The plots for v vs. AChE concentration with varying concentrations of baicalein were drawn to assess the inhibition reversibility against AChE (Figure 2B). When baicalein was added, the straight lines through the origin exhibited favourable linearity, and the slope progressively reduced, indicating that the baicalein-induced inhibition was reversible, and the binding mode was noncovalent interaction between baicalein and AChE [49]. Lineweaver–Burk plots were used to determine the inhibitory type and kinetic parameters of baicalein against AChE. The AChE activity is shown in Figure 2C at varying ATCI concentrations with varying baicalein levels. The lines of double reciprocal plots were intersected on the Y axis, which indicated a competitive inhibition, similar to chrysin's impact on xanthine oxidase [50]. The secondary plot for K_m^{app} vs. (I) (insert Figure 2C) demonstrates linear fitting, assuming a single inhibition site or a single class of inhibition sites [37]. The K_m and K_i values obtained from Equations (1) and (2) were 1.674 and (4.32 ± 0.26) µM, respectively.

3.3. Inactivation Kinetics and Rate Constants

The time courses for AChE inhibition with varying concentrations of baicalein were examined in 80 min to acquire the inactivation kinetics and rate constants (Figure 2D). The figure shows that the inhibition of AChE by baicalein was time dependent. The enzyme activity reduced rapidly in the first 20 min and then gradually became steady after 40 min. Thereafter, the enzyme activity did not change dramatically. Subsequent semilogarithmic plot analysis revealed that the baicalein-induced inactivation was a monophasic first-order process with no intermediates. When the baicalein concentration ranged between 2 µM and 16 µM, the rate constants (k) of inactivation increased from $(1.48 \pm 0.03) \times 10^{-4}\ s^{-1}$ to $(5.18 \pm 0.08) \times 10^{-4}\ s^{-1}$. By using the equation $\Delta\Delta G° = -R.T.\ \ln k$, the transition free energy change ($\Delta\Delta G°$) was measured as 21.85 kJ mol^{-1} s^{-1}, which reduced to 18.74 kJ mol^{-1} s^{-1} upon addition of baicalein. The variation in $\Delta\Delta G°$ potentially induced enzyme inactivation (Table 1) [51].

Table 1. Inactivation rate constants for AChE with baicalein.

Baicalein (µM)	Inactivation Rate Constant [a] ($\times 10^{-4}\ s^{-1}$) k	Transition Free Energy Change [b] (kJ mol^{-1} s^{-1})
2	1.48 ± 0.03	21.85
4	2.42 ± 0.05	20.63
8	4.01 ± 0.02	19.38
16	5.18 ± 0.08	18.74

[a] k is the first-order rate constant. The values of k were significantly different ($p < 0.05$) from each other; [b] Transition free-energy change, $\Delta\Delta G° = -R.T.\ \ln k$, where k is the time constant of the inactivation reaction.

3.4. Fluorescence Quenching of AChE by Baicalein

The remarkable inhibitory effect of baicalein on AChE indicated that the inhibitor was possibly bound to the enzyme [52]. Fluorescence quenching experiments were conducted to examine the binding property, binding constant, and binding site to further investigate the interactions between baicalein and AChE [53]. As shown in Figure 3A, the maximum fluorescence emission peak of AChE occurred at 343 nm after excitation at 280 nm, and baicalein exhibited no interference with AChE fluorescence since it did not display any signal under the same conditions (Curve l). The AChE fluorescence quenching was conducted by serially adding baicalein with no visible peak shift, implying that baicalein interacts with AChE.

The fluorescence quenching data, including quenching constant (K_{SV}) and bimolecular quenching constant (K_q), were assessed using the known Stern–Volmer equation to elucidate the quenching mechanism for AChE fluorescence through baicalein [54]. The Stern–Volmer plots exhibited favourable linearity at 25 °C, 31 °C, and 37 °C (Figure 3B), indicating the presence of a single quenching type alone. Additionally, the relevant K_{SV} values (shown in Table 2) reduced with an increase in temperature, proving that the fluorescence quenching mechanism through baicalein was static. Moreover, the corresponding K_q values of 10^{13} L mol^{-1} s^{-1} magnitude order were substantially higher than the quenching rate constant for maximum scattering collision (2.0×10^{10} L mol^{-1} s^{-1}) [55]. The results also indicate that the static quenching mode triggered the quenching of baicalein to form a ground-state complex rather than a dynamic collision.

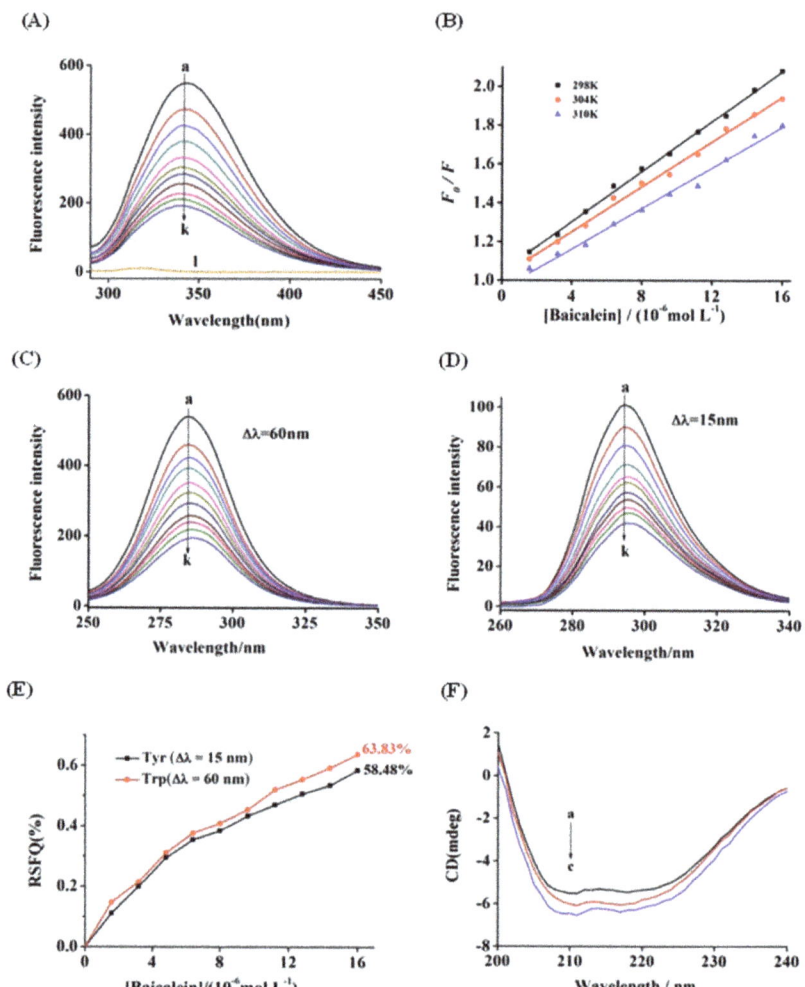

Figure 3. (**A**) Fluorescence spectra of AChE in the presence of baicalein at varying concentrations (pH 8.0, T = 25 °C). c(AChE) = 0.17 µM, c(baicalein) = 0, 1.6, 3.2, 4.8, 6.4, 8.0, 9.6, 11.2, 12.8, 14.4, and 16.0 µM (curves a→k), respectively. Curve l depicts the emission spectrum of baicalein alone. (**B**) The fluorescence quenching curve of baicalein on AChE at 25, 31, and 37 °C. Synchronous fluorescence of AChE when baicalein was added at different concentrations: (**C**) Δλ = 15 nm, (**D**) Δλ = 60 nm; c(AChE) = 0.17 µM, c(baicalein) = 0–16.0 µM for curves a→k. (**E**) Synchronous fluorescence quenching ratio (RSFQ = 1 − F/F_0) of different amino acid residues of AChE. (**F**) The CD spectra of AChE in the presence of baicalein, c(AChE) = 0.75 µM; the molar ratios of baicalein to AChE were 0:1 (a), 2:1 (b), and 6:1 (c).

Table 2. The quenching constants (K_{SV}), binding constants (K_a), and relative thermodynamic parameters for baicalein−AChE interaction under three temperature conditions.

T (°C)	K_{SV} (×10^4 L mol^{-1})	R [a]	K_a (×10^4 L mol^{-1})	R [b]	n	ΔH° (kJ mol^{-1})	ΔG° (kJ mol^{-1})	ΔS° (J mol^{-1} K^{-1})
25	6.45 ± 0.02	0.9989	6.66 ± 0.05	0.9982	0.88 ± 0.01	−17.01 ± 0.02	−22.71 ± 0.06	19.15 ± 0.09
31	5.78 ± 0.03	0.9972	5.84 ± 0.01	0.9979	0.93 ± 0.03		−22.83 ± 0.03	
37	5.22 ± 0.07	0.9946	5.11 ± 0.04	0.9967	1.13 ± 0.02		−22.94 ± 0.08	

R [a] indicates the correlation coefficient of the K_{SV} values; R [b] represents the correlation coefficient of the K_a values.

3.5. Binding Parameters and Thermodynamics

The binding constant (K_a) for the AChE–bacalein complex and the binding site number (n) were measured using Equation (3) [54]:

$$\log \frac{F_0 - F}{F} = n \log K_a - n \log \frac{1}{(Q_t) - \frac{(F_0 - F)(P_t)}{F_0}} \qquad (3)$$

(Q_t) and (P_t) denote the bacalein and AChE concentrations, respectively. The order of magnitude of K_a values was 10^4 L mol^{-1} (Table 2), which indicated that bacalein bound to AChE with a moderate affinity. The n values approached 1, indicating the presence of merely one binding site for bacalein on AChE, which is congruent with the results of the Lineweaver–Burk plot analysis [56].

In general, four noncovalent interaction forces exist between small molecules and macromolecules, namely hydrogen bonds, van der Waals forces, electrostatic forces, and hydrophobic interactions [57]. To identify the binding forces between bacalein and AChE, the thermodynamic parameters were measured following the Van't Hoff equation:

$$\log K_a = -\frac{\Delta H^\circ}{2.303RT} + \frac{\Delta S^\circ}{2.303R} \qquad (4)$$

$$\Delta G^\circ = \Delta H^\circ - T\Delta S^\circ \qquad (5)$$

where K_a denotes the binding constant, and R represents the gas constant (8.314 J mol^{-1} K^{-1}). The temperatures (T) used were 25 °C, 31 °C, and 37 °C. As shown in Table 2, ΔG° and ΔH° showed negative values, which indicated that the interplay of bacalein with AChE was spontaneous, and the binding was exothermic. Furthermore, the positive ΔS° and negative ΔH° values indicated that the hydrogen bonds and hydrophobic interactions exerted main effects on the complex formation [58].

3.6. Synchronous Fluorescence Spectroscopy

For the investigation of AChE-related conformational changes, synchronous fluorescence spectroscopy is used. The fluorescence spectra between 15 and 60 nm at the wavelength interval ($\Delta\lambda$) indicate the fluorescence characteristics for Tyr and Trp residues and can reveal the microenvironment information of the proteases. As shown in Figure 3C,D, the increase in baicalein concentration resulted in the bathochromic shifts from 294 to 295 nm for Tyr residues and 285 to 287 nm for Trp residues of AChE. The findings indicate that the hydrophobicity of Tyr and Trp residues in the milieu was increased and that baicalein triggered a conformational change in AChE and influenced the microenvironment for amino acid residues. Tyr and Trp residues were contrasted in contribution based on their ratios in synchronous fluorescence quenching (RSFQ = $1 - F/F_0$). With an increase in the concentration of baicalein, the RSFQ value for Trp residues reached 63.83%, which was higher than that for Tyr residues (58.48%) (Figure 3E). The current finding implies that the Trp residue contributed more to the interplay between baicalein and AChE than the Tyr residue because Trp residue was closer to the binding site and had more substantial binding power with baicalein [59].

3.7. CD Spectra

CD may provide information on the stereochemistry of a protein-bound drug and protein conformation through secondary structure analysis, consequently revealing information about the binding process [60]. As shown in Figure 3F, two negative CD absorption bands appeared in the AChE spectra at approximately 210 and 222 nm, due to $\pi \to \pi^*$ and $n \to \pi^*$ transitions in amide groups, which have been defined as characteristic bands of the α-helix structure [55]. Moreover, the presence of baicalein increased the negative intensity of the band at 210 nm relative to the intensity of the band at 222 nm, indicating that baicalein interacted with AChE mainly by $\pi \to \pi^*$ transition and altered the conformation of AChE. The findings of thermodynamics analysis show that the hydrophobic interactions and hydrogen bonds were the dominant forces during the interplay of baicalein with AChE. With the increase in the molar ratios of baicalein to AChE from 0:1 to 6:1, the α-helix and β-turn contents increased gradually, whereas the contents of the β-sheet and random coil decreased (Table 3). The current result is consistent with that of Manavalan et al. [61]. The observed increase in the α–helix contents indicated that the AChE structure became increasingly compact, which may have prevented the substrate from accessing the active cavity, resulting in a decrease in the AChE catalytic activity.

Table 3. Secondary structure contents of the baicalein–AChE complex of varying molar ratios.

Molar Ratio (Baicalein):(AChE)	α-Helix (%)	β-Sheet (%)	β-Turn (%)	Random Coil (%)
0:1	34.35 ± 0.74	18.62 ± 0.06	19.18 ± 0.28	28.02 ± 0.65
2:1	38.82 ± 0.08	14.57 ± 0.13	21.42 ± 0.57	25.36 ± 0.29
6:1	42.53 ± 0.35	7.73 ± 0.03	28.15 ± 0.89	21.71 ± 0.47

3.8. Cytotoxicity of Baicalein in PC12 Cells

The cytotoxicity of baicalein in PC12 cells was evaluated using the MTT assay. Figure 4 demonstrates that the viability of the PC12 cells changed slightly under different concentrations of baicalein. However, the change was not statistically significant compared with that of the control group, indicating that baicalein was nontoxic to PC12 cells at the concentration range of 1–80 μM.

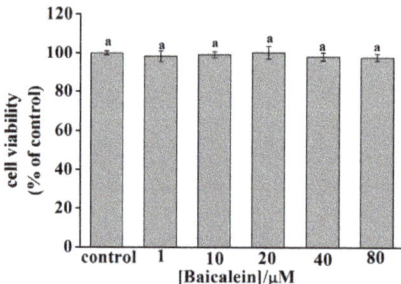

Figure 4. The cytotoxicity of baicalein in PC12 cells. Data are presented as the mean ± SEM of six independent experiments. The same letter "a" indicates no significant difference ($p > 0.05$).

3.9. Molecular Docking

The co-crystallised ligand donepezil was re-docked into the active site of AChE to validate the accuracy of the docking protocol. The outcome is depicted in Figure 5A, and the root mean square deviation (RMSD) between the docking and original co-crystallised pose was 0.844 Å.

The binding position, binding mode, and force type of the interaction between baicalein and AChE were predicted using a molecular simulation technique. According to the X-ray crystal structure analysis, the active site of AChE contains a narrow and deep gorge comprising two binding sites, namely the Ser-His-Glu catalytic site at the gorge bottom and peripheral anion site (PAS) at the gorge entrance.

As shown in Figure 5B, baicalein penetrated the active site of AChE, and the binding pattern was consistent with the competitive inhibition type of baicalein on AChE. Baicalein's lowest binding energy to AChE was −33.26 kJ mol^{-1}, close to the thermodynamic experimental data. Figure 5C,D demonstrate the baicalein binding area and the main amino acid residues interacting with baicalein in AChE in 3D and 2D modes. Baicalein formed three hydrogen bonds with AChE. The C6-hydroxyl group in the A ring of phenylchromen formed a hydrogen bond with the O atom of Trp86 (catalytic site) at the choline-binding site, with the bond length 2.04 Å. The C7-hydroxyl group formed two hydrogen bonds with the oxygen atom of Tyr72 and Asn87 residues, with bond lengths of 3.85 Å and 2.67 Å, respectively. The main phenylchromen had apparent π-π stacking, and hydrophobic interaction with the PAS active site residues Trp86, Tyr124, Phe338, and Tyr337 inhibited the AChE activity by suppressing aromatic inner surface induction. The Val73, Asp74, Tyr341, Pro88, Gln71, Ser125, Leu130, Gly126, Gly121, Gly120, and His447 residues surrounded the whole baicalein molecule by van der Waals forces. These residues were found in either the middle gorge area or the substrate-binding site, where they interacted with baicalein to decrease AChE activity. Furthermore, the PAS locus was linked to the allosteric control of AChE. [62]. Therefore, it is plausible to conclude that baicalein can simultaneously induce the allosteric structure of AChE, causing the gorge entrance to constrict and thereby preventing the combination of the substrate and enzyme active site and eventually inhibiting AChE activity.

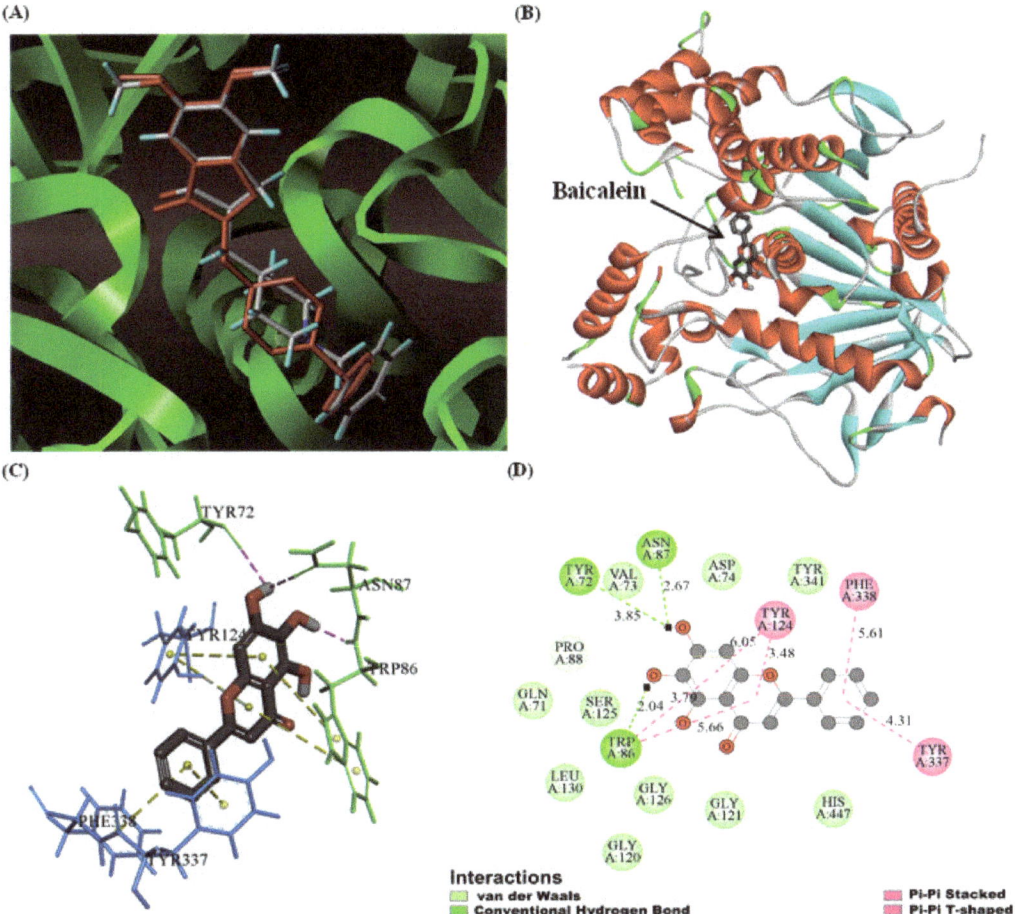

Figure 5. (**A**) Docked conformer of donepezil (white) and original pose of co-crystallised donepezil (red). The results of molecular docking and interaction of baicalein with AChE. (**B**) The 3D ribbon model of AChE (4EY7) docked with the optimal pose of baicalein. (**C**) The binding area of baicalein in AChE, with only the key residues shown. (**D**) The 2D schematic graphs of the main amino acid residues interacting with baicalein in AChE. The interaction is indicated in different colours.

3.10. M.D. Simulation

M.D., a typical computer simulation tool for studying protein stability, can visibly illustrate the dynamic changes of the protein skeleton with time and space. Understanding the link between enzyme structure and function is critical for understanding the inherent flexibility of proteins. As a result, M.D. simulations of the AChE crystal and the bacalein–AChE complex were run in 60 ns.

The findings of the RMSD were used to evaluate the dynamic changes in system stability and determine the time when the system reaches a steady state. The RMSD values for free AChE varied from 0.15 to 0.24 nm and tended to reach equilibrium at 35 ns, as shown in Figure 6A. The RMSD value of the protein skeleton fluctuated slightly in the first 35 ns, and a stable complex was formed slowly through hydrogen bonds or intermolecular forces; the RMSD value of the enzyme became stable after 35 ns. Overall, the RMSD value of the complex did not vary greatly, indicating that the binding of baicalein and AChE would only marginally alter the freedom of protein movement and that the stability of the baicalein–AChE complex was comparable to that of AChE [63].

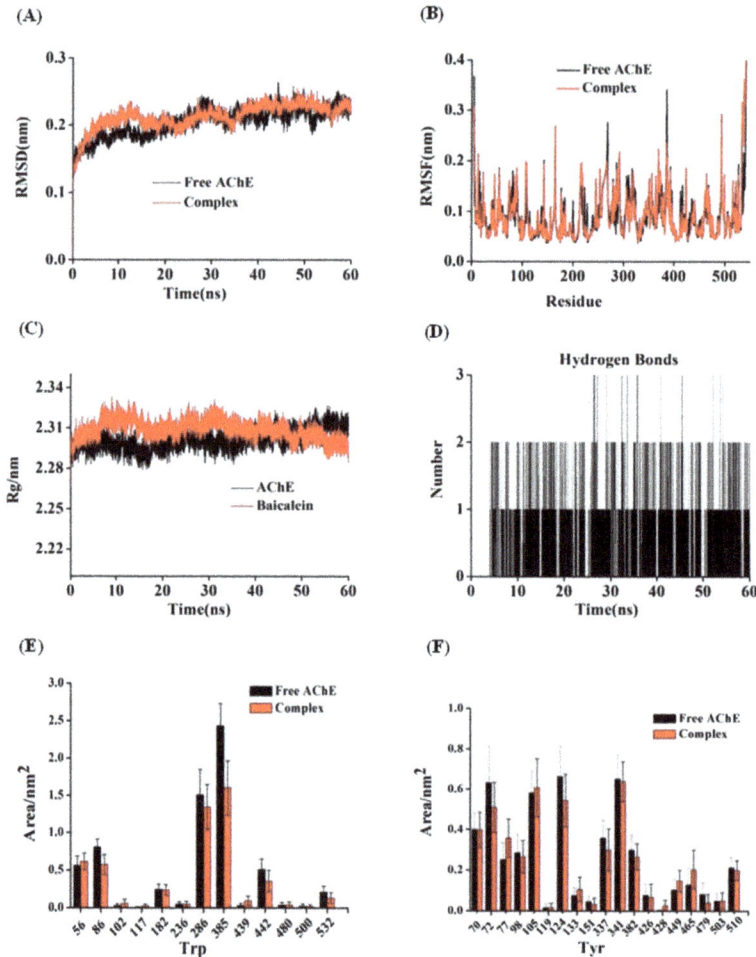

Figure 6. M.D. simulation of baicalein with AChE for 60 ns. The RMSD (**A**) and RMSF (**B**) plots, and Rg values (**C**) of the baicalein–AChE complex and free AChE backbone. (**D**) The number of hydrogen bonds between baicalein with AChE during simulation. The SASA values of the residues Trp (**E**) and Tyr (**F**).

The flexibility of amino acid residues was demonstrated by root mean square fluctuation (RMSF) during the whole simulation duration. High RMSF values of amino acids provide high flexibility during binding. The local amino acid residues of 70–100, 110–150, and 330–380 in AChE showed an obvious fluctuation, possibly due to the participation of amino acid residues at the sites in the allosteric formation of the enzyme to form stable complexes, as is consistent with the main amino acid residues during the interaction with baicalein in molecular docking. The presence of baicalein reduced the RMSF value of AChE residues (as shown in Figure 6B), indicating that baicalein limited the flexibility of the AChE structure, which might be due to the increase in the content of the rigid structure (α-helix) of AChE due to interaction [64].

The radius of gyration (Rg) is an important parameter that reflects the compactness of protein structure [65]. After 45 ns of simulation, the Rg value tended to become stable and slightly lower than that of free AChE (Figure 6C), indicating that baicalein might induce the AChE structure to become more compact. The hydrogen bonds and other forces formed during the bonding process may have rendered the steric structure of AChE more stable, which conformed to the CD analysis results. Figure 6D depicts the variation in the number of hydrogen bonds fabricated in the complex during simulation. The hydrogen bonds varied from 0 to 3, with 1–2 being the most prevalent. The solvent-accessible surface area (SASA) of a system in interaction with solvents indicates surface changes in the system. To further verify the change in the microenvironment hydrophobicity of Trp and Tyr residues, the SASA values of the Trp and Tyr residues were compared before and after stimulation (Figure 6E,F). The SASA of Trp residue decreased noticeably, implying that Trp residues' microenvironment hydrophobicity had improved. The overall SASA for Tyr residues, on the other hand, tended to decline slightly, which indicated a minor increase in the microenvironment hydrophobicity of Tyr residues [40]. The results corresponded with those of synchronous fluorescence research.

4. Conclusions

The present study investigated the inhibitory action of baicalein on AChE and the underlying mechanism at the molecular level by using several spectroscopic and computer simulation techniques. Baicalein was discovered to be a highly effective reversible competitive inhibitor of AChE. Baicalein could statically quench AChE fluorescence. The hydrogen bond and hydrophobic interaction aided the binding between baicalein and AChE. Baicalein bound to AChE at one binding site, and the order of magnitude of the binding constant was 10^4 L mol^{-1}. A cytotoxicity test showed that baicalein did not affect the activity of PC12 cells. The interaction of baicalein with PAS residues (Tyr72, Tyr124, Phe338, and Tyr337) caused an increase in the α-helix content AChE, making the structure of the enzyme more compact and increasing the surface hydrophobicity of AChE. Through hydrogen bonding, hydrophobic interaction, and van der Waals force, baicalein may embed into the active site of AChE to form a relatively stable complex. The binding caused a conformational shift in the enzyme and tightened the structure of the gorge entrance, preventing the substrate from entering and binding with the enzyme's active site, eventually inhibiting AChE activity. M.D. simulation analysis showed that the binding of baicalein with AChE might influence the stability of the protein skeleton. As fluctuating peptides, the 70–100, 110–150, and 330–380 residues might participate in causing conformational changes in the enzyme and forming stable complexes. Furthermore, a slight decrease in the Rg value reflects a moderate compact structure of AChE. Overall, the current study investigated the mechanism of AChE inhibition by baicalein and provided a new direction for the research and development of anti-AD food functional factors and drugs.

Supplementary Materials: The following supporting information can be downloaded at: https://www.mdpi.com/article/10.3390/foods11020168/s1, Figure S1: Effect of ethanol on AChE activity; Figure S2: Effect of temperature on AChE activity.

Author Contributions: Conceptualisation, Y.L. and G.Z.; data curation, Y.L. and X.H.; formal analysis, Y.L. and J.P.; funding acquisition, G.Z.; investigation, Y.L.; methodology, Y.L. and G.Z.; resources, G.Z.; software, Y.L. and X.H.; writing—review and editing, Y.L. and G.Z. All authors have read and agreed to the published version of the manuscript.

Funding: The current work was supported financially by the National Natural Science Foundation of China (22078143), Jiangxi Provincial Natural Science Foundation (20212ACB205010), and Research Project of State Key Laboratory of Food Science and Technology, Nanchang University (SKLF-ZZB−202136 and SKLF-ZZA−201912).

Institutional Review Board Statement: Not applicable.

Informed Consent Statement: Not applicable.

Conflicts of Interest: The authors declare no conflict of interest.

References

1. Scheltens, P.; Blennow, K.; Breteler, M.M.B.; de Strooper, B.; Frisoni, G.B.; Salloway, S.; Van der Flier, W.M. Alzheimer's disease. *Lancet* **2016**, *388*, 505–517. [CrossRef]
2. Goedert, M.; Spillantini, M.G. A Century of Alzheimer's Disease. *Science* **2006**, *314*, 777–781. [CrossRef] [PubMed]
3. Gauthier, S.; Rosa-Neto, P.; Morais, J.A.; Webster, C. *World Alzheimer Report 2021: Journey through the Diagnosis of Dementia*; Alzheimer's Disease International: London, UK, 2021. Available online: https://www.alzint.org/resource/world-alzheimer-report-2021/ (accessed on 12 November 2021).
4. Luo, W.; Wang, T.; Hong, C.; Yang, Y.C.; Chen, Y.; Cen, J.; Xie, S.Q.; Wang, C.J. Design, synthesis and evaluation of 4-dimethylamine flavonoid derivatives as potential multifunctional anti-Alzheimer agents. *Eur. J. Med. Chem.* **2016**, *122*, 17–26. [CrossRef] [PubMed]
5. Jin, X.F.; Wang, M.J.; Shentu, J.Y.; Huang, C.H.; Bai, Y.J.; Pan, H.B.; Zhang, D.F.; Yuan, Z.J.; Zhang, H.; Xiao, X.; et al. Inhibition of acetylcholinesterase activity and beta-amyloid oligomer formation by 6-bromotryptamine A, a multi-target anti-Alzheimer's molecule. *Oncol. Lett.* **2020**, *19*, 1593–1601. [CrossRef] [PubMed]
6. Ashford, J.W. Treatment of Alzheimer's Disease: The Legacy of the Cholinergic Hypothesis, Neuroplasticity, and Future Directions. *J. Alzheimer's Dis.* **2015**, *47*, 149–156. [CrossRef] [PubMed]
7. Hardy, J.; Allsop, D. Amyloici deposition as the central event in the aetiology of Alzheimer's disease. *Trends Pharmacol. Sci.* **1991**, *12*, 383–388. [CrossRef]
8. Bartus, R.T.; Bernard Beer, R.L.; Lippa, A.S. The Cholinergic Hypothesis of Geriatric Memory Dysfunction. *Science* **1982**, *217*, 408–417. [CrossRef] [PubMed]
9. Hampel, H.; Mesulam, M.M.; Cuello, A.C.; Farlow, M.R.; Giacobini, E.; Grossberg, G.T.; Khachaturian, A.S.; Vergallo, A.; Cavedo, E.; Snyder, P.J.; et al. The cholinergic system in the pathophysiology and treatment of Alzheimer's disease. *Brain* **2018**, *141*, 1917–1933. [CrossRef]
10. Xu, Y.C.; Cheng, S.M.; Sussman, J.L.; Silman, I.; Jiang, H.L. Computational Studies on Acetylcholinesterases. *Molecules* **2017**, *22*, 1324. [CrossRef] [PubMed]
11. Sussman, J.L.; Silman, I. Computational studies on cholinesterases: Strengthening our understanding of the integration of structure, dynamics and function. *Neuropharmacology* **2020**, *179*, 108265. [CrossRef]
12. Craig, L.A.; Hong, N.S.; McDonald, R.J. Revisiting the cholinergic hypothesis in the development of Alzheimer's disease. *Neurosci. Biobehav. Rev.* **2011**, *35*, 1397–1409. [CrossRef] [PubMed]
13. Islam, B.U.; Tabrez, S. Management of Alzheimer's disease—An insight of the enzymatic and other novel potential targets. *Int. J. Biol. Macromol.* **2017**, *97*, 700–709. [CrossRef]
14. Woodruff-Pak, D.S.; Lander, C.; Geerts, H. Nicotinic Cholinergic Modulation Galantamine as a Prototype. *CNS Drug Rev.* **2002**, *8*, 405–426. [CrossRef]
15. Sharma, K. Cholinesterase inhibitors as Alzheimer's therapeutics. *Mol. Med. Rep.* **2019**, *20*, 1479–1487. [CrossRef]
16. Birsan, R.I.; Wilde, P.; Waldron, K.W.; Rai, D.K. Anticholinesterase activities of different solvent extracts of Brewer's Spent Grain. *Foods* **2021**, *10*, 930. [CrossRef]
17. Jeon, S.G.; Song, E.J.; Lee, D.; Park, J.; Nam, Y.; Kim, J.I.; Moon, M. Traditional oriental medicines and Alzheimer's disease. *Aging Dis.* **2019**, *10*, 307–328. [CrossRef]
18. Khan, H.; Marya; Amin, S.; Kamal, M.A.; Patel, S. Flavonoids as acetylcholinesterase inhibitors: Current therapeutic standing and future prospects. *Biomed. Pharmacother.* **2018**, *101*, 860–870. [CrossRef]
19. Beking, K.; Vieira, A. Flavonoid intake and disability-adjusted life years due to Alzheimer's and related dementias: A population-based study involving twenty-three developed countries. *Public Health Nutr.* **2010**, *13*, 1403–1409. [CrossRef] [PubMed]
20. Liu, J.Y.H.; Sun, M.Y.Y.; Sommerville, N.; Ngan, M.P.; Ponomarev, E.D.; Lin, G.; Rudd, J.A. Soy flavonoids prevent cognitive deficits induced by intra-gastrointestinal administration of beta-amyloid. *Food Chem. Toxicol.* **2020**, *141*, 111396. [CrossRef] [PubMed]

21. Zhou, L.; Tan, S.; Shan, Y.L.; Wang, Y.G.; Cai, W.; Huang, X.H.; Liao, X.Y.; Li, H.Y.; Zhang, L.; Zhang, B.J.; et al. Baicalein improves behavioral dysfunction induced by Alzheimer's disease in rats. *Neuropsychiatr. Dis. Treat.* **2016**, *12*, 3145–3152. [CrossRef] [PubMed]
22. Sahu, B.D.; Mahesh Kumar, J.; Sistla, R. Baicalein, a bioflavonoid, prevents cisplatin-induced acute kidney injury by up-regulating antioxidant defenses and down-regulating the MAPKs and NF-κB Pathways. *PLoS ONE* **2015**, *10*, e0134139. [CrossRef]
23. Wang, Y.H.; Yu, H.T.; Pu, X.P.; Du, G.H. Baicalein prevents 6-hydroxydopamine-induced mitochondrial dysfunction in SH-SY5Y cells via inhibition of mitochondrial oxidation and up-regulation of DJ-1 protein expression. *Molecules* **2013**, *18*, 14726–14738. [CrossRef]
24. Patwardhan, R.S.; Sharma, D.; Thoh, M.; Checker, R.; Sandur, S.K. Baicalein exhibits anti-inflammatory effects via inhibition of NF-κB transactivation. *Biochem. Pharmacol.* **2016**, *108*, 75–89. [CrossRef] [PubMed]
25. Goc, A.; Niedzwiecki, A.; Rath, M. In vitro evaluation of antibacterial activity of phytochemicals and micronutrients against *Borrelia burgdorferi* and *Borrelia garinii*. *J. Appl. Microbiol.* **2015**, *119*, 1561–1572. [CrossRef] [PubMed]
26. Han, Z.Q.; Zhu, S.M.; Han, X.; Wang, Z.; Wu, S.W.; Zheng, R.S. Baicalein inhibits hepatocellular carcinoma cells through suppressing the expression of CD24. *Int. Immunopharmacol.* **2015**, *29*, 416–422. [CrossRef] [PubMed]
27. Zhang, Z.; Cui, W.; Li, G.; Yuan, S.; Xu, D.; Hoi, M.P.; Lin, Z.; Dou, J.; Han, Y.; Lee, S.M. Baicalein protects against 6-OHDA-induced neurotoxicity through activation of Keap1/Nrf2/HO-1 and involving PKCalpha and PI3K/AKT signaling pathways. *J. Agric. Food Chem.* **2012**, *60*, 8171–8182. [CrossRef]
28. Li, Y.; Zhao, J.; Holscher, C. Therapeutic Potential of Baicalein in Alzheimer's Disease and Parkinson's Disease. *CNS Drugs* **2017**, *31*, 639–652. [CrossRef] [PubMed]
29. Li, C.; Lin, G.; Zuo, Z. Pharmacological effects and pharmacokinetics properties of Radix Scutellariae and its bioactive flavones. *Biopharm. Drug Dispos.* **2011**, *32*, 427–445. [CrossRef]
30. Gao, L.; Li, C.; Yang, R.Y.; Lian, W.W.; Fang, J.S.; Pang, X.C.; Qin, X.M.; Liu, A.L.; Du, G.H. Ameliorative effects of baicalein in MPTP-induced mouse model of Parkinson's disease: A microarray study. *Pharmacol. Biochem. Behav.* **2015**, *133*, 155–163. [CrossRef] [PubMed]
31. Gu, X.H.; Xu, L.J.; Liu, Z.Q.; Wei, B.; Yang, Y.J.; Xu, G.G.; Yin, X.P.; Wang, W. The flavonoid baicalein rescues synaptic plasticity and memory deficits in a mouse model of Alzheimer's disease. *Behav. Brain Res.* **2016**, *311*, 309–321. [CrossRef] [PubMed]
32. Wei, D.F.; Tang, J.F.; Bai, W.G.; Wang, Y.Y.; Zhang, Z.J. Ameliorative effects of baicalein on an amyloid-β induced Alzheimer's disease rat model: A proteomics study. *Curr. Alzheimer Res.* **2014**, *11*, 869–881. [CrossRef] [PubMed]
33. Han, J.; Ji, Y.; Youn, K.; Lim, G.; Lee, J.; Kim, D.H.; Jun, M. Baicalein as a Potential Inhibitor against BACE1 and AChE: Mechanistic Comprehension through In Vitro and Computational Approaches. *Nutrients* **2019**, *11*, 2694. [CrossRef] [PubMed]
34. Xie, Y.; Yang, W.; Chen, X.; Xiao, J. Inhibition of flavonoids on acetylcholine esterase: Binding and structure-activity relationship. *Food Funct.* **2014**, *5*, 2582–2589. [CrossRef] [PubMed]
35. Ellman, G.L.; Courtney, K.D.; Andres, V.J.; Featherstone, J.M. A new and rapid colorimetric determination of acetylcholinesterase activity. *Biochem. Pharmacol.* **1961**, *7*, 88–95. [CrossRef]
36. Ding, H.F.; Wu, X.Q.; Pan, J.H.; Hu, X.; Gong, D.M.; Zhang, G.W. New insights into the inhibition mechanism of betulinic acid on alpha-glucosidase. *J. Agric. Food Chem.* **2018**, *66*, 7065–7075. [CrossRef]
37. Wang, Y.J.; Zhang, G.W.; Yan, J.K.; Gong, D.M. Inhibitory effect of morin on tyrosinase: Insights from spectroscopic and molecular docking studies. *Food Chem.* **2014**, *163*, 226–233. [CrossRef]
38. Zeng, L.; Ding, H.F.; Hu, X.; Zhang, G.W.; Gong, D.M. Galangin inhibits α-glucosidase activity and formation of non-enzymatic glycation products. *Food Chem.* **2019**, *271*, 70–79. [CrossRef]
39. Fan, M.H.; Zhang, G.W.; Hu, X.; Xu, X.M.; Gong, D.M. Quercetin as a tyrosinasm inhibitor: Inhibitory activity, conformational change and mechanism. *Food Res. Int.* **2017**, *100*, 226–233. [CrossRef]
40. Ni, M.T.; Hu, X.; Gong, D.M.; Zhang, G.W. Inhibitory mechanism of vitexin on α-glucosidase and its synergy with acarbose. *Food Hydrocoll.* **2020**, *105*, 105824. [CrossRef]
41. Cheung, J.; Rudolph, M.J.; Burshteyn, F.; Cassidy, M.S.; Gary, E.N.; Love, J.; Franklin, M.C.; Height, J.J. Structures of human acetylcholinesterase in complex with pharmacologically important ligands. *J. Med. Chem.* **2012**, *55*, 10282–10286. [CrossRef]
42. Hornak, V.; Abel, R.; Okur, A.; Strockbine, B.; Roitberg, A.; Simmerling, C. Comparison of multiple Amber force fields and development of improved protein backbone parameters. *Proteins* **2006**, *65*, 712–725. [CrossRef] [PubMed]
43. Lindorff-Larsen, K.; Piana, S.; Palmo, K.; Maragakis, P.; Klepeis, J.L.; Dror, R.O.; Shaw, D.E. Improved side-chain torsion potentials for the Amber ff99SB protein force field. *Proteins* **2010**, *78*, 1950–1958. [CrossRef] [PubMed]
44. Chandar, N.B.; Efremenko, I.; Silman, I.; Martin, J.M.L.; Sussman, J.L. Molecular dynamics simulations of the interaction of Mouse and Torpedo acetylcholinesterase with covalent inhibitors explain their differential reactivity: Implications for drug design. *Chem. Biol. Interact.* **2019**, *310*, 108715. [CrossRef] [PubMed]
45. Song, X.; Hu, X.; Zhang, Y.; Pan, J.H.; Gong, D.M.; Zhang, G.W. Inhibitory mechanism of epicatechin gallate on tyrosinase: Inhibitory interaction, conformational change and computational simulation. *Food Funct.* **2020**, *11*, 4892–4902. [CrossRef] [PubMed]
46. Xiao, J.B.; Cao, H.; Wang, Y.F.; Zhao, J.Y.; Wei, X.L. Glycosylation of dietary flavonoids decreases the affinities for plasma protein. *J. Agric. Food Chem.* **2009**, *57*, 6642–6648. [CrossRef]

47. Cao, H.; Liu, X.J.; Ulrih, N.P.; Sengupta, P.K.; Xiao, J.B. Plasma protein binding of dietary polyphenols to human serum albumin: A high performance affinity chromatography approach. *Food Chem.* **2019**, *270*, 257–263. [CrossRef]
48. Katalinic, M.; Rusak, G.; Domacinovic Barovic, J.; Sinko, G.; Jelic, D.; Antolovic, R.; Kovarik, Z. Structural aspects of flavonoids as inhibitors of human butyrylcholinesterase. *Eur. J. Med. Chem.* **2010**, *45*, 186–192. [CrossRef]
49. Han, L.; Fang, C.; Zhu, R.; Peng, Q.; Li, D.; Wang, M. Inhibitory effect of phloretin on α-glucosidase: Kinetics, interaction mechanism and molecular docking. *Int. J. Biol. Macromol.* **2017**, *95*, 520–527. [CrossRef]
50. Lin, S.Y.; Zhang, G.W.; Liao, Y.J.; Pan, J.H. Inhibition of chrysin on xanthine oxidase activity and its inhibition mechanism. *Int. J. Biol. Macromol.* **2015**, *81*, 274–282. [CrossRef]
51. Jin, Q.X.; Yin, S.J.; Wang, W.; Wang, Z.J.; Yang, J.M.; Qian, G.Y.; Si, Y.X.; Park, Y.D. The effect of Zn^{2+} on Euphausia superba arginine kinase: Unfolding and aggregation studies. *Process Biochem.* **2014**, *49*, 821–829. [CrossRef]
52. Yan, J.K.; Zhang, G.W.; Pan, J.H.; Wang, Y.J. α-Glucosidase inhibition by luteolin: Kinetics, interaction and molecular docking. *Int. J. Biol. Macromol.* **2014**, *64*, 213–223. [CrossRef]
53. Huang, Y.; Wu, P.; Ying, J.; Dong, Z.; Chen, X.D. Mechanistic study on inhibition of porcine pancreatic α-amylase using the flavonoids from dandelion. *Food Chem.* **2021**, *344*, 128610. [CrossRef]
54. Zeng, L.; Zhang, G.W.; Lin, S.Y.; Gong, D.M. Inhibitory mechanism of apigenin on α-glucosidase and synergy analysis of flavonoids. *J. Agric. Food Chem.* **2016**, *64*, 6939–6949. [CrossRef]
55. Danesh, N.; Navaee Sedighi, Z.; Beigoli, S.; Sharifi-Rad, A.; Saberi, M.R.; Chamani, J. Determining the binding site and binding affinity of estradiol to human serum albumin and holo-transferrin: Fluorescence spectroscopic, isothermal titration calorimetry and molecular modeling approaches. *J. Biomol. Struct. Dyn.* **2018**, *36*, 1747–1763. [CrossRef] [PubMed]
56. Chai, W.M.; Lin, M.Z.; Wang, Y.X.; Xu, K.L.; Huang, W.Y.; Pan, D.D.; Zou, Z.R.; Peng, Y.Y. Inhibition of tyrosinase by cherimoya pericarp proanthocyanidins: Structural characterization, inhibitory activity and mechanism. *Food Res. Int.* **2017**, *100*, 731–739. [CrossRef] [PubMed]
57. He, X.M.; Carter, D.C. Atomic structure and chemistry of human serum albumin. *Nature* **1992**, *358*, 209–214. [CrossRef]
58. Dai, T.T.; Chen, J.; McClements, D.J.; Li, T.; Liu, C.M. Investigation the interaction between procyanidin dimer and α-glucosidase: Spectroscopic analyses and molecular docking simulation. *Int. J. Biol. Macromol.* **2019**, *130*, 315–322. [CrossRef]
59. Wu, X.Q.; Ding, H.F.; Hu, X.; Pan, J.H.; Liao, Y.J.; Gong, D.M.; Zhang, G.W. Exploring inhibitory mechanism of gallocatechin gallate on α-amylase and α-glucosidase relevant to postprandial hyperglycemia. *J. Funct. Foods* **2018**, *48*, 200–209. [CrossRef]
60. Bertucci, C.; Pistolozzi, M.; De Simone, A. Circular dichroism in drug discovery and development: An abridged review. *Anal. Bioanal. Chem.* **2010**, *398*, 155–166. [CrossRef]
61. Manavalan, P.; Taylor, P.; Johnson, W.C., Jr. Circular dichroism studies of acetylcholinesterase conformation. Comparison of the 11 S and 5.6 S species and the differences induced by inhibitory ligands. *Biochim. Biophys. Acta* **1985**, *829*, 365–370. [CrossRef]
62. Golicnik, M.; Fournier, D.; Stojan, J. Interaction of drosophila acetylcholinesterases with d-Tubocurarine: An explanation of the activation by an inhibitor. *Biochemistry* **2001**, *40*, 1214–1219. [CrossRef]
63. Zhu, M.; Pan, J.H.; Hu, X.; Zhang, G.W. Epicatechin gallate as xanthine oxidase inhibitor: Inhibitory kinetics, binding characteristics, synergistic inhibition, and action mechanism. *Foods* **2021**, *10*, 2191. [CrossRef] [PubMed]
64. Song, X.; Ni, M.T.; Zhang, Y.; Zhang, G.W.; Pan, J.H.; Gong, D.M. Comparing the inhibitory abilities of epigallocatechin-3-gallate and gallocatechin gallate against tyrosinase and their combined effects with kojic acid. *Food Chem.* **2021**, *349*, 129172. [CrossRef] [PubMed]
65. Sahihi, M.; Ghayeb, Y. An investigation of molecular dynamics simulation and molecular docking: Interaction of citrus flavonoids and bovine beta-lactoglobulin in focus. *Comput. Biol. Med.* **2014**, *51*, 44–50. [CrossRef] [PubMed]

Article

Mechanistic Insights into Biological Activities of Polyphenolic Compounds from Rosemary Obtained by Inverse Molecular Docking

Samo Lešnik [1] and Urban Bren [1,2,*]

[1] Laboratory of Physical Chemistry and Chemical Thermodynamics, Faculty of Chemistry and Chemical Engineering, University of Maribor, Smetanova 17, SI-2000 Maribor, Slovenia; samo.lesnik@um.si
[2] Faculty of Mathematics, Natural Sciences and Information Technologies, University of Primorska, Glagoljaška 8, SI-6000 Koper, Slovenia
* Correspondence: urban.bren@um.si; Tel.: +386-2-2294-421

Abstract: Rosemary (*Rosmarinus officinalis* L.) represents a medicinal plant known for its various health-promoting properties. Its extracts and essential oils exhibit antioxidative, anti-inflammatory, anticarcinogenic, and antimicrobial activities. The main compounds responsible for these effects are the diterpenes carnosic acid, carnosol, and rosmanol, as well as the phenolic acid ester rosmarinic acid. However, surprisingly little is known about the molecular mechanisms responsible for the pharmacological activities of rosemary and its compounds. To discern these mechanisms, we performed a large-scale inverse molecular docking study to identify their potential protein targets. Listed compounds were separately docked into predicted binding sites of all non-redundant holo proteins from the Protein Data Bank and those with the top scores were further examined. We focused on proteins directly related to human health, including human and mammalian proteins as well as proteins from pathogenic bacteria, viruses, and parasites. The observed interactions of rosemary compounds indeed confirm the beforementioned activities, whereas we also identified their potential for anticoagulant and antiparasitic actions. The obtained results were carefully checked against the existing experimental findings from the scientific literature as well as further validated using both redocking procedures and retrospective metrics.

Keywords: rosemary; inverse molecular docking; carnosol; carnosic acid; rosmanol; rosmarinic acid

1. Introduction

Rosemary (*Rosmarinus officinalis* L.), which belongs to the Lamiaceae family, represents an evergreen, perennial, branched shrub that can grow up to three feet tall. It grows fragrant, needle-like, dark green leaves with curved margins and tiny white, pink, purple, or blue flowers [1,2]. The plant is native to the Mediterranean region and its leaves are used extensively in Mediterranean cuisine, mainly as a spice.

Rosemary has been found to possess several bioactive compounds that exert various pharmacological activities, particularly antioxidative [3], anti-inflammatory [4], antidiabetic [5], and antibacterial [6], effects. Moreover, rosemary extracts exhibit promising anticarcinogenic activities in several in vitro [7–9] as well as in vivo studies [10,11].

Carnosic acid (Figure 1a), carnosol (Figure 1b), rosmanol (Figure 1c), and rosmarinic acid (Figure 1d) are most frequently cited in relation to the beneficial pharmacological activities of compounds found in rosemary [12]. Carnosol, carnosic acid, and rosmanol represent polyphenolic diterpenes with similar structures. They consist of the main abietane scaffold, a fused six-membered tricyclic ring system, with one of these rings being aromatic. Carnosic acid represents the major constituent of rosemary and constitutes up to 4% of the dried leaves [13]. However, it is not very stable and, once isolated, undergoes oxidation leading to the formation of the γ-lactone carnosol, which causes it to lose the acidic properties [14]. Oxidation of carnosic acid can alternatively lead to rosmanol, which differs from carnosol in that it has a free hydroxyl group at the C-7 atom and that the γ-lactone is formed via C-20 and C-6 atoms. The three diterpenes form a very effective oxidation cascade, which is vital for the rosemary's potent antioxidative activity. When carnosic acid is oxidized by free radicals, it forms a quinone derivative. This substance can then undergo isomerization, producing carnosol, or a redox reaction, yielding rosmanol. Thus, carnosic acid, while itself a potent antioxidant, can form two additional substances that also exhibit potent antioxidative activities. This mechanism probably represents the main reason behind the extraordinary antioxidative properties of rosemary [15]. Moreover, these compounds also exhibit antibacterial [16], antiviral [17,18], anti-inflammatory [19,20], antiproliferative [7,8,21–28], and antidepressant [29,30] effects. The study by Romo Vaquero, et al. [31] in rats showed that after oral intake, the glucuronide derivatives of these compounds can be found in plasma as early as 25 min after administration, indicating a good bioavailability. Moreover, carnosic acid was also found in the brain tissue of rats, suggesting that it is able to cross the blood–brain barrier, giving credence to a number of studies in which various positive neuroprotective and cognitive effects were established [32,33].

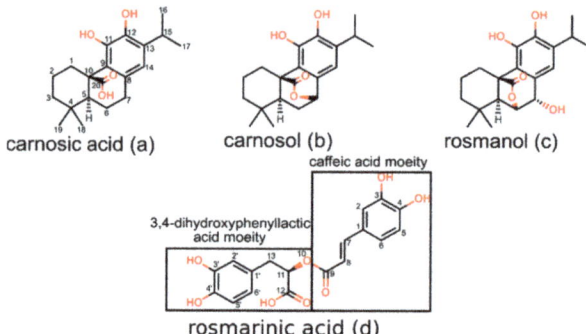

Figure 1. Molecular structures and atom numbering of compounds investigated in the inverse docking study (**a**) molecular structure of carnosic acid, (**b**) carnosol, (**c**) rosmanol, and (**d**) rosmarinic acid.

Rosmarinic acid represents an ester of caffeic acid and 3,4-dihydroxyphenyllactic acid [34]. The structure contains two electroactive catechol moieties that can neutralize free radicals through the electron/proton donor mechanism. Examination of the steps reveals that rosmarinic acid is first oxidized at the caffeic acid moiety of the molecule, while the second step corresponds to the oxidation of the 3,4-dihydroxyphenylic acid moiety. Moreover, the hydroxyl and carboxylic oxygens form a system that exerts good metal chelating properties [35]. Rosmarinic acid can also insert itself into lipid membranes where it effectively inhibits lipid peroxidation [36]. Numerous studies describe that rosmarinic acid exhibits also anti-inflammatory [37,38], antimicrobial [39], anticarcinogenic [7,40,41], and neuroprotective effects [42]. However, unlike the diterpenes in rosemary, the oral bioavailability of rosmarinic acid is poor and amounts to only about 1% in rats [43]. This highlights the need to develop novel delivery systems, such as nanoparticles, to improve the poor pharmacokinetic properties of rosmarinic acid [44,45].

Our aim is to identify potential protein targets of carnosic acid, carnosol, rosmanol, and rosmarinic acid using the inverse docking methodology [46], in which a ligand is docked to a multitude of protein binding sites. The method is typically applied to discover new potential protein targets for small molecule drugs [47] or natural products [48–50] and to explain their mechanisms of action in various diseases. To the best of our knowledge such an investigation has never been performed for the major rosemary compounds.

2. Materials and Methods

2.1. Starting Coordinates of Rosemary Compounds

The initial coordinates of carnosic acid, carnosol, rosmanol, and rosmarinic acid were obtained from the ZINC15 database [51], using ZINC IDs ZINC000003984016, ZINC000003871891, ZINC000031157853, and ZINC0000899870, respectively. Prior to performing inverse molecular docking, all molecules were subjected to a quantum mechanical geometry optimization procedure using the MP2/6-31G* level of theory/basis set combination. This optimization was performed in Gaussian 16 [52].

2.2. In Silico Determination of ADME Properties

In silico determined ADME/Tox profiles provide a useful tool for predicting the pharmacological and toxicological properties of investigated molecules [53]. To provide a more detailed prediction of the pharmacokinetic properties of carnosic acid, carnosol, rosmanol, and rosmarinic acid, which would complement the known experimental data, we implemented the SwissADME web server [54]. SwissADME represents a freely available tool that enables robust predictions of absorption, distribution, metabolism, and extraction, based on the two-dimensional data of the molecule. In addition, it yields predictions on drug-likeness based on well-established metrics.

All compounds were inputted on the SwissADME webpage (http://www.swissadme.ch/ date accessed: 20 December 2021) using the Simplified Molecular-Input Line-Entry System (SMILES) strings.

2.3. Inverse Molecular Docking

Our goal was to gain mechanistic insight into the potential mechanism of pharmacological actions of the investigated rosemary compounds using CANDOCK (Chemical Atomic Network based Docking) [55] inverse molecular docking on more than 65,000 protein structures potentially associated with human pathologies. Protein binding sites for small molecules were obtained from the ProBiS-Dock Database [56]. The main advantage of defining binding sites in this way is that multiple spherical centroids are defined in advance to describe a very accurate 3D shape that can be used in conjunction with the CANDOCK algorithm. Moreover, binding sites at the interface of multiple protein chains are also considered for docking.

For docking, the CANDOCK algorithm applies a hierarchical approach to reconstruct small molecules from the atomic lattice using graph theory, while applying a generalized statistical potential function for scoring. The docking scores represent approximations of the relative binding free energies and are expressed in arbitrary units. Specifically, CANDOCK finds the best-docked poses of small-molecule fragments and applies a fast-maximum-clique algorithm [57] to link them together. In the molecular reconstruction, the algorithm uses iterative dynamics for better placement of the ligand in the binding pocket. After the initial docking and reconnection procedure is completed, a minimization procedure based on the Chemistry at Harvard Macromolecular Mechanics (CHARMM) force field [58] is performed to model the induced fit of the ligand binding to the protein binding site.

2.4. Method Validation

To retrospectively validate our inverse molecular docking procedure, we applied receiver operating characteristic curves (ROC) [59], enrichment curves [60], and predictiveness curves (PC) [61]. Briefly, the ROC metric plot shows a correlation between the true-positive fraction (TPF) on the y-axis and the false-positive fraction (FPF) on the x-axis. In our case, the TPF represents experimentally confirmed protein targets of rosmarinic acid from the ChEMBL database [62] with the corresponding PDB entries, while the FPF represents all other protein targets from the ProBiS-Dock database. We did not perform an analogous validation for diterpenes as only a small number of confirmed targets is available for them. The area under the ROC curve (ROC AUC) represents a simple measure to evaluate the overall performance of the inverse molecular docking method. The larger the ROC AUC, the more effective is the method at discriminating true from false targets. The enrichment curve represents the early quantification of target proteins from the TPF. Moreover, PC also provides the early detection quantification of target proteins from the TPF, but in addition, it can be used to define the threshold for potential targets from the inverse molecular docking to be tested experimentally. Contrary to ROC, PC can describe the dispersion of the inverse docking scores well. To quantify the early detection, we applied the enrichment factor of 1% of the compounds tested (EF) [63], the Boltzmann-enhanced discrimination of ROC (BEDROC) [59], and the robust initial enrichment (RIE) [63] measures as well. Using PC, the standardized total gain (TG) [61] was also determined, which summarizes the contribution of the inverse molecular docking scores in explaining the probability of targets over the entire protein dataset. To calculate all of the listed measures, the Screening Explorer web server [64] was implemented.

3. Results

3.1. Inverse Molecular Docking of Diterpenes

Because of their similar structure and good agreement, the docking results for carnosic acid, carnosol, and rosmanol were combined and analyzed together: the diterpene ligand with the best score for the individual protein was considered. The 0.05% (3.5σ) top scoring proteins from the entire docked database were selected (Figure 2) and among them, those with implications for human health were chosen. Human and mammalian proteins as well as proteins from pathogenic bacteria, viruses, and parasites were considered. Moreover, mammalian proteins were considered in order to increase the protein space available for docking, where we assumed that within the class of mammals, analogous proteins and their binding sites are similar enough so that our findings from non-human mammals are transferable to human proteins.

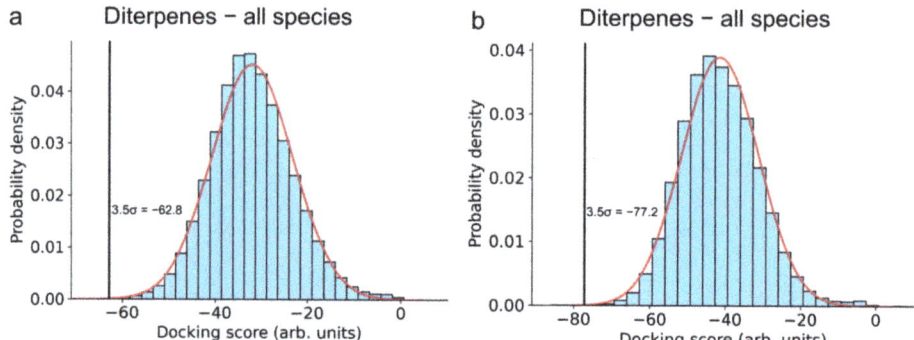

Figure 2. Normal distribution fit of the inverse docking scores. (**a**) Combined distribution of docking scores obtained by inversely docking the rosemary diterpenes carnosic acid, carnosol, and rosmanol to the whole ProBiS-Dock Database. (**b**) Distribution of docking scores for rosmarinic acid. A cut-off criterion of 3.5 σ was used to select the most promising protein–ligand complexes to be further investigated in more detail.

In Table 1, we present the highest-scoring protein–ligand complexes based on the cut-off criterion of 3.5 σ. Moreover, where data were available, we redocked ligands/drugs that are known to bind to the presented targets using an analogous procedure as the one applied for inverse docking. These results, presented in Table S1, show that in all cases except for K-Ras G12C and enhanced intracellular survival protein, the docking scores of the known ligand/drugs are worse than the ones of the rosemary diterpenes. This indicates an already strong binding affinity of the rosemary compounds, although they have not yet been rationally optimized for these specific protein targets.

Table 1. Best scoring human, mammalian, and pathogen protein targets of rosemary diterpenes carnosic acid, carnosol, and rosmanol. Docking scores independent of the organism or type of protein are collected in Supplementary Materials in Table S2.

Rank	PDB ID with Chain	Ligand	Predicted Ligand Docking Score (arb. Units)	Protein Name	Organism	Protein Function and Disease Correlation	Reported Experimental Correlation of Protein and Ligand
1	4lucB	Carnosic acid	−69.9	K-Ras G12C	Homo sapiens	Controls cell proliferation and differentiation. Its gene is a proto-oncogene.	[65]
2	3oojA	Carnosic acid	−68.2	Glucosamine-fructose-6-phosphate aminotransferase	Escherichia coli	Catalyzes the first step in hexosamine metabolism and is needed for E. coli growth and infection spread.	[66–68]
3	3srdD	Carnosic acid	−68.1	Pyruvate kinase M2	Homo sapiens	Catalyzes the last step in the glycolysis. Important in providing ATP to cancer cells.	
4	1kenA	Carnosic acid	−66.9	Hemagglutinin HA1	Influenza A virus	Enables viral entry into cells causing the flu.	
5	2hpeA	Carnosic acid	−65.0	HIV-2 protease	Human immunodeficiency virus 2	Hydrolyzes peptide bonds leading to functional proteins essential for HIV infectivity.	[69]
6	4jd6C	Carnosic acid	−64.8	Enhanced intracellular survival protein	Mycobacterium tuberculosis	Acetylates amine groups in aminoglycoside drugs, thus preventing the binding to the ribosome, leading to M. tuberculosis resistance.	
7	5u46A	Carnosic acid	−64.7	Peroxisome proliferator activated receptor delta	Homo sapiens	Regulates lipid catabolism and its transport and storage and is also associated with insulin secretion and resistance. It is implicated in metabolic disorders and cancer.	γ isoform [70]
8	3mt7A	Carnosic acid	−64.5	Glycogen phosphorylase	Oryctolagus cuniculus	Breaks the non-reducing ends in the chain of glycogen that enables glucose production. Its inhibition can manage type II diabetes.	
9	3rycC	Carnosic acid	−64.2	Tubulin	Rattus norvegicus	Involved in cell division as it forms microtubules which in turn form mitotic spindles that pull chromosomes apart during cell division. Tubulin targeting is used in cancer treatment.	
10	2j9kB	Carnosic acid	−63.5	HIV-1 protease	Human immunodeficiency virus 1	Hydrolyzes peptide bonds leading to functional proteins essential to HIV infectivity.	[69]
11	1fxfB	Carnosol	−63.3	Phospholipase A2	Sus scrofa	Catalyzes the hydrolysis of glycerophospholipids thus releasing free fatty acids, including arachidonic acid. Its action is implicated in several inflammation-based diseases such as arthritis, coronary artery disease, Alzheimer's and cancer.	
12	3ogpA	Carnosic acid	−63.3	FIV Protease	Feline immunodeficiency virus	Hydrolyzes peptide bonds leading to functional proteins essential to FIV infectivity in cats.	
13	2p2hA	Carnosic acid	−63.1	Vascular endothelial growth factor receptor 2	Homo sapiens	Signal protein crucial in angiogenesis. Its inhibition is used in cancer treatment.	Negative: [71]

Table 1. Cont.

Rank	PDB ID with Chain	Ligand	Predicted Ligand Docking Score (arb. Units)	Protein Name	Organism	Protein Function and Disease Correlation	Reported Experimental Correlation of Protein and Ligand
14	5ilqC	Carnosic acid	−63.0	Aspartate carbamoyltransferase	*Plasmodium falciparum*	Enzyme involved in pyrimidine biosynthesis, crucial for *Plasmodium falciparum* (causative agent of malaria) survival and replication.	
15	4iv5A	Carnosic acid	−62.8	Aspartate carbamoyltransferase	*Trypanosoma cruzi*	Enzyme involved in pyrimidine biosynthesis, crucial for *Trypanosoma cruzi* (causative agent of Chagas disease) survival and replication	

3.1.1. K-Ras

K-Ras is a GTPase responsible for relaying signals from outside the cell to the nucleus. It represents a part of the rat sarcoma/mitogen-activated protein kinase (RAS/MAPK) pathway, and K-Ras signaling leads to cell growth, proliferation, and differentiation. K-Ras is of utmost clinical importance as it represents the most frequently mutated oncogene in pancreatic, colon, and lung cancers [72]. Numerous attempts have been made to develop compounds that inhibit the function of K-Ras, but with limited success only [73]. The non-druggability of K-Ras is mainly due to the lack of a well-defined binding pocket, as well as the high affinity for guanosine triphosphate (GTP), with which alternative drug molecules have difficulty competing. Nevertheless, progress has been made in recent years in modulating K-Ras with small-molecule ligands. Fell, et al. [74] developed a potent inhibitor of the oncogenic K-Ras G12C mutant that induces the formation of a new binding pocket near the nucleotide (GTP) binding site (Figure 3a). Binding to this new pocket results in signal inhibition by arresting the enzyme in its inactive state. Interestingly, this induced binding pocket was ranked most favorable of all the protein binding sites tested by our method for carnosic acid (Table 1). Carnosic acid docks at this induced binding site where it forms two hydrogen bonds with Thr58 side chain and two hydrogen bonds to the backbone atoms of Ala59 and Gly60 (Figure 3b, Table S4). A strong salt bridge with a distance of 4.1 Å is additionally created between the carboxylate of carnosic acid and Arg68. Finally, the relatively large hydrophobic ring system of carnosic acid forms hydrophobic interactions with Glu62, Tyr96, and Gln99. Although none of the diterpenes have been previously reported to bind directly to K-Ras, rosemary extracts have indeed been shown to lead to the down-regulation of K-Ras expression in colon cancer cells [75]. This suggests an interesting potential of carnosic acid for a two-pronged attack on the protein by down-regulating its expression and by inhibiting it directly.

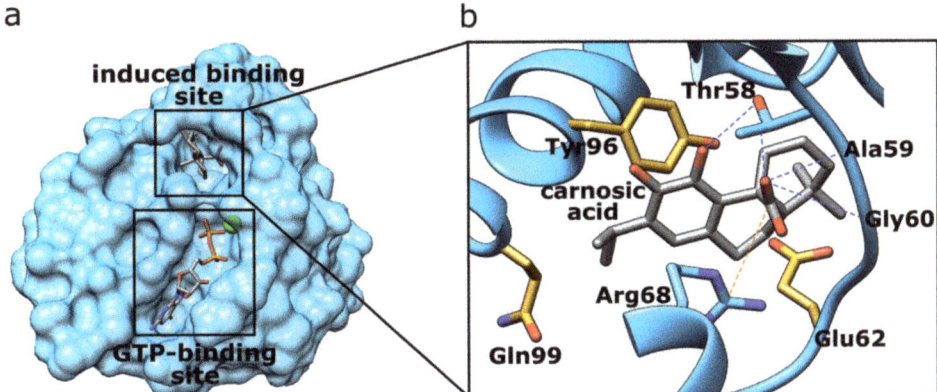

Figure 3. (**a**) K-Ras protein structure highlighting the GTP- and induced-binding site. (**b**) The induced binding site of the K-Ras protein (blue) with docked carnosic acid (carbon atoms depicted in grey). Orange dotted lines represent salt–bridge interactions, and blue dotted lines H-bonding interactions. Amino acid residues forming hydrophobic interactions are denoted with yellow sticks.

3.1.2. Glucosamine/Fructose-6-Phosphate Aminotransferase

In humans, infection with pathogenic strains of *Escherichia coli* leads to various diseases such as gastroenteritis, septic shock, and urinary tract infections. In addition, some strains have been linked to colon cancer because they can synthesize substances that damage DNA [76]. While most *Escherichia coli* infections can be treated with existing antibiotics, such as fluoroquinolones, the proliferation of multidrug-resistant strains produces the need to identify new compounds with antimicrobial activity. Although specific binding of rosemary diterpenes to glucosamine/fructose-6-phosphate aminotransferase (GlmS) is not reported in the scientific literature, a number of studies shows that rosemary compounds indeed exhibit activity against *Escherichia coli* [66–68]. Since no mechanism of this inhibition has yet been reported, we speculate that carnosic acid may bind to GlmS, which catalyzes the first step in hexosamine metabolism by converting fructose-6P to glucosamine-6P using glutamine as a nitrogen source [77], yielding N-acetylglucosamine an essential building block of bacterial cell walls. Therefore, targeting this enzyme could lead to the inhibition of bacterial growth [78]. Predicted interactions between carnosic acid and GlmS are presented in Table S5.

3.1.3. Pyruvate Kinase 2–Muscle Isoform

Cancer cells often rely on glycolysis to meet their high energy demands, whereas normal cells derive most of their energy from oxidative phosphorylation [79]. This difference in cell metabolism can be, therefore, exploited to target cancer cells. The muscle isoform of pyruvate kinase 2 (PKM2) is universally expressed in cancer cells and catalyzes the final step of glycolysis by transferring a phosphate group from phosphoenolpyruvate (PEP) to adenosine diphosphate (ADP), resulting in one molecule of pyruvate and one molecule of adenosine triphosphate (ATP). On the other hand, the remaining isozymes of pyruvate kinase are expressed in most normal tissues, so targeting PKM2 represents a viable way to selectively inhibit glucose metabolism in cancer cells [80]. Carnosic acid binds at the site where variations in two amino acid residues are present compared to PKM1, namely Ile389Met and Gln393Lys (Table S6). These variations result in a significant decrease in docking score as the best PKM1 isoform scores −65.1 A.U compared to −68.1 for the M2 isoform (Table 1), which may indicate that carnosic acid is indeed selective towards PKM2.

3.1.4. Hemagglutinin HA1

Influenza virus hemagglutinin (HA) represents a surface glycoprotein that is critical for viral infectivity. It has multifunctional activity, allowing entry of the virus by binding to sialic acid at the surface of host cells, while also being responsible for the fusion of the viral envelope to the endosomal membrane [81]. Due to its importance, this protein forms a key target for neutralizing antibodies [82]. However, it is also possible to target it with small molecules such as arbidol [83]. Carnosic acid docks to a cavity in the HA trimer stem at the interface between the three protomers. This binding site is separate from the conserved epitope targeted by the neutralizing antibodies. The drug arbidol is known to stabilize the conformation of HA, thereby preventing the large conformational changes required for membrane fusion. This could potentially also be the case with carnosic acid, as it forms three hydrogen bonds, one with each protomer, and could thus act as a so-called molecular glue that binds the protomers together, making them nonfunctional (Figure 4, Table S7).

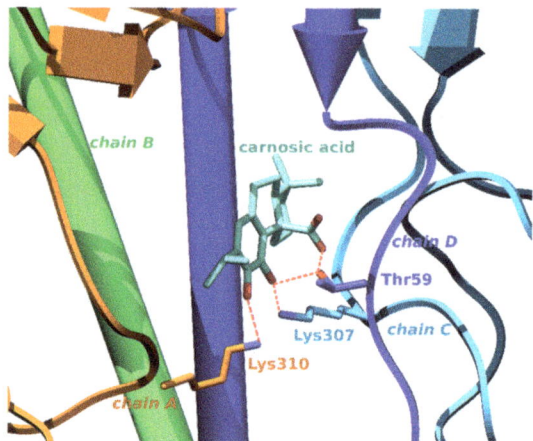

Figure 4. Carnosic acid glues together chains A, C and D of the HA glycoprotein. Carbons of carnosic acid are displayed as teal sticks, chain A in orange, chain B in green, chain C in sky blue and chain D in dark blue pipes and planks. Amino acids forming hydrogen bonds (denoted with red dotted lines) are displayed as sticks of matching colors.

3.1.5. HIV-1 and HIV-2 Protease

Human immunodeficiency viruses (HIV) protease is a retroviral aspartyl protease involved in the hydrolysis of several peptide bonds, which is essential for the life cycle and replication of HIV [84]. Small molecule inhibitors of HIV protease play a critical role in the effective treatment of acquired immunodeficiency syndrome AIDS, as they represent part of the highly active antiretroviral therapy (HAART). While HIV-1, carrier of the HIV-1 protease isoform, forms the most common subtype worldwide, HIV-2 remains mainly confined to West Africa and is also spreading in India [85,86]. However, the treatment of HIV-2 is more difficult than that of HIV-1, as most antiviral drugs have been developed for the HIV-1 isoform. HIV-2 proteases have also been found more resistant to small-molecule inhibition [87]. Moreover, dual infection with both isoforms is possible as well [88]. Consequently, novel inhibitors for both HIV proteases would be of great benefit. It has been shown that carnosic acid exhibits potent inhibition of the HIV-1 protease isoenzyme with an IC_{90} = 0.08 µg/mL [69]. Inhibition has not yet been experimentally demonstrated for the HIV-2 isoform; however, our studies suggest that carnosic acid is also capable of inhibiting this isoform. This finding can also be corroborated by the fact that the binding sites of both isoforms are very similar, with a sequence identity close to 70% and the ProBiS Z-score of 3.76 [58]. ProBiS Z-score measures the statistical and structural significance of local binding site similarity. Binding site alignments leading to ProBiS Z-Scores higher than 2 are considered to be very similar. In addition, all the equivalent binding site amino acid residues are of the same charge and polarity type. Overall, carnosic acid could prove to be a valuable starting point for the development of antivirals that would be effective against both strains of HIV.

Interestingly, the inverse docking results also suggest a high binding ability of carnosic acid to the HIV-related feline immunodeficiency virus (FIV) protease, which causes an AIDS-like syndrome in cats. HIV-2 and FIV proteases possess a binding site similarity of 1.90, expressed by the ProBiS Z-score, and a general sequence similarity of 26%. The relatively different binding sites result in different binding positions of carnosic acid in HIV-2 and FIV proteases. In HIV-2, the ligand is positioned deeper in the major binding site, which is located between the two protomers (Figure 5, Tables S8, S13 and S15).

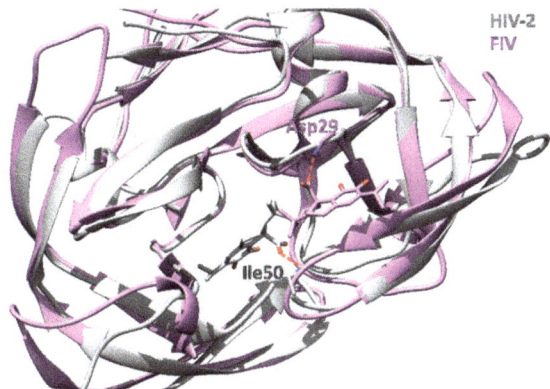

Figure 5. Comparison of HIV-2 (grey cartoon) and FIV (purple cartoon) protease. Carnosic acid in HIV-2 binds deep into the protease binding site and forms a hydrogen bond with the backbone nitrogen of Ile50 (red dotted line). On the other hand, carnosic acid is located closer to the protease surface in FIV and forms a hydrogen bond H-bond with the backbone nitrogen of Asp29 (red dotted line).

3.1.6. Enhanced Intra-Cellular Survival Protein

Tuberculosis represents the leading cause of infectious death worldwide, primarily due to the emergence of multidrug-resistant tuberculosis and due to extensively drug-resistant strains of *Mycobacterium tuberculosis* [89]. Up-regulation of the enhanced intra-cellular survival (Eis) protein was found to be the sole cause of resistance to the aminoglycoside of last resort-kanamycin in approximately one-third of *Mycobacterium tuberculosis* isolates. Specifically, Eis represents an acetyltransferase responsible for *Mycobacterium tuberculosis* resistance to multiple aminoglycoside drugs. A distinctive property of Eis is that it acetylates the aminoglycoside drugs at multiple amine functional groups, preventing them from binding to their target, the ribosome. The simultaneous use of Eis inhibitors with anti-tuberculosis drugs may therefore provide a way to combat this resistance by restoring aminoglycoside drug activity [90]. Carnosic acid docks to the aminoglycoside binding pocket formed by the N-terminal domain to which also tobramycin binds, thereby suggesting the possibility of competitive inhibition of Eis by carnosic acid (Table S9) [91].

3.1.7. Peroxisome Proliferator-Activated Receptor δ

The peroxisome proliferator-activated receptor (PPARδ) functions as a sensor for dietary and endogenous fats [92]. It regulates the transcription of genes associated with lipid and glucose metabolism. Specifically, it controls lipid degradation, transport, and storage, while also being associated with insulin secretion and resistance. PPARδ agonists have been shown beneficial in models of metabolic disorders in primates and may thus possess therapeutic potential in hyperlipidemia, atherosclerosis, obesity, and diabetes [93,94]. PPARδ is also associated with cancer by promoting chronic inflammation through increasing cyclooxygenase-2 (COX-2) expression and prostaglandin E2 production, leading to an increase in proinflammatory cytokine concentrations. Moreover, the ability of PPARδ to promote the use of fatty acids as the energy source may enhance cell survival and proliferation under harsh metabolic conditions often found in tumors. Therefore, PPARδ agonists may be useful in treating metabolic disorders, while antagonists may reduce inflammation-related disorders and slow down cancer progression.

Whereas there are no experimental data that carnosic acid, carnosol, or rosmanol bind to PPARδ, it is known that both carnosol and carnosic acid represents agonists of the PPARγ isoform with half maximal effective concentration (EC_{50})values of 41 and 20 μM, respectively [95]. Carnosic acid docks to PPARδ in the same Ω-pocket where serotonin binds to PPARγ which also acts as agonists at PPARδ. The binding site possesses 62% amino acid sequence identity and a ProBiS Z-score of 3.36. From the superposition of PPARδ (with docked carnosic acid) and PPARγ (with serotonin) Ω-pockets, we observe that PPARγ produce steric clashes with carnosic acid (Figure 6, Table S10). However, due to experimental evidence, that carnosic acid indeed binds to PPARγ, we can predict that induced fitting effects play an important role. Because PPARδ possesses a smaller threonine in this place and because the overall binding site is similar, we can hypothesize that carnosic acid could bind even stronger to the PPARδ isotype as preliminary induced fitting would not be required.

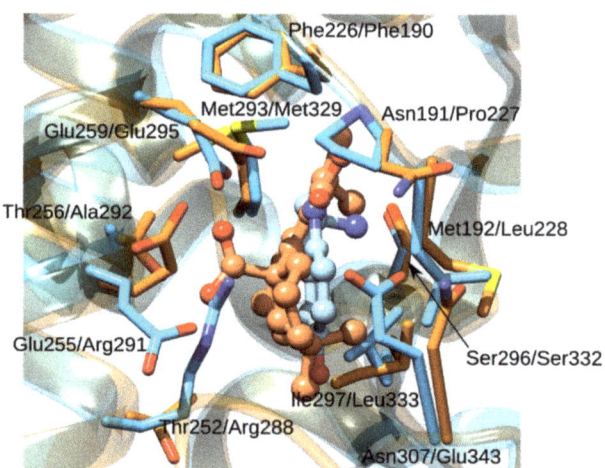

Figure 6. The Ω-pocket superimposition between PPARδ (orange cartoon and sticks) and PPARγ (blue cartoon and sticks). The first amino acid residue numbering corresponds to PPARδ, and the second to PPARγ. Serotonin is displayed using blue balls and sticks and the docked carnosic acid using orange balls-and-sticks. We emphasize the difference in amino acid residues Thr252 versus Arg288. Compared to Thr252 in PPARδ, the large Arg288 in PPARγ would lead to stearic clashes with carnosic acid.

3.1.8. Glycogen Phosphorylase

Glycogen phosphorylase (GP) is an enzyme that cleaves the non-reducing ends in the chain of glycogen to produce glucose-1-phosphate monomers which can be further converted to free glucose [96]. Because glycogen is an important source of blood glucose, GP represents a promising target for the treatment of type II diabetes, and its inhibitors have been shown effective in controlling blood glucose concentrations in animal studies [97].

GP can exist in two different forms that bind different regulatory molecules: the active phosphorylated (on Ser14) GPa and the non-phosphorylated GPb form. In addition, GP has been reported to bind compounds at four different binding sites, identified as: (a) the catalytic, (b) the allosteric (indole), (c) the novel allosteric, and (d) the inhibitory site (caffeine) [98]. In our study, carnosic acid docked to the catalytic site (a) of the GPb form, specifically to the α-D-glucose binding site, therefore, it might act as a competitive inhibitor with respect to glucose-1-phosphate (Figure 7) [96,99]. Glucose-1-phosphate forms hydrogen bonds with Glu672, Asn284, Ser674, His337, and Asn484, while the docked pose of carnosic acid binds to the cofactor pyridoxal phosphate with two hydrogen bonds, but also forming hydrogen bonds with Lys574 and Thr676 (Table S11).

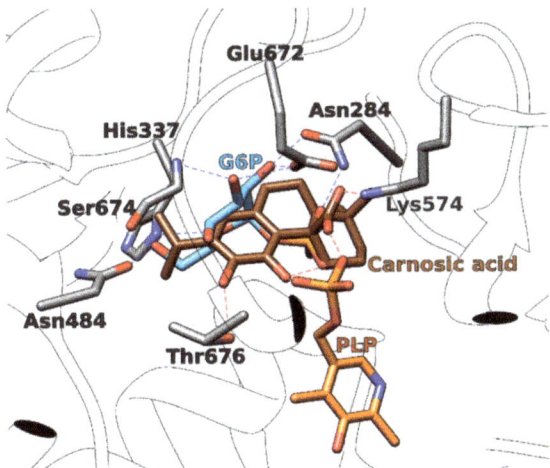

Figure 7. Glucose-6-phosphate (G6P, light blue sticks) overlapped with carnosic acid (brown sticks) in the catalytic binding site of glycogen phosphatase. Important amino acid residues are shown in grey sticks. Hydrogen bonds formed by carnosic acid are presented with red dotted lines, while the hydrogen bonds formed by glucose-6-phosphate are shown with blue dotted lines. The cofactor pyridoxal phosphate (PLP) is presented in orange sticks.

3.1.9. Tubulin

Tubulins represent protein monomers of microtubules, which form an essential component of the eukaryotic cytoskeleton [100,101]. They are involved in cell division as microtubules form mitotic spindles that are used by the cell to pull the chromosomes apart. Microtubules are produced by the polymerization of dimers of α- and β-tubulin that join together to form long hollow tubes called microtubules. Microtubule targeting agents such as chemotherapeutics vinblastine, colchicine, and paclitaxel bind to tubulin and disrupt microtubule dynamics, leading to a loss of function and to subsequent cell arrest or apoptosis. They can be classified into subgroups based on their binding site within the tubulin dimer: (a) the paclitaxel site at the β-tubulin in the microtubule lumen; (b) the vinblastine site at the interdimeric interface of two heterodimers; and (c) the colchicine site at the β-tubulin at the intra-subunit interface of a heterodimer. In our study, carnosic acid docked to the colchicine-binding site (c) (Table S12) and could therefore, like colchicine, potentially lead to microtubule depolymerization.

3.1.10. Phosholipase A2

Phospholipases A2 (PLA2) represent enzymes that catalyze the hydrolysis of glycerophospholipids at the sn-2 position, releasing free fatty acids, including arachidonic acid. The action of PLA2 forms a crucial upstream step that increases free arachidonic acid levels and triggers the storm of eicosanoids, especially after inflammatory cell activation. Due to their involvement in the inflammatory response, PLA2 are thought to be associated with various diseases such as arthritis [102], cancer [103], coronary heart disease [104], and neurological disorders such as Alzheimer's disease and multiple sclerosis [105]. In our study, carnosol docks between the two subunits of the dimer and forms a large hydrophobic and desolvated surface that is buried. Most of the carnosol molecule is located within the B subunit. (Figure 8, Table S14). Binding to identical active site as the alkyl portion of the tetrahedral mimic inhibitor MJ33.

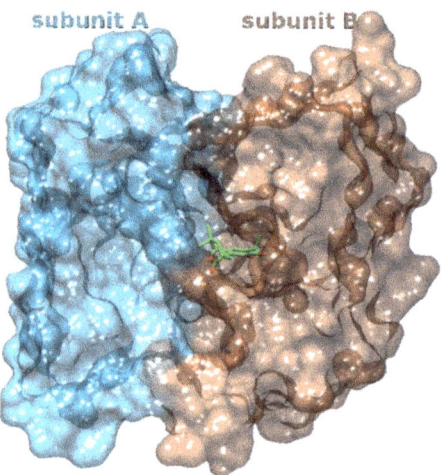

Figure 8. Carnosol (carbons denoted with green sticks) docks between subunits A (blue surface) and B (orange surface) of the phospholipase A2.

3.1.11. Vascular Endothelial Growth Factor Receptor 2

Vascular endothelial growth factor receptors (VEGFR) represent tyrosine kinase receptors for vascular endothelial growth factor (VEGF), a signaling protein critical in angiogenesis [106]. Because solid cancer tumors require an adequate blood supply to grow and metastasize, the inhibition of VEGFR signaling with small molecule drugs such as sorafenib or pazopanib is used as a well-established treatment in various cancers, since tumors cannot grow more than 2 mm without angiogenesis. VEGFR-2 plays an important role in cell migration and proliferation-two crucial steps of angiogenesis. Carnosic acid docks to the binding site representative of type II kinase inhibitors. In general, type II inhibitors, such as sorafenib and lenvatinib, are often more specific than those targeting only the ATP binding site [107,108]. They represent a class of compounds that capture kinases in an inactive form and occupy both the adenine region (of ATP) as well as a hydrophobic pocket adjacent to the ATP binding site [109]. However, due to the small size of carnosic acid, only the hydrophobic binding site is actually occupied (Figure 9, Table S16). This is consistent with the experimental finding that carnosic acid or carnosol actually do not possess a measurable inhibitory activity against VEGFR2 [109]. However, given the strong interaction measured between carnosic acid and VEGFR2 applied in the inverse docking method, carnosic acid could potentially serve as a base compound to which a specific ring system region would be added to also target the adenine binding site.

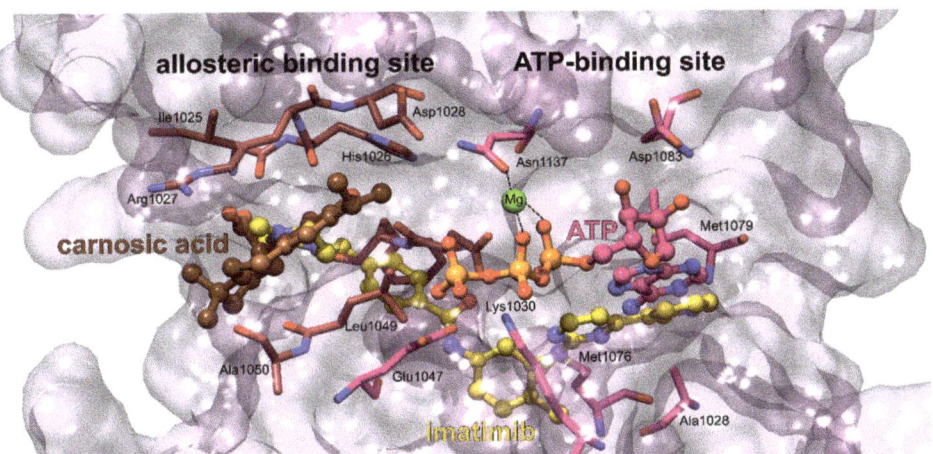

Figure 9. VEGFR2 binding sites. The VEGFR2 enzyme is presented with pink surface; the amino acid residues of the adenosine binding site are shown in pink sticks and of the lipophilic allosteric binding site in brown sticks. The docked carnosic acid located in the allosteric site displayed in brown balls and sticks and the ATP molecule in pink balls and sticks. The typical type II inhibitor imatimib, binding to both sites concurrently is depicted in yellow balls-and-sticks.

3.1.12. Aspartate Carbamoyltransferase

Plasmodium falciparum and *Trypanosoma cruzi* represent parasites that cause malaria and Chagas disease, respectively [110]. Aspartate carbamoyltransferase is an enzyme involved in pyrimidine biosynthesis that catalyzes the formation of phosphate and N-carbamoyl L-aspartate from carbamoyl phosphate and L-aspartate. Reproduction of both parasites requires a sufficient supply of purines, as they form the building blocks of nucleic acid molecules. Recent studies in *Plasmodium falciparum* have shown that aspartate carbamoyltransferase represents a suitable drug target, as its inhibition leads to a reduction in parasite growth [111]. Carnosic acid docks in the aspartate carbamoyltransferase of both *Plasmodium falciparum* and *Trypanosoma cruzi* at the interface between the protomers in the carbamoyl phosphate domain, where the carbamoyl phosphate substrate binds (Tables S17 and S18) [112].

3.2. Inverse Docking of Rosmarinic Acid

Table 2 lists the top-scoring protein–ligand complexes based on the cut-off criterion of 3.5 σ. We focus only on protein targets related to human health, i.e., we present only proteins from humans and mammals, as well as proteins from pathogenic microorganisms. As before, where data were available, we redocked ligands/drugs known to bind to the presented protein targets using a procedure analogous to inverse docking. These results, presented in Table S1, show that the docking scores of known ligand/drugs are in all cases worse than those of rosemarinic acid. Again, this may indicate an already strong binding affinity of rosmarinic acid, although it has not yet been rationally optimized for these specific protein targets.

Table 2. Best scoring mammalian, human, and pathogen protein targets of rosmarinic acid. Docking scores independent of the organism or type of protein are collected in Supplementary Materials in Table S3.

Rank	PDB ID with Chain	Predicted Ligand Docking Score (arb. Units)	Protein Name	Organism	Protein Function and Disease Correlation	Reported Experimental Correlation of Protein and Ligand
1	2d1jA	−86.1	Coagulation factor X	Homo sapiens	Serine endopeptidase is involved in the coagulation cascade. Its deficiency leads to a bleeding disorder. Its inhibitors are popular anticoagulants.	
2	1fxfB	−84.8	Phospholipase A2	Sus scrofa	Catalyzes the hydrolysis of glycerophospholipids thus releasing free fatty acids, including arachidonic acid. Its action is implicated in several inflammation-based disease such as arthritis, coronary artery disease, Alzheimer's and cancer.	[113]
3	2jt5A	−84.5	Matrix metalloproteinase-3	Homo sapiens	Zinc-dependent endopeptidase which is involved in the remodeling of the extracellular matrix. Involved in arthritis, multiple sclerosis, aneurysms, and the spread of metastatic cancer. After traumatic brain injury, matrix metalloproteinase-3 (MMP-3) concentrations increase and lead to additional damage to the blood–brain barrier.	
4	4jzbA	−83.2	Farnesyl pyrophosphate synthase	Leishmania major	Farnesyl pyrophosphate synthase (FPPS) is an essential enzyme involved in the biosynthesis of ergosterol in leishmania parasites, the causative agents of leishmaniasis.	[114]
5	3qmuB	−80.2	Glutamate dehydrogenase 1	Bos Taurus	Part of the glutaminolysis pathway, playing a crucial role in nitrogen and carbon metabolism. Inhibition leads to in vivo and in vitro reduced viability of cancer cells.	
6	5fi6A	−77.6	Glutaminase	Homo sapiens		

3.2.1. Coagulation Factor X

Factor X represents an enzyme involved in the coagulation cascade that, when activated by the hydrolysis of factor Xa, claves prothrombin to the active thrombin, which in turn converts soluble fibrinogen to insoluble fibrin strands [115]. The role of factor X is particularly important because it is the first enzyme where the intrinsic and extrinsic coagulation pathways converge. Drug manipulation of the coagulation cascade is extremely important in modern medicine, since reducing excessive coagulation is critical for preventing diseases such as myocardial infarction and ischemic stroke, which belong among the leading causes of death and disability in the Western world [116,117]. Oral inhibitors of factor X, such as rivaroxaban, are already successfully used in clinical practice [118]. Rosmarinic acid docks in an analogous manner to a number of sulfonamide factor X inhibitors (Table S19) [119]. Its caffeic acid ring binds to the S1 pocket, while its 3,4-dihydroxyphenyllactic acid moiety binds to the S4 pocket (Figure 10).

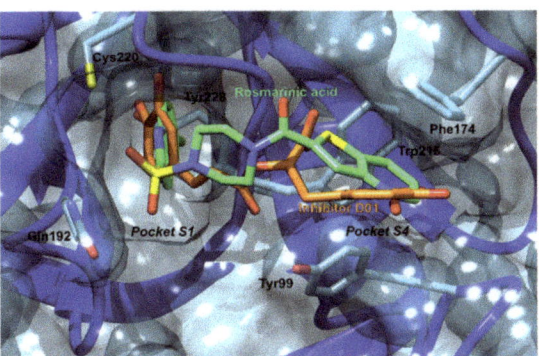

Figure 10. Factor Xa binding site. Factor Xa is shown in blue ribbons and surface, its important amino acid residues in blue sticks. Rosmarinic acid (green carbons) is docked in the same binding site as the one occupied by a known inhibitor (orange sticks) with a PDB ID: D01. The caffeic acid part of rosmarinic acid docks to the S1 pocket, and the 3,4-dihydroxyphenyllactic acid moiety to the S4 pocket.

3.2.2. Phospholipase A2

Similar to the case of carnosol, our inverse docking algorithm also detected a strong binding to the enzyme phospholipase A2. This is consistent with existing experimental evidence, as the PDB contains a snake toxin phospholipase A2 homolog (PDB ID: 3QNL) bound with rosmarinic acid, and this complex was also applied later on in our study to validate the inverse docking algorithm by redocking. Compared to the main active site, its binding site is located in a different region between the dimer site where the MJ33 inhibitor was reported to bind and where carnosol was docked in this study (Table S20). We have here an interesting case where two rosemary compounds potentially inhibit the same enzyme.

3.2.3. Matrix Metalloproteinase-3

Matrix metalloproteinase-3 (MMP-3) represents a zinc-dependent endopeptidase Matrix metalloproteinase-3 (MMP-3) represents a zinc-dependent endopeptidase involved in extracellular matrix remodeling [120]. It is, therefore, required for physiological processes such as embryonic development and reproduction and is also involved in various pathological processes. Moreover, MMP-3 can also activate other metalloproteinases, enter cell nuclei, and control gene expression. Excessive activation of MMPs can lead to excessive degradation of the extracellular matrix and to numerous pathological conditions such as arthritis, multiple sclerosis, aneurysms, as well as the spread of metastatic cancer. Furthermore, it has been shown that following a traumatic brain injury, MMP-3 levels can also increase and cause additional damage to the blood–brain barrier [70]. The discovery of novel small molecule inhibitors of MMP-3 is, therefore, of great importance for the treatment of numerous diseases [120]. The 3,4-dihydroxyphenyllactic moiety of rosmarinic acid docks to the catalytic region, but it is too far from the catalytic zinc ion to form direct interactions with it (Figure 11, Table S21). The caffeic acid moiety docks to the S1' pocket that delimits the active site. The S1' pocket is known to confer the selectivity of compounds towards different matrix metalloproteinases. Therefore, compounds that interact within the S1' pocket and not with the catalytic zinc could selectively inhibit one particular MMP without affecting the activities of the remaining ones.

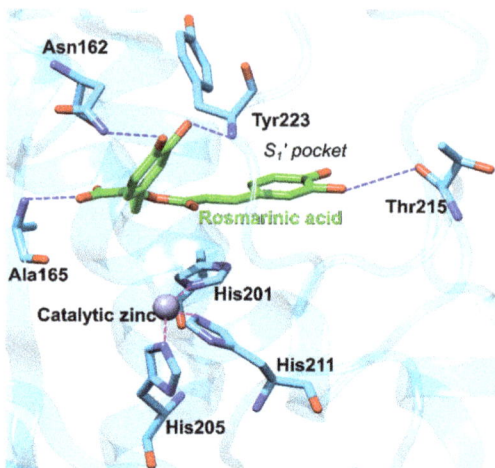

Figure 11. Rosmarinic acid docked in MMP3 (blue ribbons). The rosmarinic acid docks near the catalytic zinc ion and one of the catechol groups positions inside the $S_{1'}$ selectivity pocket. Important amino acid residues are shown in blue sticks, hydrogen bonds are denoted with dotted blue lines and coordinative bonds with dotted purple lines.

3.2.4. Farnesyl Pyrophosphate Synthase

Leishmania major represents an intracellular, pathogenic, parasitic organism that causes cutaneous leishmaniasis. The World Health Organization stated that leishmaniasis is one of the most neglected diseases. Moreover, 350 million people are considered at risk of contracting the disease, approximately 12 million people are infected worldwide, and an estimated two million new cases occur each year [121]. Farnesyl pyrophosphate synthase (FPPS) represents an important enzyme involved in the biosynthesis of ergosterol in Leishmania parasites. Antiparasitic compounds targeting the ergosterol biosynthesis play an important role in the treatment of leishmaniasis, and the inhibition of FPPS has been shown largely effective against the related *Leishmania donovani* [122]. Interestingly, a study [114] showed that carnosic acid and carnosol form potent inhibitors of human FPPS, with IC_{50} values of 20.0 and 13.3 µM, respectively. It also demonstrated that inhibition of the human form of the enzyme leads to the induction of apoptosis in pancreatic cell lines by downregulating RAS prenylation. *Leishmania major* FPPS is not among the 0.05% best scoring proteins of rosemary diterpenes and is not listed in Table 1. However, it still scored extremely high with carnosol (−60.4), which is within the 3.0σ. Thus, as with phospholipase A2, we have yet another interesting case of two rosemary polyphenols potentially inhibiting the same enzyme. Both rosmarinic acid and carnosol bind approximately to the same protein space, with portions of the ligands occupying the same region as the reported *Leismania minor* FPPS inhibitor 1-(2-hydroxy-2,2-diphosphonoethyl)-3-phenylpyridinium (300B) (Figure 12, Table S22). Part of the rosmarinic acid enters the substrate-binding region where the substrate isopentenyl pyrophosphate is present in an uninhibited enzyme.

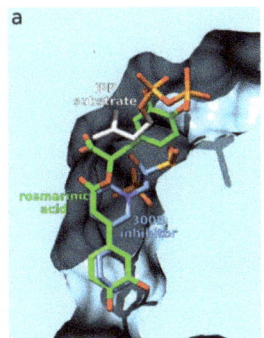 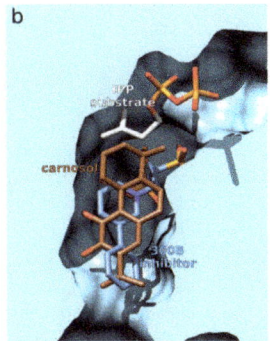

Figure 12. Comparison of ligand binding to farnesyl pyrophosphate synthase (blue surfaces). (**a**) Comparison between the crystal ligand 300B (blue carbons) and rosmarinic acid (green carbons) binding. (**b**) Comparison between the crystal ligand 300B (blue carbons) and carnosol (brown carbons) binding. The enzyme substrate isopentenyl pyrophosphate (IPP) was not present during the inverse docking but is shown for comparison purposes using white carbons.

3.2.5. Glutamate Dehydrogenase 1 and Glutaminase

Glutaminase and glutamate dehydrogenase 1 (GDH1) represent enzymes that are both part of the glutaminolysis pathway. Glutaminolysis begins with the conversion of glutamine to glutamate by glutaminase, while the next step is catalyzed by GDH, which converts glutamate to 2-oxoglutarate. The two enzymes play a crucial role in nitrogen and carbon metabolism, as the product 2-oxoglutarate feeds the citric acid cycle. Numerous cancer cells rely on increased glutaminolysis to meet their energy requirements. It has thus been shown that the inhibition of glutaminase and GDH1 by small molecules leads to a decreased viability of cancer cells in vivo and in vitro. Consequently, they form promising targets for cancer treatment [123,124]. It has been already shown that the plant compounds from green tea epigallocatechin gallate and epicatechin gallate strongly inhibit GDH [125–127]. According to our inverse docking procedure, rosmarinic acid is located at a different binding site than the green tea compounds. It binds at hexameric 2-fold axes between the dimers of the GDH subunits, where known inhibitors such as bithionol are also located (Table S23) [127].

Rosmarinic acid binds to the allosteric pocket formed at the interface between the two dimers of glutaminase (Figure 13). In numerous crystal structures of glutaminase in the PDB, co-crystallized inhibitors have occupied this binding site, e.g., 3UO9, 3VOZ, and 3VP1 (Table S24) [128,129].

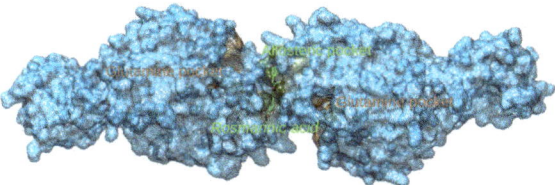

Figure 13. Glutaminase binding sites. Two protomers forming glutaminase are shown on blue surfaces. The main glutamine substrate-binding pockets are highlighted in orange surfaces, whereas rosmarinic acid (carbons denoted with green sticks) docks into the allosteric binding site (green surfaces) formed between the two glutamase protomers.

3.3. Method Validation

3.3.1. Redocking Procedure

To validate the inverse molecular docking procedure, a redocking study was performed using all available protein complexes from the PDB containing rosmarinic acid (PDB structures: 6MQD, 3QNL, and 4PWI). An analogous redocking procedure using the investigated diterpene structures could not be performed because protein structures containing carnosic acid, carnosol, or rosmanol do not yet exist in the PDB. Redocking of rosmarinic acid was performed by first removing the ligand from the binding site. Then, the CANDOCK algorithm was used with identical settings for inverse molecular docking to bind rosmarinic acid to the binding site defined by the crystal structure. The actual binding site definition was again identical to the one found in the ProBiS-Dock Database. To evaluate the success of the redocking procedure, the root-mean-square deviation (RMSD) of all heavy atoms between the co-crystallized and the redocked rosmarinic acid was measured.

From a molecular docking perspective, rosmarinic acid represents a problematic molecule, because it contains a high number, namely seven, rotatable bonds, which makes it difficult for the docking algorithms to consistently identify the correct conformer of this molecule. This problem is reflected in the fact that we obtained a low RMSD value of 1.3 Å only with the PDB structure 3QNL, which is a snake venom-derived phospholipase A2 structure [113], compared to the original crystal structure (Figure 14). The redocking procedure was not successful for 4PWI or 6MQD structures with significantly larger RMSD values (not shown), implying that the correct pose was not detected with the CANDOCK docking algorithm. However, based on a successful redock with 3QNL and on the fact that the docking algorithm identified numerous targets that have already been also experimentally confirmed for both rosemary diterpenes as well as rosmarinic acid, we are confident that the method is capable of recognizing correct protein targets to large extent.

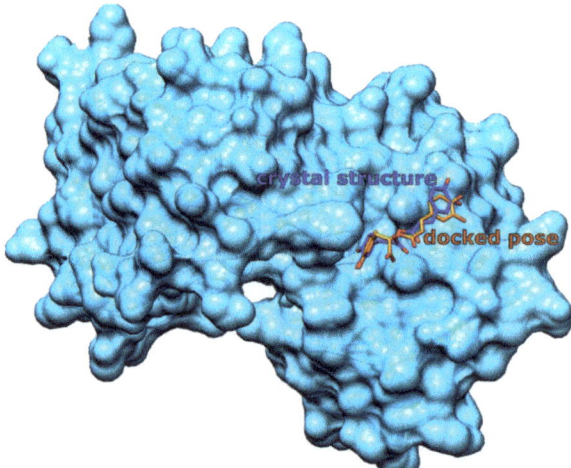

Figure 14. A successful redocking of rosmarinic acid to the crystal structure of phospholipase A2 (PDB ID: 3QNL) from the snake venom (depicted in blue ribbons and transparent surfaces). The stick structure of rosmarinic acid with blue carbons represents the native ligand position found in the crystal structure, while the structure with orange carbons displays the redocked structure. Hydrogen atoms are not shown for clarity. The RMSD between the two rosmarinic acid structures is 1.3 Å.

3.3.2. Validation Using ROC, EF, and PC

We performed the inverse molecular docking using the CANDOCK algorithm on all proteins from the ProBiS-Dock database, including 206 experimentally confirmed targets of rosmarinic acid, whose measured IC_{50} values were < 10 μM [62]. The ability of our protocol to distinguish the confirmed protein targets from proteins that reportedly do not bind rosmarinic acid was evaluated using the metrics established in the virtual screening community (Figure 15). It was successful in discriminating between the true and false targets of rosmarinic acid (ROC AUC of 0.627). The early detection of protein targets was assessed by the BEDROC of 0.071, by the RIE of 1.403, and by the EF 1% of 1.46, which is satisfactory. The inverse molecular docking protocol based on the CANDOCK algorithm resulted in score variations for the detection of true target proteins (TG determined from PC has a value of 0.171), which in combination with ROC AUC above 0.6 indicates that the protocol provides good results in agreement with the experiments [61]. Moreover, our inverse molecular docking protocol has been already extensively validated by Fine and Konc et al. [55], Furlan et al. [50,130], Kores et al. [49,131], and Jukič et al. [132] using different molecules of interest.

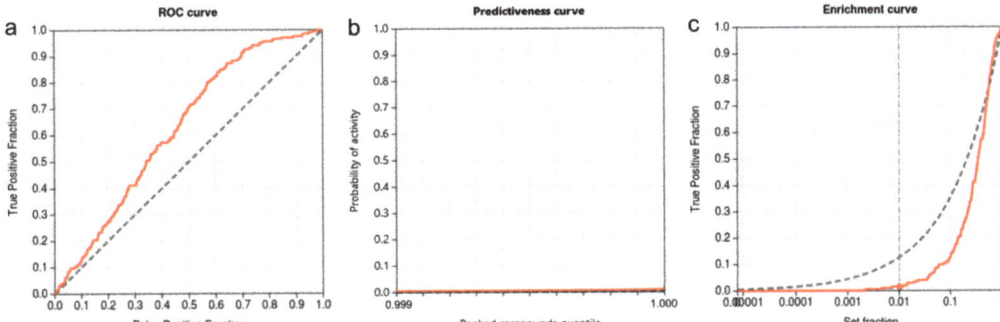

Figure 15. Validation of the inverse molecular docking protocol using rosmarinic acid: (**a**) the ROC curve, with the ROC AUC of 0.627; (**b**) predictiveness curve, from which the TG of 0.171 is determined; (**c**) enrichment curve.

3.4. In Silico Prediction of Pharmacokinetic Properties

In concurrence with experimental findings [31], the SwissADME web server [54] indeed predicts that carnosic acid, carnosol, and rosmanol all exhibit high gastrointestinal absorption (data shown in Supplementary Materials). On the contrary, the server predicts low gastrointestinal absorption for rosmarinic acid, which is again in line with the available in vivo data [43]. All compounds are predicted to be moderately soluble. Diterpenes are overall predicted as quite lipophilic, with a consensus score of logP above 3.5 for carnosic acid and carnosol, and 2.9 for rosmanol. Rosmarinic acid, as expected, due to the large number of polar functional groups, exhibits a much lower logP value of 1.2. Interestingly, carnosol is the only molecule predicted to penetrate the blood–brain barrier, however experimental studies on rat animal models show that carnosic acid also effectively penetrates the blood–brain barrier [32]. The SwissADME output is presented in its entirety in the Supplementary Materials.

4. Discussion

Natural plant-based compounds play an important role in the development of novel drugs as they may possess several advantages over conventional synthetic compounds, namely, fewer side effects, lower long-term toxicity, and versatile biological effects [130]. We report the potential targets of the major compounds from *Rosmarinus officinalis*, including the diterpenes carnosic acid, carnosol, and rosmanol, as well as the polyphenolic ester rosmarinic acid. Their targets were identified in silico using an inverse molecular docking approach. All four compounds were individually docked to all non-redundant holo-proteins available in the PDB. To identify the binding sites of each protein in advance, we applied the recently developed ProBiS-Dock Database—a freely available repository of binding sites between small ligands and proteins. Thereby, the docking procedure was limited to binding sites already known to bind at least one drug-like small-molecule ligand or to binding sites exhibiting a high similarity with the already known binding sites. Moreover, we used the novel CANDOCK algorithm, which employs a fragment-based docking approach with maximum clique and a knowledge-based scoring function.

Due to the similar molecular structure and docking/scoring values, we combined the results of all three investigated diterpenes into a single set (Table 1). We identified numerous human/mammalian proteins that could explain the observed anticarcinogenic activities of rosemary diterpenes. The best docking score was obtained for the complex between carnosic acid and the proto-oncogene K-Ras G12C. Moreover, the anticarcinogenic activities can also be explained by the potential binding of rosemary diterpenes to pyruvate kinase, PPARδ, tubulin, VEGFR2, and phospholipase A2. In general, phospholipase A2 has also been strongly implicated in inflammation-related disorders, so its inhibition may be likewise beneficial in arthritis, coronary artery disease, or dementia. Due to the identification of potential binding of the investigated diterpenes to glycogen phosphorylase, which facilitates glucose production, these compounds may be also useful in the treatment of type II diabetes. Furthermore, rosemary diterpenes exhibit antiviral activities.

From previous experimental studies, it is known that carnosol strongly inhibits HIV-1 protease. However, we also found out that rosemary diterpenes may bind strongly to the HIV-2 enzyme isotype. These compounds therefore likely represent a good starting point for the development of drugs against AIDS that could treat concurrent infections with HIV-1 and HIV-2. Interestingly, all diterpenes also yield good docking scores when bound to the feline immunodeficiency virus (FIV) protease, which is strongly related to HIV proteases, suggesting their potential utility in veterinary medicine. Finally, we have also found out that these compounds can bind to HA1 of the influenza A virus, potentially reducing its infectivity.

The antibacterial activity of investigated diterpenes can be explained by our discovery that they can bind to the enzyme glucosamine-fructose-6-phosphate aminotransferase in *Escherichia coli*, which is critical for the first step of hexosamine metabolism responsible for bacterial growth. Encouragingly, we have also found out that they can bind to the Eis protein of *Mycobacterium tuberculosis*, which confers resistance to aminoglycoside drugs, rendering them inactive. Therefore, the inhibition of this enzyme in conjunction with tuberculosis treatment could be beneficial in reducing the bacterial resistance to these drugs.

The investigated diterpenes also displayed binding to aspartate carbamoyltransferase of two different pathogenic parasites-*P. falciparum* and *T. cruzi*. *P. falciparum* represents the causative agent of malaria, while *T. cruzi* causes Chagas disease. Inhibition of this enzyme results in the inability of the two parasites to produce pyrimidines, limiting their biosynthesis of new nucleic acids.

Like diterpenes, rosmarinic acid also shows binding to proteins involved in carcinogenesis, namely matrix metalloproteinase-3 and phospholipase A2. Interestingly, all four compounds display very favorable binding scores for the enzyme phospholipase A2, which could provide a possible explanation for the strong anti-inflammatory effects of rosemary. According to our results, rosmarinic acid may also interfere with the glutaminolysis pathway, as it forms top-scoring complexes with two related enzymes—glutaminase and glutamate dehydrogenase. Inhibition of this pathway by small-molecule drugs has been indeed shown to reduce cancer cell viability. Moreover, the complex between rosmarinic acid and coagulation factor X yielded the best scoring result. Regulating blood clotting with drugs is of utmost importance, as reducing excessive blood clotting is crucial in preventing diseases such as heart attacks and ischemic strokes, which belong among the leading causes of death and disability in the Western world. Furthermore, rosmarinic acid might also possess antiparasitic activity as its binding to farnesyl pyrophosphate synthase (FPPS) of *Leishmania major* obtained a favorable docking score. This parasite causes zoonotic cutaneous leishmaniasis, and inhibition of the FPPS prevents the biosynthesis of ergosterol.

The results of this study will facilitate future molecular dynamics studies. Therein, we plan to investigate the dynamic binding patterns of prior parameterized rosemary compounds to the notable protein targets identified here. The molecular dynamics observations will be extended with the linear interaction energy as well as linear response approximation calculations to obtain the binding free energy values, which will then be compared with drug ligands already known to bind to the protein targets described here.

5. Conclusions

Using an in silico inverse molecular docking procedure, we identified protein targets that could explain the observed pharmacological activities of rosemary or its major polyphenolic constituents. By identifying protein structures to which carnosic acid, carnosol, rosmanol, and rosmarinic acid can bind, we provide possible explanations for the observed anticarcinogenic, anti-inflammatory, antidiabetic, antiviral, and antibacterial activities of rosemary. In addition, using this methodology we were able to predict new effects of these compounds that have not yet been reported, namely their anticoagulant and antiparasitic activities. Lastly, we believe that our research can form the basis for the development of novel drugs, where the rosemary compounds studied here could serve as a starting point for efficient drug design.

Supplementary Materials: The following are available online at https://www.mdpi.com/article/10.3390/foods11010067/s1. Table S1: Best docking scores of ligands/drugs known to bind to a specific target. Table S2: Best docking scores for carnosol, carnosic acid and rosmanol. Table S3: Best docking scores for rosmarinic acid. Table S4–S24: Interaction of compounds with respective targets presented in Table 1; Table 2. The SwissADME output file can be found in the file swissadme_output.xlsx supplied as part of the Supplementary Materials.

Author Contributions: Conceptualization, S.L. and U.B.; methodology, S.L. and U.B.; validation, S.L.; formal analysis, S.L.; investigation, S.L.; resources, U.B.; Writing—Original draft preparation, S.L.; Writing—review and editing, S.L. and U.B.; visualization, S.L.; supervision, U.B. All authors have read and agreed to the published version of the manuscript.

Funding: Financial support through Slovenian Research Agency (ARRS) grants project and programme J1-2471 and P2-0046 as well as through Slovenian Ministry of Education, Science and Sports project grants C3330-19-952021 and AB FREE is gratefully acknowledged.

Institutional Review Board Statement: Not applicable.

Informed Consent Statement: Not applicable.

Data Availability Statement: All data generated or analyzed during this study are included in the published article.

Conflicts of Interest: The authors declare no conflict of interest.

Abbreviations

SMILES	Simplified Molecular-Input Line-Entry System
CANDOCK	Chemical Atomic Network based Docking
CHARMM	Chemistry at Harvard Macromolecular Mechanics
ROC	Receiver operating characteristics curve
PC	Predictiveness curve
TPF	True positive fraction
FPF	False positive fraction
PDB	Protein Data Bank
ROC AUC	Area under the receiver operating characteristics curve
EF	Enrichment factor
BEDROC	Boltzmann-enhanced discrimination of ROC
RIE	Robust initial enhancement
TG	Total gain

Eis	Enhanced intracellular survival
HIV	Human immunodeficiency virus
FIV	Feline immunodeficiency virus
RAS/MAPK	Rat sarcoma/mitogen-activated protein kinase
GTP	Guanosine triphosphate
GlmS	Glucosamine/fructose-6-phosphate aminotransferase
PKM	Pyruvate kinase M
PEP	Phosphoenolpyruvate
ADP	Adenosine diphosphate
ATP	Adenosine triphosphate
HA	Hemagglutinin
AIDS	Acquired immunodeficiency syndrome
HAART	Highly active antiretroviral therapy
PPAR	Peroxisome proliferator-activated receptor
COX-2	Cyclooxygenase-2
EC_{50}	Half maximal effective concentration
GP	Glycogen phosphorylase
PLA2	Phospholipase A2
VEGFR	Vascular endothelial growth factor receptor
VEGF	Vascular endothelial growth factor
MMP	Matrix metalloproteinase
FPPS	Farnesyl pyrophosphate synthase
IPP	Isopentenyl pyrophosphate
GDH1	Glutamate dehydrogenase 1
RMSD	Root-mean-square deviation

References

1. Begum, A.; Sandhya, S.; Ali, S.S.; Vinod, K.R.; Reddy, S.; Banji, D. An in-depth review on the medicinal flora *Rosmarinus officinalis* (Lamiaceae). *Acta Sci. Pol. Technol. Aliment.* **2013**, *12*, 61–74.
2. Al-Sereiti, M.R.; Abu-Amer, K.M.; Sena, P. Pharmacology of rosemary (*Rosmarinus officinalis* Linn.) and its therapeutic potentials. *IJEB* **1999**, *37*, 124–130.
3. Perez-Fons, L.; Garzon, M.T.; Micol, V. Relationship between the antioxidant capacity and effect of rosemary (*Rosmarinus officinalis* L.) polyphenols on membrane phospholipid order. *J. Agric. Food Chem.* **2009**, *58*, 161–171. [CrossRef]
4. Yu, M.-H.; Choi, J.-H.; Chae, I.-G.; Im, H.-G.; Yang, S.-A.; More, K.; Lee, I.-S.; Lee, J. Suppression of LPS-induced inflammatory activities by *Rosmarinus officinalis* L. *Food Chem.* **2013**, *136*, 1047–1054. [CrossRef]
5. Bakırel, T.; Bakırel, U.; Keleş, O.Ü.; Ülgen, S.G.; Yardibi, H. In vivo assessment of antidiabetic and antioxidant activities of rosemary (*Rosmarinus officinalis*) in alloxan-diabetic rabbits. *J. Ethnopharmacol.* **2008**, *116*, 64–73. [CrossRef] [PubMed]
6. Bozin, B.; Mimica-Dukic, N.; Samojlik, I.; Jovin, E. Antimicrobial and antioxidant properties of rosemary and sage (*Rosmarinus officinalis* L. and *Salvia officinalis* L., Lamiaceae) essential oils. *J. Agric. Food Chem.* **2007**, *55*, 7879–7885. [CrossRef]
7. Tai, J.; Cheung, S.; Wu, M.; Hasman, D. Antiproliferation effect of Rosemary (*Rosmarinus officinalis*) on human ovarian cancer cells in vitro. *Phytomedicine* **2012**, *19*, 436–443. [CrossRef] [PubMed]
8. Valdés, A.; García-Cañas, V.; Rocamora-Reverte, L.; Gómez-Martínez, Á.; Ferragut, J.A.; Cifuentes, A. Effect of rosemary polyphenols on human colon cancer cells: Transcriptomic profiling and functional enrichment analysis. *Genes Nutr.* **2013**, *8*, 43–60. [CrossRef]
9. Yesil-Celiktas, O.; Sevimli, C.; Bedir, E.; Vardar-Sukan, F. Inhibitory effects of rosemary extracts, carnosic acid and rosmarinic acid on the growth of various human cancer cell lines. *Plant Foods Hum. Nutr.* **2010**, *65*, 158–163. [CrossRef]
10. Singletary, K.; MacDonald, C.; Wallig, M. Inhibition by rosemary and carnosol of 7,12-dimethylbenz[a]anthracene (DMBA)-induced rat mammary tumorigenesis and in vivo DMBA-DNA adduct formation. *Cancer Lett.* **1996**, *104*, 43–48. [CrossRef]
11. Huang, M.-T.; Ho, C.-T.; Wang, Z.Y.; Ferraro, T.; Lou, Y.-R.; Stauber, K.; Ma, W.; Georgiadis, C.; Laskin, J.D.; Conney, A.H. Inhibition of Skin Tumorigenesis by Rosemary and Its Constituents Carnosol and Ursolic Acid. *Cancer Res.* **1994**, *54*, 701–708.
12. Lešnik, S.; Furlan, V.; Bren, U. Rosemary (*Rosmarinus officinalis* L.): Extraction techniques, analytical methods and health-promoting biological effects. *Phytochem. Rev.* **2021**, *20*, 1273–1328. [CrossRef]
13. Okamura, N.; Fujimoto, Y.; Kuwabara, S.; Yagi, A. High-performance liquid chromatographic determination of carnosic acid and carnosol in *Rosmarinus officinalis* and *Salvia officinalis*. *J. Chromatogr. A* **1994**, *679*, 381–386. [CrossRef]
14. Johnson, J.J. Carnosol: A promising anti-cancer and anti-inflammatory agent. *Cancer Lett.* **2011**, *305*, 1–7. [CrossRef] [PubMed]
15. Masuda, T.; Inaba, Y.; Maekawa, T.; Takeda, Y.; Tamura, H.; Yamaguchi, H. Recovery Mechanism of the Antioxidant Activity from Carnosic Acid Quinone, an Oxidized Sage and Rosemary Antioxidant. *J. Agric. Food Chem.* **2002**, *50*, 5863–5869. [CrossRef]

16. Collins, M.A.; Charles, H.P. Antimicrobial activity of Carnosol and Ursolic acid: Two anti-oxidant constituents of *Rosmarinus officinalis* L. *Food Microbiol.* **1987**, *4*, 311–315. [CrossRef]
17. Shin, H.-B.; Choi, M.-S.; Ryu, B.; Lee, N.-R.; Kim, H.-I.; Choi, H.-E.; Chang, J.; Lee, K.-T.; Jang, D.S.; Inn, K.-S. Antiviral activity of carnosic acid against respiratory syncytial virus. *Virol. J.* **2013**, *10*, 303. [CrossRef]
18. Pukl, M.; Umek, A.; Pariš, A.; Štrukelf, B.; Renko, M.; Korant, B.D.; Turk, V. Inhibitory effect of carnosolic acid on HIV-1 protease. *Planta Med.* **1992**, *58*, 632. [CrossRef]
19. Bai, N.; He, K.; Roller, M.; Lai, C.-S.; Shao, X.; Pan, M.-H.; Ho, C.-T. Flavonoids and Phenolic Compounds from Rosmarinus officinalis. *J. Agric. Food Chem.* **2010**, *58*, 5363–5367. [CrossRef]
20. Kuhlmann, A.; Röhl, C. Phenolic Antioxidant Compounds Produced by in Vitro. Cultures of Rosemary (*Rosmarinus officinalis*) and Their Anti-inflammatory Effect on Lipopolysaccharide-Activated Microglia. *Pharm. Biol.* **2006**, *44*, 401–410. [CrossRef]
21. Visanji, J.M.; Thompson, D.G.; Padfield, P.J. Induction of G2/M phase cell cycle arrest by carnosol and carnosic acid is associated with alteration of cyclin A and cyclin B1 levels. *Cancer Lett.* **2006**, *237*, 130–136. [CrossRef] [PubMed]
22. González-Vallinas, M.; Molina, S.; Vicente, G.; Zarza, V.; Martín-Hernández, R.; García-Risco, M.R.; Fornari, T.; Reglero, G.; De Molina, A.R. Expression of MicroRNA-15b and the Glycosyltransferase GCNT3 Correlates with Antitumor Efficacy of Rosemary Diterpenes in Colon and Pancreatic Cancer. *PLoS ONE* **2014**, *9*, e98556. [CrossRef] [PubMed]
23. Xiang, Q.; Ma, Y.; Dong, J.; Shen, R. Carnosic acid induces apoptosis associated with mitochondrial dysfunction and Akt inactivation in HepG2 cells. *Int. J. Food Sci. Nutr.* **2015**, *66*, 76–84. [CrossRef]
24. Kar, S.; Palit, S.; Ball, W.B.; Das, P.K. Carnosic acid modulates Akt/IKK/NF-κB signaling by PP2A and induces intrinsic and extrinsic pathway mediated apoptosis in human prostate carcinoma PC-3 cells. *Apoptosis* **2012**, *17*, 735–747. [CrossRef]
25. Barni, M.V.; Carlini, M.J.; Cafferata, E.G.; Puricelli, L.; Moreno, S. Carnosic acid inhibits the proliferation and migration capacity of human colorectal cancer cells. *Oncol. Rep.* **2012**, *27*, 1041–1048. [CrossRef] [PubMed]
26. Aliebrahimi, S.; Kouhsari, S.M.; Arab, S.S.; Shadboorestan, A.; Ostad, S.N. Phytochemicals, withaferin A and carnosol, overcome pancreatic cancer stem cells as c-Met inhibitors. *Biomed. Pharmacother.* **2018**, *106*, 1527–1536. [CrossRef]
27. Lo, A.-H.; Liang, Y.-C.; Lin-Shiau, S.-Y.; Ho, C.-T.; Lin, J.-K. Carnosol, an antioxidant in rosemary, suppresses inducible nitric oxide synthase through down-regulating nuclear factor-κB in mouse macrophages. *Carcinogenesis* **2002**, *23*, 983–991. [CrossRef]
28. Cheng, A.-C.; Lee, M.-F.; Tsai, M.-L.; Lai, C.-S.; Lee, J.H.; Ho, C.-T.; Pan, M.-H. Rosmanol potently induces apoptosis through both the mitochondrial apoptotic pathway and death receptor pathway in human colon adenocarcinoma COLO 205 cells. *Food Chem. Toxicol.* **2011**, *49*, 485–493. [CrossRef]
29. Machado, D.G.; Cunha, M.P.; Neis, V.B.; Balen, G.O.; Colla, A.; Bettio, L.E.B.; Oliveira, Á.; Pazini, F.L.; Dalmarco, J.B.; Simionatto, E.L.; et al. Antidepressant-like effects of fractions, essential oil, carnosol and betulinic acid isolated from Rosmarinus officinalis L. *Food Chem.* **2013**, *136*, 999–1005. [CrossRef]
30. Sasaki, K.; El Omri, A.; Kondo, S.; Han, J.; Isoda, H. *Rosmarinus officinalis* polyphenols produce anti-depressant like effect through monoaminergic and cholinergic functions modulation. *Behav. Brain Res.* **2013**, *238*, 86–94. [CrossRef]
31. Romo Vaquero, M.; Garcia Villalba, R.; Larrosa, M.; Yáñez-Gascón, M.J.; Fromentin, E.; Flanagan, J.; Roller, M.; Tomás-Barberán, F.A.; Espín, J.C.; García-Conesa, M.-T. Bioavailability of the major bioactive diterpenoids in a rosemary extract: Metabolic profile in the intestine, liver, plasma, and brain of Zucker rats. *Mol. Nutr. Food Res.* **2013**, *57*, 1834–1846. [CrossRef] [PubMed]
32. de Oliveira, M.R. The dietary components carnosic acid and carnosol as neuroprotective agents: A mechanistic view. *Mol. Neurobiol.* **2016**, *53*, 6155–6168. [CrossRef]
33. Rasoolijazi, H.; Azad, N.; Joghataei, M.; Kerdari, M.; Nikbakht, F.; Soleimani, M. The protective role of carnosic acid against beta-amyloid toxicity in rats. *Sci. World J.* **2013**, *2013*, 1–5. [CrossRef] [PubMed]
34. Petersen, M. Rosmarinic acid: New aspects. *Phytochem. Rev.* **2013**, *12*, 207–227. [CrossRef]
35. De Souza Gil, E.; Adrian Enache, J.; Maria Oliveira-Brett, A. Redox behaviour of verbascoside and rosmarinic acid. *Comb. Chem. High Throughput Screen.* **2013**, *16*, 92–97. [CrossRef]
36. Fadel, O.; El Kirat, K.; Morandat, S. The natural antioxidant rosmarinic acid spontaneously penetrates membranes to inhibit lipid peroxidation in situ. *Biochim. Biophys. Acta* **2011**, *1808*, 2973–2980. [CrossRef]
37. Kimura, Y.; Okuda, H.; Okuda, T.; Hatano, T.; Arichi, S. Studies on the activities of tannins and related compounds, X. Effects of caffeetannins and related compounds on arachidonate metabolism in human polymorphonuclear leukocytes. *J. Nat. Prod.* **1987**, *50*, 392–399. [CrossRef]
38. Lucarini, R.; Bernardes, W.A.; Ferreira, D.S.; Tozatti, M.G.; Furtado, R.; Bastos, J.K.; Pauletti, P.M.; Januário, A.H.; Silva, M.L.A.; Cunha, W.R. In vivo analgesic and anti-inflammatory activities of *Rosmarinus officinalis* aqueous extracts, rosmarinic acid and its acetyl ester derivative. *Pharm. Biol.* **2013**, *51*, 1087–1090. [CrossRef]
39. Amaral, G.P.; Mizdal, C.R.; Stefanello, S.T.; Mendez, A.S.L.; Puntel, R.L.; de Campos, M.M.A.; Soares, F.A.A.; Fachinetto, R. Antibacterial and antioxidant effects of *Rosmarinus officinalis* L. extract and its fractions. *J. Tradit. Complement. Med.* **2019**, *9*, 383–392. [CrossRef]
40. Radziejewska, I.; Supruniuk, K.; Nazaruk, J.; Karna, E.; Poplawska, B.; Bielawska, A.; Galicka, A. Rosmarinic acid influences collagen, MMPs, TIMPs, glycosylation and MUC1 in CRL-1739 gastric cancer cell line. *Biomed. Pharmacother.* **2018**, *107*, 397–407. [CrossRef] [PubMed]

41. Ma, Z.-J.; Yan, H.; Wang, Y.-J.; Yang, Y.; Li, X.-B.; Shi, A.-C.; Jing-Wen, X.; Yu-Bao, L.; Li, L.; Wang, X.-X. Proteomics analysis demonstrating rosmarinic acid suppresses cell growth by blocking the glycolytic pathway in human HepG2 cells. *Biomed. Pharmacother.* **2018**, *105*, 334–349. [CrossRef]
42. Cui, H.-Y.; Zhang, X.-J.; Yang, Y.; Zhang, C.; Zhu, C.-H.; Miao, J.-Y.; Chen, R. Rosmarinic acid elicits neuroprotection in ischemic stroke via Nrf2 and heme oxygenase 1 signaling. *Neural Regen. Res.* **2018**, *13*, 2119–2128.
43. Wang, J.; Li, G.; Rui, T.; Kang, A.; Li, G.; Fu, T.; Li, J.; Di, L.; Cai, B. Pharmacokinetics of rosmarinic acid in rats by LC-MS/MS: Absolute bioavailability and dose proportionality. *RSC Adv.* **2017**, *7*, 9057–9063. [CrossRef]
44. da Silva, S.B.; Amorim, M.; Fonte, P.; Madureira, R.; Ferreira, D.; Pintado, M.; Sarmento, B. Natural extracts into chitosan nanocarriers for rosmarinic acid drug delivery. *Pharm. Biol.* **2015**, *53*, 642–652. [CrossRef]
45. Madureira, A.R.; Campos, D.A.; Fonte, P.; Nunes, S.; Reis, F.; Gomes, A.M.; Sarmento, B.; Pintado, M.M. Characterization of solid lipid nanoparticles produced with carnauba wax for rosmarinic acid oral delivery. *RSC Adv.* **2015**, *5*, 22665–22673. [CrossRef]
46. Xu, X.; Huang, M.; Zou, X. Docking-based inverse virtual screening: Methods, applications, and challenges. *Biochem. Biophys. Rep.* **2018**, *4*, 1–16. [CrossRef]
47. Warrier, S.B.; Kharkar, P.S. Inverse Virtual Screening in Drug Repositioning: Detailed Investigation and Case Studies. In *Crystallizing Ideas–The Role of Chemistry*; Springer: Berlin/Heidelberg, Germany, 2016; pp. 71–83.
48. Konc, J. Identification of neurological disease targets of natural products by computational screening. *Neural Regen. Res.* **2019**, *14*, 2075. [CrossRef]
49. Kores, K.; Lešnik, S.; Bren, U.; Janežič, D.; Konc, J. Discovery of novel potential human targets of resveratrol by inverse molecular docking. *J. Chem. Inf. Model.* **2019**, *59*, 2467–2478. [CrossRef] [PubMed]
50. Furlan, V.; Konc, J.; Bren, U. Inverse molecular docking as a novel approach to study anticarcinogenic and anti-neuroinflammatory effects of curcumin. *Molecules* **2018**, *23*, 3351. [CrossRef] [PubMed]
51. Sterling, T.; Irwin, J.J. ZINC 15–ligand discovery for everyone. *J. Chem. Inf. Model.* **2015**, *55*, 2324–2337. [CrossRef]
52. Frisch, M.; Trucks, G.; Schlegel, H.; Scuseria, G.; Robb, M.; Cheeseman, J.; Scalmani, G.; Barone, V.; Petersson, G.; Nakatsuji, H. *Gaussian 16*; Gaussian, Inc.: Wallingford, CT, USA, 2016.
53. Duraán-Iturbide, N.A.; Díaz-Eufracio, B.r.I.; Medina-Franco, J.L. In silico ADME/Tox profiling of natural products: A focus on BIOFACQUIM. *ACS Omega* **2020**, *5*, 16076–16084. [CrossRef]
54. Daina, A.; Michielin, O.; Zoete, V. SwissADME: A free web tool to evaluate pharmacokinetics, drug-likeness and medicinal chemistry friendliness of small molecules. *Sci. Rep.* **2017**, *7*, 1–13. [CrossRef]
55. Fine, J.; Konc, J.; Samudrala, R.; Chopra, G. Candock: Chemical atomic network-based hierarchical flexible docking algorithm using generalized statistical potentials. *J. Chem. Inf. Model.* **2020**, *60*, 1509–1527. [CrossRef]
56. Konc, J.; Lešnik, S.; Škrlj, B.; Janežič, D. ProBiS-Dock Database: A Web Server and Interactive Web Repository of Small Ligand–Protein Binding Sites for Drug Design. *J. Chem. Inf. Model.* **2021**, *61*, 4097–4107. [CrossRef]
57. Konc, J.; Janežič, D. An improved branch and bound algorithm for the maximum clique problem. *MATCH Commun. Math. Comput. Chem.* **2007**, *4*, 569–590.
58. Konc, J.; Miller, B.T.; Štular, T.; Lešnik, S.; Woodcock, H.L.; Brooks, B.R.; Janežič, D. ProBiS-CHARMMing: Web interface for prediction and optimization of ligands in protein binding sites. *J. Chem. Inf. Model.* **2015**, *55*, 2308–2314. [CrossRef] [PubMed]
59. Triballeau, N.; Acher, F.; Brabet, I.; Pin, J.-P.; Bertrand, H.-O. Virtual screening workflow development guided by the "receiver operating characteristic" curve approach. Application to high-throughput docking on metabotropic glutamate receptor subtype 4. *J. Med. Chem.* **2005**, *48*, 2534–2547. [CrossRef] [PubMed]
60. Truchon, J.-F.; Bayly, C.I. Evaluating virtual screening methods: Good and bad metrics for the "early recognition" problem. *J. Chem. Inf. Model.* **2007**, *47*, 488–508. [CrossRef] [PubMed]
61. Empereur-Mot, C.; Guillemain, H.; Latouche, A.; Zagury, J.-F.; Viallon, V.; Montes, M. Predictiveness curves in virtual screening. *J. Cheminform.* **2015**, *7*, 1–17. [CrossRef]
62. Gaulton, A.; Bellis, L.J.; Bento, A.P.; Chambers, J.; Davies, M.; Hersey, A.; Light, Y.; McGlinchey, S.; Michalovich, D.; Al-Lazikani, B. ChEMBL: A large-scale bioactivity database for drug discovery. *Nucleic Acids Res.* **2012**, *40*, D1100–D1107. [CrossRef]
63. Sheridan, R.P.; Singh, S.B.; Fluder, E.M.; Kearsley, S.K. Protocols for bridging the peptide to nonpeptide gap in topological similarity searches. *J. Chem. Inf. Comput. Sci.* **2001**, *41*, 1395–1406. [CrossRef]
64. Empereur-Mot, C.; Zagury, J.-F.; Montes, M. Screening explorer–An interactive tool for the analysis of screening results. *J. Chem. Inf. Model.* **2016**, *56*, 2281–2286. [CrossRef]
65. Ahmad, H.H.; Hamza, A.H.; Hassan, A.Z.; Sayed, A.H. Promising therapeutic role of Rosmarinus officinalis successive methanolic fraction against colorectal cancer. *Int. J. Pharm. Pharm. Sci* **2013**, *5*, 164–170.
66. Sacco, C.; Bellumori, M.; Santomauro, F.; Donato, R.; Capei, R.; Innocenti, M.; Mulinacci, N. An in vitro evaluation of the antibacterial activity of the non-volatile phenolic fraction from rosemary leaves. *Nat. Prod. Res.* **2015**, *29*, 1537–1544. [CrossRef]
67. Moreno, S.; Scheyer, T.; Romano, C.S.; Vojnov, A.A. Antioxidant and antimicrobial activities of rosemary extracts linked to their polyphenol composition. *Free Radic. Res.* **2006**, *40*, 223–231. [CrossRef] [PubMed]
68. Pavić, V.; Jakovljević, M.; Molnar, M.; Jokić, S. Extraction of carnosic acid and carnosol from sage (*Salvia officinalis* L.) leaves by supercritical fluid extraction and their antioxidant and antibacterial activity. *Plants* **2019**, *8*, 16. [CrossRef] [PubMed]
69. Pariš, A.; Štrukelj, B.; Renko, M.; Turk, V.; Pukl, M.; Umek, A.; Korant, B.D. Inhibitory Effect of Carnosolic Acid on HIV-1 Protease in Cell-Free Assays. *J. Nat. Prod.* **1993**, *56*, 1426–1430. [CrossRef]

70. Falo, M.; Fillmore, H.; Reeves, T.; Phillips, L. Matrix metalloproteinase-3 expression profile differentiates adaptive and maladaptive synaptic plasticity induced by traumatic brain injury. *J. Neurosci. Res.* **2006**, *84*, 768–781. [CrossRef] [PubMed]
71. López-Jiménez, A.; García-Caballero, M.; Medina, M.Á.; Quesada, A.R. Anti-angiogenic properties of carnosol and carnosic acid, two major dietary compounds from rosemary. *Eur. J. Nutr.* **2013**, *52*, 85–95. [CrossRef]
72. Kranenburg, O. The KRAS oncogene: Past, present, and future. *Biochim. Biophys. Acta* **2005**, *1756*, 81–82. [CrossRef]
73. Matikas, A.; Mistriotis, D.; Georgoulias, V.; Kotsakis, A. Targeting KRAS mutated non-small cell lung cancer: A history of failures and a future of hope for a diverse entity. *Crit. Rev. Oncol. Hematol.* **2017**, *110*, 1–12. [CrossRef]
74. Fell, J.B.; Fischer, J.P.; Baer, B.R.; Blake, J.F.; Bouhana, K.; Briere, D.M.; Brown, K.D.; Burgess, L.E.; Burns, A.C.; Burkard, M.R.; et al. Identification of the clinical development candidate MRTX849, a covalent KRASG12C inhibitor for the treatment of cancer. *J. Med. Chem.* **2020**, *63*, 6679–6693. [CrossRef]
75. Ahmed, Z.; Abdeslam-Hassan, M.; Ouassila, L.; Danielle, B. Extraction and modeling of Algerian rosemary essential oil using supercritical CO_2: Effect of pressure and temperature. *Energy Procedia* **2012**, *18*, 1038–1046. [CrossRef]
76. Balskus, E.P. Colibactin: Understanding an elusive gut bacterial genotoxin. *Nat. Prod. Rep.* **2015**, *32*, 1534–1540. [CrossRef]
77. Teplyakov, A.; Obmolova, G.; Badet-Denisot, M.-A.; Badet, B.; Polikarpov, I. Involvement of the C terminus in intramolecular nitrogen channeling in glucosamine 6-phosphate synthase: Evidence from a 1.6\AA crystal structure of the isomerase domain. *Structure* **1998**, *6*, 1047–1055. [CrossRef]
78. Bearne, S.L.; Blouin, C. Inhibition of Escherichia coli Glucosamine-6-phosphate Synthase by Reactive Intermediate Analogues: The Role of the 2-amino function in Catalysis. *J. Biol. Chem.* **2000**, *275*, 135–140. [CrossRef]
79. Fadaka, A.; Ajiboye, B.; Ojo, O.; Adewale, O.; Olayide, I.; Emuowhochere, R. Biology of glucose metabolization in cancer cells. *J. Oncol. Sci.* **2017**, *3*, 45–51. [CrossRef]
80. Vander Heiden, M.G.; Christofk, H.R.; Schuman, E.; Subtelny, A.O.; Sharfi, H.; Harlow, E.E.; Xian, J.; Cantley, L.C. Identification of small molecule inhibitors of pyruvate kinase M2. *Biochem. Pharmacol.* **2010**, *79*, 1118–1124. [CrossRef] [PubMed]
81. Skehel, J.J.; Wiley, D.C. Receptor binding and membrane fusion in virus entry: The influenza hemagglutinin. *Annu. Rev. Biochem.* **2000**, *69*, 531–569. [CrossRef] [PubMed]
82. Wong, S.-S.; Webby, R.J. Traditional and new influenza vaccines. *Clin. Microbiol. Rev.* **2013**, *26*, 476–492. [CrossRef]
83. Kadam, R.U.; Wilson, I.A. Structural basis of influenza virus fusion inhibition by the antiviral drug Arbidol. *Proc. Natl. Acad. Sci. USA* **2017**, *114*, 206–214. [CrossRef]
84. Lv, Z.; Chu, Y.; Wang, Y. HIV protease inhibitors: A review of molecular selectivity and toxicity. *HIV AIDS* **2015**, *7*, 95.
85. Malhotra, S.; Dhundial, R.; Bhatia, N.; Duggal, N. HIV-2 Infections from a Tertiary Care Hospital in India-A Case Report. *J. Hum. Virol. Retrovirol.* **2018**, *5*, 2–6.
86. De Silva, T.; Weiss, R.A. HIV-2 goes global: An unaddressed issue in Indian anti-retroviral programmes. *Indian J. Med. Res.* **2010**, *132*, 660.
87. Visseaux, B.; Damond, F.; Matheron, S.; Descamps, D.; Charpentier, C. HIV-2 molecular epidemiology. *Infect. Genet. Evol.* **2016**, *46*, 233–240. [CrossRef]
88. De Silva, T.I.; van Tienen, C.; Rowland-Jones, S.L.; Cotten, M. Dual infection with HIV-1 and HIV-2: Double trouble or destructive interference? *HIV Ther.* **2010**, *4*, 305–323.
89. Annabel, B.; Anna, D.; Hannah, M. *Global Tuberculosis Report 2019*; World Health Organization: Geneva, Switzerland, 2019.
90. Garzan, A.; Willby, M.J.; Green, K.D.; Tsodikov, O.V.; Posey, J.E.; Garneau-Tsodikova, S. Discovery and optimization of two Eis inhibitor families as kanamycin adjuvants against drug-resistant M. tuberculosis. *ACS Med. Chem. Lett.* **2016**, *7*, 1219–1221. [CrossRef]
91. Houghton, J.L.; Biswas, T.; Chen, W.; Tsodikov, O.V.; Garneau-Tsodikova, S. Chemical and structural insights into the regioversatility of the aminoglycoside acetyltransferase Eis. *Chembiochem* **2013**, *14*, 2127. [CrossRef] [PubMed]
92. Wu, C.-C.; Baiga, T.J.; Downes, M.; La Clair, J.J.; Atkins, A.R.; Richard, S.B.; Fan, W.; Stockley-Noel, T.A.; Bowman, M.E.; Noel, J.P.; et al. Structural basis for specific ligation of the peroxisome proliferator-activated receptor δ. *Proc. Natl. Acad. Sci. USA* **2017**, *114*, E2563–E2570. [CrossRef] [PubMed]
93. Dressel, U.; Allen, T.L.; Pippal, J.B.; Rohde, P.R.; Lau, P.; Muscat, G.E. The peroxisome proliferator-activated receptor β/δ agonist, GW501516, regulates the expression of genes involved in lipid catabolism and energy uncoupling in skeletal muscle cells. *Mol. Endocrinol.* **2003**, *17*, 2477–2493. [CrossRef] [PubMed]
94. Liu, Y.; Colby, J.K.; Zuo, X.; Jaoude, J.; Wei, D.; Shureiqi, I. The role of PPAR-δ in metabolism, inflammation, and cancer: Many characters of a critical transcription factor. *Int. J. Mol. Sci.* **2018**, *19*, 3339. [CrossRef]
95. Waku, T.; Shiraki, T.; Oyama, T.; Maebara, K.; Nakamori, R.; Morikawa, K. The nuclear receptor PPARγ individually responds to serotonin-and fatty acid-metabolites. *EMBO J.* **2010**, *29*, 3395–3407. [CrossRef]
96. Alexacou, K.-M.; Tenchiu, A.-C.; Chrysina, E.D.; Charavgi, M.-D.; Kostas, I.D.; Zographos, S.E.; Oikonomakos, N.G.; Leonidas, D.D. The binding of β-d-glucopyranosyl-thiosemicarbazone derivatives to glycogen phosphorylase: A new class of inhibitors. *Biorg. Med. Chem.* **2010**, *18*, 7911–7922. [CrossRef]
97. Treadway, J.L.; Mendys, P.; Hoover, D.J. Glycogen phosphorylase inhibitors for treatment of type 2 diabetes mellitus. *Expert Opin. Investig. Drugs* **2001**, *10*, 439–454. [CrossRef]
98. Spasov, A.; Chepljaeva, N.; Vorob'ev, E. Glycogen phosphorylase inhibitors in the regulation of carbohydrate metabolism in type 2 diabetes. *Russ. J. Bioorganic Chem.* **2016**, *42*, 133–142. [CrossRef]

99. Martin, J.L.; Johnson, L.N.; Withers, S.G. Comparison of the binding of glucose and glucose 1-phosphate derivatives to T-state glycogen phosphorylase b. *Biochemistry* **1990**, *29*, 10745–10757. [CrossRef]
100. Barbier, P.; Tsvetkov, P.O.; Breuzard, G.; Devred, F. Deciphering the molecular mechanisms of anti-tubulin plant derived drugs. *Phytochem. Rev.* **2014**, *13*, 157–169. [CrossRef]
101. Nawrotek, A.; Knossow, M.; Gigant, B. The determinants that govern microtubule assembly from the atomic structure of GTP-tubulin. *J. Mol. Biol.* **2011**, *412*, 35–42. [CrossRef] [PubMed]
102. Bomalaski, J.S.; Clark, M.A. Phospholipase A2 and arthritis. *Arthritis Rheum.* **1993**, *36*, 190–198. [CrossRef] [PubMed]
103. Cummings, B.S. Phospholipase A2 as targets for anti-cancer drugs. *Biochem. Pharmacol.* **2007**, *74*, 949–959. [CrossRef]
104. Mallat, Z.; Lambeau, G.; Tedgui, A. Lipoprotein-associated and secreted phospholipases A2 in cardiovascular disease: Roles as biological effectors and biomarkers. *Circulation* **2010**, *122*, 2183–2200. [CrossRef] [PubMed]
105. Ong, W.-Y.; Farooqui, T.; Kokotos, G.; Farooqui, A.A. Synthetic and natural inhibitors of phospholipases A2: Their importance for understanding and treatment of neurological disorders. *ACS Chem. Neurosci.* **2015**, *6*, 814–831. [CrossRef]
106. Ivy, S.P.; Wick, J.Y.; Kaufman, B.M. An overview of small-molecule inhibitors of VEGFR signaling. *Nat. Rev. Clin. Oncol.* **2009**, *6*, 569–579. [CrossRef]
107. Kufareva, I.; Abagyan, R. Type-II kinase inhibitor docking, screening, and profiling using modified structures of active kinase states. *J. Med. Chem.* **2008**, *51*, 7921–7932. [CrossRef] [PubMed]
108. Vásquez, A.F.; Reyes Munoz, A.; Duitama, J.; González Barrios, A. Discovery of new potential CDK2/VEGFR2 type II inhibitors by fragmentation and virtual screening of natural products. *J. Biomol. Struct. Dyn.* **2021**, *39*, 3285–3299. [CrossRef]
109. Rathi, E.; Kumar, A.; Kini, S.G. Molecular dynamics guided insight, binding free energy calculations and pharmacophore-based virtual screening for the identification of potential VEGFR2 inhibitors. *J. Recept. Signal Transduct.* **2019**, *39*, 415–433. [CrossRef] [PubMed]
110. Banerjee, A.K.; Arora, N.; Murty, U.S.N. Aspartate carbamoyltransferase of *Plasmodium falciparum* as a potential drug target for designing anti-malarial chemotherapeutic agents. *Med. Chem. Res.* **2012**, *21*, 2480–2493. [CrossRef]
111. Bosch, S.S.; Lunev, S.; Batista, F.A.; Linzke, M.; Kronenberger, T.; Dömling, A.S.; Groves, M.R.; Wrenger, C. Molecular Target Validation of Aspartate Transcarbamoylase from *Plasmodium falciparum* by Torin 2. *ACS Infect. Dis.* **2020**, *6*, 986–999. [CrossRef]
112. Wang, J.; Stieglitz, K.A.; Cardia, J.P.; Kantrowitz, E.R. Structural basis for ordered substrate binding and cooperativity in aspartate transcarbamoylase. *Proc. Natl. Acad. Sci. USA* **2005**, *102*, 8881–8886. [CrossRef] [PubMed]
113. Dos Santos, J.I.; Cardoso, F.F.; Soares, A.M.; dal Pai Silva, M.; Gallacci, M.; Fontes, M.R. Structural and functional studies of a bothropic myotoxin complexed to rosmarinic acid: New insights into Lys49-PLA2 inhibition. *PLoS ONE* **2011**, *6*, e28521. [CrossRef] [PubMed]
114. Han, S.; Li, X.; Xia, Y.; Yu, Z.; Cai, N.; Malwal, S.R.; Han, X.; Oldfield, E.; Zhang, Y. Farnesyl Pyrophosphate Synthase as a Target for Drug Development: Discovery of Natural-Product-Derived Inhibitors and Their Activity in Pancreatic Cancer Cells. *J. Med. Chem.* **2019**, *62*, 10867–10896. [CrossRef] [PubMed]
115. Hoffman, M.; Monroe, D.; Oliver, J.; Roberts, H. Factors IXa and Xa play distinct roles in tissue factor-dependent initiation of coagulation. *Blood* **1995**, *86*, 1794–1801. [CrossRef]
116. Chen, X.; Zhou, L.; Zhang, Y.; Yi, D.; Liu, L.; Rao, W.; Wu, Y.; Ma, D.; Liu, X.; Zhou, X.-H.A.; et al. Risk factors of stroke in Western and Asian countries: A systematic review and meta-analysis of prospective cohort studies. *BMC Public Health* **2014**, *14*, 776. [CrossRef]
117. Li, S.; Peng, Y.; Wang, X.; Qian, Y.; Xiang, P.; Wade, S.W.; Guo, H.; Lopez, J.A.G.; Herzog, C.A.; Handelsman, Y. Cardiovascular events and death after myocardial infarction or ischemic stroke in an older Medicare population. *Clin. Cardiol.* **2019**, *42*, 391–399. [CrossRef] [PubMed]
118. Perzborn, E.; Roehrig, S.; Straub, A.; Kubitza, D.; Mueck, W.; Laux, V. Rivaroxaban: A new oral factor Xa inhibitor. *Atertio. Thromb. Vasc. Biol.* **2010**, *30*, 376–381. [CrossRef]
119. Komoriya, S.; Kobayashi, S.; Osanai, K.; Yoshino, T.; Nagata, T.; Haginoya, N.; Nakamoto, Y.; Mochizuki, A.; Nagahara, T.; Suzuki, M. Design, synthesis, and biological activity of novel factor Xa inhibitors: Improving metabolic stability by S1 and S4 ligand modification. *Biorg. Med. Chem.* **2006**, *14*, 1309–1330. [CrossRef] [PubMed]
120. Alcaraz, L.A.; Banci, L.; Bertini, I.; Cantini, F.; Donaire, A.; Gonnelli, L. Matrix metalloproteinase–inhibitor interaction: The solution structure of the catalytic domain of human matrix metalloproteinase-3 with different inhibitors. *J. Biol. Inorg. Chem.* **2007**, *12*, 1197–1206. [CrossRef]
121. World Helath Organization. *Control of the Leishmaniases WHO Technical Report Series 949*; World Health Organization: Geneva, Switzerland, 2010.
122. Martin, M.B.; Grimley, J.S.; Lewis, J.C.; Heath, H.T.; Bailey, B.N.; Kendrick, H.; Yardley, V.; Caldera, A.; Lira, R.; Urbina, J.A.; et al. Bisphosphonates Inhibit the Growth of Trypanosoma b rucei, Trypanosoma c ruzi, Leishmania d onovani, Toxoplasma g ondii, and Plasmodium f alciparum: A Potential Route to Chemotherapy. *J. Med. Chem.* **2001**, *44*, 909–916. [CrossRef] [PubMed]
123. Jin, L.; Li, D.; Alesi, G.N.; Fan, J.; Kang, H.-B.; Lu, Z.; Boggon, T.J.; Jin, P.; Yi, H.; Wright, E.R.; et al. Glutamate dehydrogenase 1 signals through antioxidant glutathione peroxidase 1 to regulate redox homeostasis and tumor growth. *Cancer Cell* **2015**, *27*, 257–270. [CrossRef]
124. McDermott, L.A.; Iyer, P.; Vernetti, L.; Rimer, S.; Sun, J.; Boby, M.; Yang, T.; Fioravanti, M.; O'Neill, J.; Wang, L.; et al. Design and evaluation of novel glutaminase inhibitors. *Biorg. Med. Chem.* **2016**, *24*, 1819–1839. [CrossRef]

125. Li, C.; Allen, A.; Kwagh, J.; Doliba, N.M.; Qin, W.; Najafi, H.; Collins, H.W.; Matschinsky, F.M.; Stanley, C.A.; Smith, T.J. Green tea polyphenols modulate insulin secretion by inhibiting glutamate dehydrogenase. *J. Biol. Chem.* **2006**, *281*, 10214–10221. [CrossRef] [PubMed]
126. Li, C.; Li, M.; Chen, P.; Narayan, S.; Matschinsky, F.M.; Bennett, M.J.; Stanley, C.A.; Smith, T.J. Green tea polyphenols control dysregulated glutamate dehydrogenase in transgenic mice by hijacking the ADP activation site. *J. Biol. Chem.* **2011**, *286*, 34164–34174. [CrossRef]
127. Li, M.; Smith, C.J.; Walker, M.T.; Smith, T.J. Novel Inhibitors Complexed with Glutamate Dehydrogenase Allosteric Regulation by Control of Protein Dynamics. *J. Biol. Chem.* **2009**, *284*, 22988–23000. [CrossRef] [PubMed]
128. DeLaBarre, B.; Gross, S.; Fang, C.; Gao, Y.; Jha, A.; Jiang, F.; Song, J.J.; Wei, W.; Hurov, J.B. Full-length human glutaminase in complex with an allosteric inhibitor. *Biochemistry* **2011**, *50*, 10764–10770. [CrossRef] [PubMed]
129. Thangavelu, K.; Pan, C.Q.; Karlberg, T.; Balaji, G.; Uttamchandani, M.; Suresh, V.; Schüler, H.; Low, B.C.; Sivaraman, J. Structural basis for the allosteric inhibitory mechanism of human kidney-type glutaminase (KGA) and its regulation by Raf-Mek-Erk signaling in cancer cell metabolism. *Proc. Natl. Acad. Sci. USA* **2012**, *109*, 7705–7710. [CrossRef]
130. Furlan, V.; Bren, U. Insight into Inhibitory Mechanism of PDE4D by Dietary Polyphenols Using Molecular Dynamics Simulations and Free Energy Calculations. *Biomolecules* **2021**, *11*, 479. [CrossRef]
131. Kores, K.; Konc, J.; Bren, U. Mechanistic insights into side effects of troglitazone and rosiglitazone using a novel inverse molecular docking protocol. *Pharmaceutics* **2021**, *13*, 315. [CrossRef]
132. Jukič, M.; Kores, K.; Janežič, D.; Bren, U. Repurposing of Drugs for SARS-CoV-2 Using Inverse Docking Fingerprints. *Front. Chem.* **2021**, *9*, 757826. [CrossRef]

Article

Inhibition of Biofilm Formation of Foodborne *Staphylococcus aureus* by the Citrus Flavonoid Naringenin

Qing-Hui Wen [1,2], Rui Wang [1,2], Si-Qi Zhao [1,2], Bo-Ru Chen [1,2] and Xin-An Zeng [1,3,*]

1. School of Food Science and Engineering, South China University of Technology, Guangzhou 510641, China; 201610103987@mail.scut.edu.cn (Q.-H.W.); 201910105089@mail.scut.edu.cn (R.W.); 202120126516@mail.scut.edu.cn (S.-Q.Z.); 201811006830@mail.scut.edu.cn (B.-R.C.)
2. Overseas Expertise Introduction Center for Discipline Innovation of Food Nutrition and Human Health (111 Center), Guangzhou 510641, China
3. School of Food Science and Engineering, Foshan University, Foshan 528000, China
* Correspondence: xazeng@scut.edu.cn; Tel.: +86-208-7112-894

Abstract: Taking into consideration the importance of biofilms in food deterioration and the potential risks of antiseptic compounds, antimicrobial agents that naturally occurring are a more acceptable choice for preventing biofilm formation and in attempts to improve antibacterial effects and efficacy. Citrus flavonoids possess a variety of biological activities, including antimicrobial properties. Therefore, the anti-biofilm formation properties of the citrus flavonoid naringenin on the *Staphylococcus aureus* ATCC 6538 (*S. aureus*) were investigated using subminimum inhibitory concentrations (sub-MICs) of 5~60 mg/L. The results were confirmed using laser and scanning electron microscopy techniques, which revealed that the thick coating of *S. aureus* biofilms became thinner and finally separated into individual colonies when exposed to naringenin. The decreased biofilm formation of *S. aureus* cells may be due to a decrease in cell surface hydrophobicity and exopolysaccharide production, which is involved in the adherence or maturation of biofilms. Moreover, transcriptional results show that there was a downregulation in the expression of biofilm-related genes and alternative sigma factor *sigB* induced by naringenin. This work provides insight into the anti-biofilm mechanism of naringenin in *S. aureus* and suggests the possibility of naringenin being used in the industrial food industry for the prevention of biofilm formation.

Keywords: naringenin; biofilm formation; cell surface hydrophobicity; confocal laser scanning microscopy; biofilm-related genes

1. Introduction

Staphylococcus aureus is a common pathogen and is responsible for food poisoning through the production of thermally stable enterotoxins in various kinds of food [1,2]. A microbial biofilm is an aggregation of bacteria that is composed of extracellular polymeric substances, which are attached on the surface of microorganisms [3]. The most common feature of microbial lifestyles is attachment onto a surface by biofilm formation. Notably, *S. aureus* can form biofilms on different surfaces in food processing plants and is very adaptable to various environmental stressors including acids, salts, antibiotics, and detergents [4–6]. The presence of *S. aureus* biofilm on food contact surfaces creates serious problems for the food industry because it can lead to food spoilage and disease transmission [7,8]. Therefore, it is important to inhibit the formation of *S. aureus* biofilms on food contact to surfaces ensure the manufacture of safe food products.

Flavonoids from fruits and vegetables have been shown to has a range of biological activities [9,10]. For example, the citrus flavonoid naringenin have beneficial effects on human health by preventing various diseases, including diabetes, hypertension and cancer [11–13]. Moreover, naringenin has wide antibacterial activity and can prevent the growth of numerous microorganisms [14,15]. Specifically, naringenin from bergamot peel inhibits *Escherichia coli*, *Lactococcus lactis*, *Salmonella* Enteritidis, and *Pseudomonas putida* with minimum inhibitory concentration (MIC) values ranging from 0.25 to 1.0 g/L [9]. A small number of studies have reported that naringenin inhibits the biofilm formation of microorganisms (*Escherichia coli*, *Vibrio harveyi* and *Streptococcus mutans*) by affecting the expression of bacteria related genes and surface hydrophobicity [16,17]. In our earlier study, we determined that naringenin has strong antibacterial activity against *S. aureus* via such mechanisms of action as disrupting the bacterial cytoplasmic membrane and binding to its genomic DNA [18]. Moreover, we also found that naringenin has a strong effect in suppressing the biofilm formation of S. aureus on the surface of glass and plastic well plates. However, to the best of our knowledge, research into biofilm inhibition by naringenin is limited, and its anti-biofilm mechanism is also unclear.

Hence, we aimed to study the effect of naringenin on the inhibition of the biofilm formation of *S. aureus* at different temperatures (25 and 37 °C) using confocal laser and scanning electron microscopy techniques and exopolysaccharide production (EPS) and hydrophobicity assays. Furthermore, our study also investigated the genes (*sigB*) related to *S. aureus* biofilms using RT-qPCR, which is the main regulator of gene transcription and expression under the stress conditions induced by naringenin.

2. Materials and Methods

2.1. Bacterial Strain and Biofilm Formation

The foodborne strain *S. aureus* was obtained from the Microbial Culture Collection Center of Guangdong Institute of Microbiology (Guangzhou, China) and activated by culturing twice in 100 mL of sterile tryptic soy broth (TSB, Beijing Aoboxing Biotechnology Co., Ltd., Beijing, China) at 37 °C for 24 h. The effect of naringenin (purity ≥ 98%, Aladdin Chemical Co., Shanghai, China) on *S. aureus* growth was evaluated by transferring pre-cultured *S. aureus* cells into fresh TSB liquid medium ($OD_{600\ nm} \approx 0.08$) and cultivating in 96-well polystyrene plates at different temperature (25 and 37 °C) with gentle shaking. In order to measure the absorbance value of *S. aureus* growth, a FilterMax F5 multifunctional microplate reader (American molecular devices company, Sunnyvale, CA, USA) was used.

The biofilm assay was performed under similar conditions without shaking. The volume of DMSO, that was used to dissolve naringenin was equal in all of the samples, while the final concentration of naringenin varied from 0 to 60 mg/L. The crystal violet staining method was used to quantify *S. aureus* biofilm according to a relevant publication [19].

2.2. Cell Surface Hydrophobicity of S. aureus

The effects of naringenin on *S. aureus* cell surface hydrophobicity were evaluated at 25 and 37 °C by analyzing cells adhesion to xylene, as previously described [20]. After cultivation to the stationary-phase (48 h for *S. aureus* at 25 °C, and 12 h for *S. aureus* at 37 °C), *S. aureus* cells were collected by refrigerated centrifugation at 4000× *g* for 5 min. The *S. aureus* pellet was washed twice using distilled water and resuspended in 3 mL of a 0.85% NaCl solution ($OD_{600\ nm} \approx 0.3$), which defined as A_1. Xylene (1 mL) was added to a 3 mL suspension of *S. aureus* and then incubated for 15 min at 25 °C. After vortexing for two minutes and then incubating for 15 min, the mixture separated into a xylene/water bilayer system. The $OD_{600\ nm}$ of the aqueous phase of the bilayer was recorded as A_2. The index of cell surface hydrophobicity (I) was determined using Equation (1):

$$I = (1 - A_2/A_1) \times 100\% \tag{1}$$

2.3. Quantification EPS Production of S. aureus

After growing in the various subminimum inhibitory concentrations of naringenin, the *S. aureus* cells were centrifuged (12,000× *g* for 15 min at 4 °C) and the supernatant of *S. aureus* was then filtered through glass fiber filters. An equal volume of absolute ethanol was then added to this supernatant of *S. aureus* and incubated overnight at 4 °C to precipitate EPS. The precipitate was resuspended in water with gentle heating (50 °C) and then quantified using a phenol–sulfuric acid procedure [21]. The percentage of EPS reduction upon exposure to naringenin was evaluated using Equation (2):

$$\text{Reduction of EPS (\%)} = (1 - E_2/E_1) \times 100\% \qquad (2)$$

where E_1 and E_2 are the absorbances at 490 nm for *S. aureus* cells grown in the absence and presence of naringenin, respectively.

2.4. CLSM and SEM of S. aureus Biofilms

A 3 mL aliquot of inoculum ($OD_{600\,nm} \approx 0.1$) was transferred into the wells of a 6-well plate containing 13 mm-diameter sterile glass coverslips. After incubation at 25 and 37 °C for 48 h and 12 h respectively, the medium with free-floating *S. aureus* cells was removed and the coverslips washed thrice in sterile 0.85% saline solution. *S. aureus* biofilms on glass coverslips were then stained for 20 min in the dark at room temperature with diluted 5(6)-carboxy fluorescein diacetate succinimidyl ester (Aladdin Chemical Co., Shanghai, China). The stained biofilms were adjusted and photographed using a confocal laser scanning microscope (Leica, Wetzlar, Germany). At least ten pictures were taken from different locations for each sample, and the image data were then processed and analyzed.

S. aureus biofilms were prepared for SEM analysis as previously described, with minor modifications [18]. Glutaraldehyde (2.5% in 0.01 M phosphate buffer, pH 7.2) was added to the samples and incubated overnight at 4 °C and then dehydrated using a series of ethanol solutions (20 min each time) of increasing concentration (30~100%). The dehydrated biofilms were then incubated in tertiary butanol twice for 20 min each, followed by air-drying overnight. After gold-coating by ion sputtering (Jeol JFC-1100, Tokyo, Japan), *S. aureus* biofilms were photographed by scanning electron microscopy (SEM, Zeiss EVO18, Germany) with operation at 10.0 kV.

2.5. RNA Extraction and Real-Time Quantitative PCR (RT-qPCR) Analysis

TRIzol reagent (Invitrogen, CA, USA) was used according to the kit instructions to extract the *S. aureus* RNA. To check the concentration and purity, the RNA was measure at OD_{260} and OD_{280} using an 1800 UV spectrophotometer (Shimadzu Corporation, Kyoto, Japan). cDNA was reverse transcribed from 800 ng RNA with 4.0 µL of 5 × reaction buffer, 0.5 µL Thermo Scientific RiboLock RNase Inhibitor (20 U) and 1.0 µL RevertAid Premium Reverse Transcriptase (200 U), following the protocol of RevertAid Premium First Strand cDNA Synthesis Kit (Thermo Scientific™ EP0733, Thermo Fisher Scientific, Waltham, MA, USA).

RT-qPCRs was performed on an Applied Biosystems StepOne Plus™ thermocycler (Life Technologies Inc., Milano, Italy) using the SybrGreen qPCR Master Mix, following the kit instructions. Reactions were carried out in a system which was composed of 10 µL Master Mix, 0.4 µL of 0.25 µM solutions of each primer (Table 1), and 2 µL cDNA, diluted to a final volume of 20 µL using double-distilled water (DNase-free). The following thermal profile was used: a holding step for 3 min at 95 °C, followed by a cycling step consisting of 45 cycles at 95 °C for 7 s (to melt), 57 °C for 10 s (to anneal) and 72 °C for 15 s (to extend). The endogenous reference gene of 16S rRNA was used to evaluate the changes in transcriptional levels of the *S. aureus* RNA.

2.6. Statistical Analysis

Results are expressed as means ± standard deviation (SD), and data graphics were drawn using OriginPro 7.0 (OriginLab, Northampton, MA, USA). SPSS software (IBM, Armonk, NY, USA) was used to analyze the variance (ANOVA) by Tukey's test, and $p < 0.05$ was represented for significant difference.

Table 1. Sequences of the primers used for RT-qPCR.

Gene	Primer
cidA	Forward 5′-AGCGTAATTTCGGAAGCAACATCCA-3′
	Reverse 5′-CCCTTAGCCGGCAGTATTGTTGGTC-3′
icaA	Forward 5′-CTG GCG CAG TCA ATA CTA TTT CGG GTG TCT-3′
	Reverse 5′-GAC CTC CCA ATG TTT CTG GAA CCA ACA TCC-3′
dltB	Forward 5′-GTGGACATCAGATTCACTTCC-3′
	Reverse 5′-ATAGAACCATCACGAATTTCC-3′
agrA	Forward 5′-TGATAATCCTTATGAGGTGCTT-3′
	Reverse 5′-CACTGTGACTCGTAACGAAAA-3′
sortaseA	Forward 5′-AAACCACATATCGATAATTATC-3′
	Reverse 5′-TTATTTGACTTCTGTAGCTACAA-3′
sarA	Forward 5′-CAAACAACCACAAGTTGTTAAAGC-3′
	Reverse 5′-TGTTTGCTTCAGTGATTCGTTT-3′
sigB	Forward 5′-AAGTGATTCGTAAGGACGTCT-3′
	Reverse 5′-TCGATAACTATAACCAAAGCCT-3′
16S rRNA	Forward 5′-CGGTGAATACGTTCYCGG-3′
	Reverse 5′-GGWTACCTTGTTACGACTT-3′

3. Results

3.1. Effects of Naringenin on S. aureus at Different Growth Temperatures

Figure 1 shows the effect of naringenin on cell growth profiles of *S. aureus* at different temperatures as reflected by the optical density (OD) at 600 nm. Subminimum inhibitory concentrations (MIC for 0.5 g/L) [22] values in the range of (5~60 mg/L) did not decrease the cell density of *S. aureus* at 25 °C (Figure 1a) and 37 °C (Figure 1c). However, the time taken to reach the stationary phase for *S. aureus* was significantly affected by temperatures. *S. aureus* cultured at 37 °C takes 12 h to reach the stationary phase, as compared to 48 h for *S. aureus* grown at 25 °C.

The effect of naringenin on the biofilms formed at different temperatures was measured by crystal violet staining, expressed as OD_{570} nm. Due to different growth rates, *S. aureus* at 25 and 37 °C were incubated for 48 and 12 h, respectively. *S. aureus* biofilm formation (measured at OD_{570} nm) decreased with increasing concentrations of naringenin. For example, the OD_{570} nm of *S. aureus* grown at 37 °C was reduced by 0.94 (32.3%) with 10 mg/mL and 0.61 (56.1%) with 20 mg/mL naringenin ($p < 0.05$). A further decrease was observed after the addition of 30 mg/mL or higher naringenin concentrations (Figure 1d). By contrast, *S. aureus* cultivated at 25 °C showed a larger decrease ($p < 0.05$) in biofilm formation (as measured by OD_{570} nm) in the presence of naringenin. For example, OD_{570} nm decreased from 1.21 to 0.22 (81.8%) after exposure to naringenin at a concentration of 20 mg/L (Figure 1b).

3.2. Changes in Cell Surface Hydrophobicity and EPS Production of S. aureus

Surface hydrophobicity of *S. aureus* cells was determined and expressed as hydrophobicity index (I). Figure 2a shows the significant dose-related reduction ($p < 0.05$) in cell hydrophobicity of *S. aureus* with increasing naringenin concentration at both 25 and 37 °C. Naringenin at a concentration of 10 mg/L dramatically reduced the surface hydrophobicity of *S. aureus* cells grown at 25 and 37 °C by 40.6% and 57.2% ($p < 0.05$), respectively. The respective values further decreased to 14.4% and 21.2% ($p < 0.05$) when the concentration of naringenin was increased to 40 mg/L.

The effect of naringenin on the EPS production of *S. aureus* was also investigated. The results reveal that *S. aureus* treated with various concentrations of naringenin (20, 40, and 60 mg/L) show a significant reduction ($p < 0.05$) in EPS compared to that of the control (Figure 2b). For *S. aureus* grown at 25°C, naringenin at 20 and 60 mg/mL reduced EPS by 59% and 5%, respectively. At 37 °C, the same concentrations of naringenin reduced EPS by 67% and 18%, respectively.

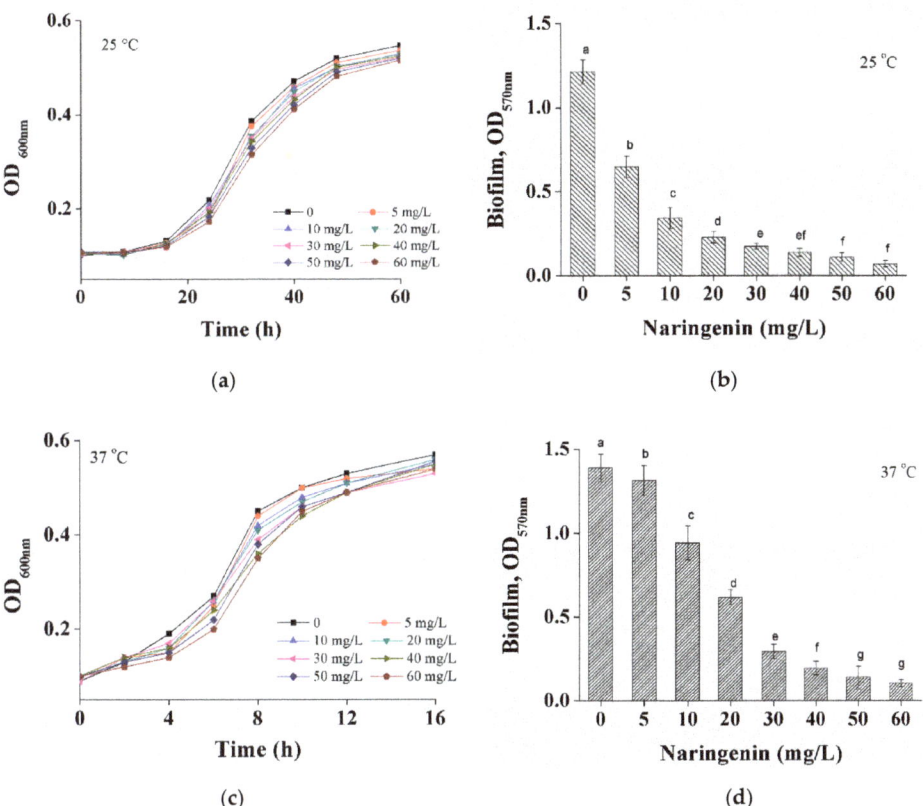

Figure 1. Cell growth of *S. aureus* in the presence of naringenin (0~60 mg/mL) was measured at $OD_{600\ nm}$ in 96-well plates at 25 °C (**a**) or 37 °C (**c**). Biofilm formations of *S. aureus* with naringenin concentration of 0~60 mg/mL at 25 °C (**b**) or 37 °C (**d**) for 48 and 12 h in 96-well plates, respectively. Biofilm OD values are processed as mean ± SD and a–g indicate significant differences between different columns ($p < 0.05$).

3.3. Microscopic Observations of S. aureus Biofilm

Direct visual information, including the surface coverage and thickness of *S. aureus* biofilms, were obtained by CLSM analyses. As shown in Figure 3a,b, *S. aureus* formed thick and compact biofilms covering the surface of glass coverslips at 25 and 37 °C when grown in the absence of naringenin. The confocal stack images show that the thick coating of *S. aureus* biofilms represented by cell aggregations became thinner and looser on the surfaces in the presence of 20 mg/L naringenin (Figure 3d). At 40 mg/L of naringenin, there was a visible reduction in the numbers of microcolonies for *S. aureus* cells grown at 37 °C (Figure 3f). Compared to 37 °C, the cells grown at 25 °C had a more obvious decrease associated with naringenin exposure, and the bacterial density was significantly decreased (Figure 3c,e). These results were further confirmed by SEM images.

The SEM images show that naringenin inhibited the bacterial growth of *S. aureus* at subminimum inhibitory concentrations (sub-MICs) values of 0, 20 and 40 mg/L. As the concentration of naringenin increases, the total number of bacteria obviously decreased, especially at high concentrations (Figure 4e,f) [17]. Compared to the incubation temperature of 37 °C (Figure 4b,d,f), the total number of *S. aureus* cultivated at 25 °C (Figure 4a,c,e) showed a greater decrease, demonstrating that naringenin has a significant effect in suppressing the biofilm formation of bacteria.

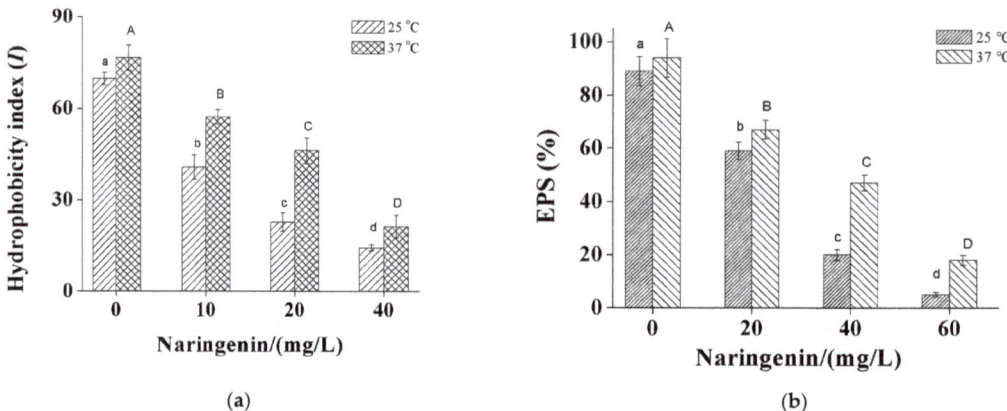

Figure 2. Effects of naringenin in the concentrations of 0, 20 and 40 mg/L on cell surface hydrophobicity (**a**) and EPS production (**b**) of *S. aureus*. Values are mean ± SD and there are significant differences between the values of columns marked with different letters (a–d) and (A–D), as indicated ($p < 0.05$).

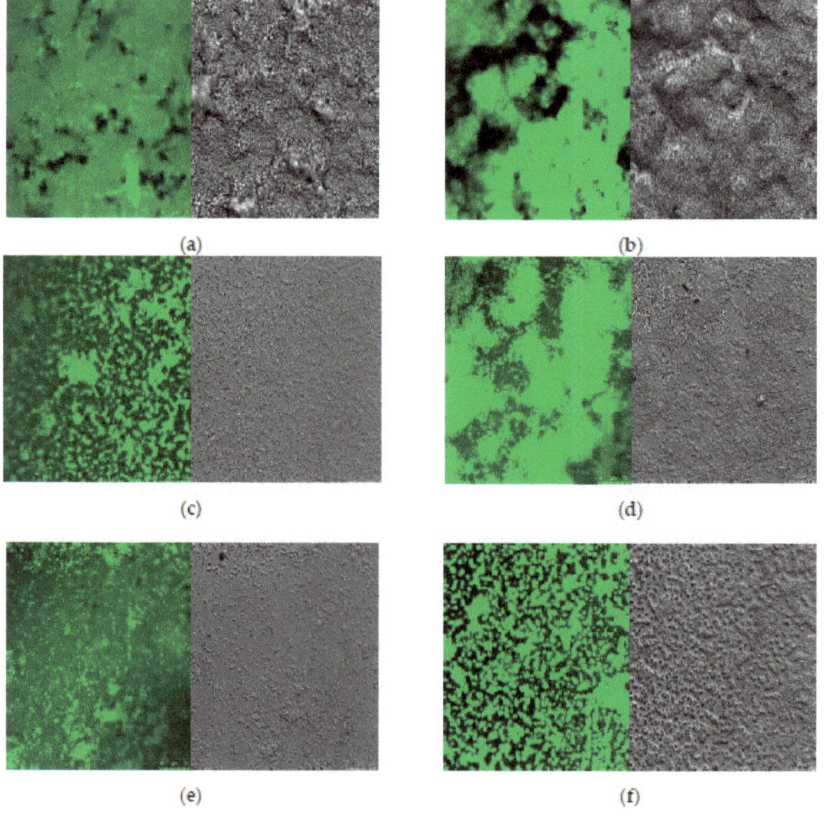

Figure 3. Confocal laser scanning microscopy (CLSM) analysis of biofilms formed by *S. aureus* incubated with different concentrations of naringenin. (**a,c,e**) for *S. aureus* cells were grown at 25 °C with naringenin at 0, 20 and 40 mg/L, respectively; (**b,d,f**) for *S. aureus* cells were grown at 37 °C with naringenin at 0, 20 and 40 mg/L, respectively.

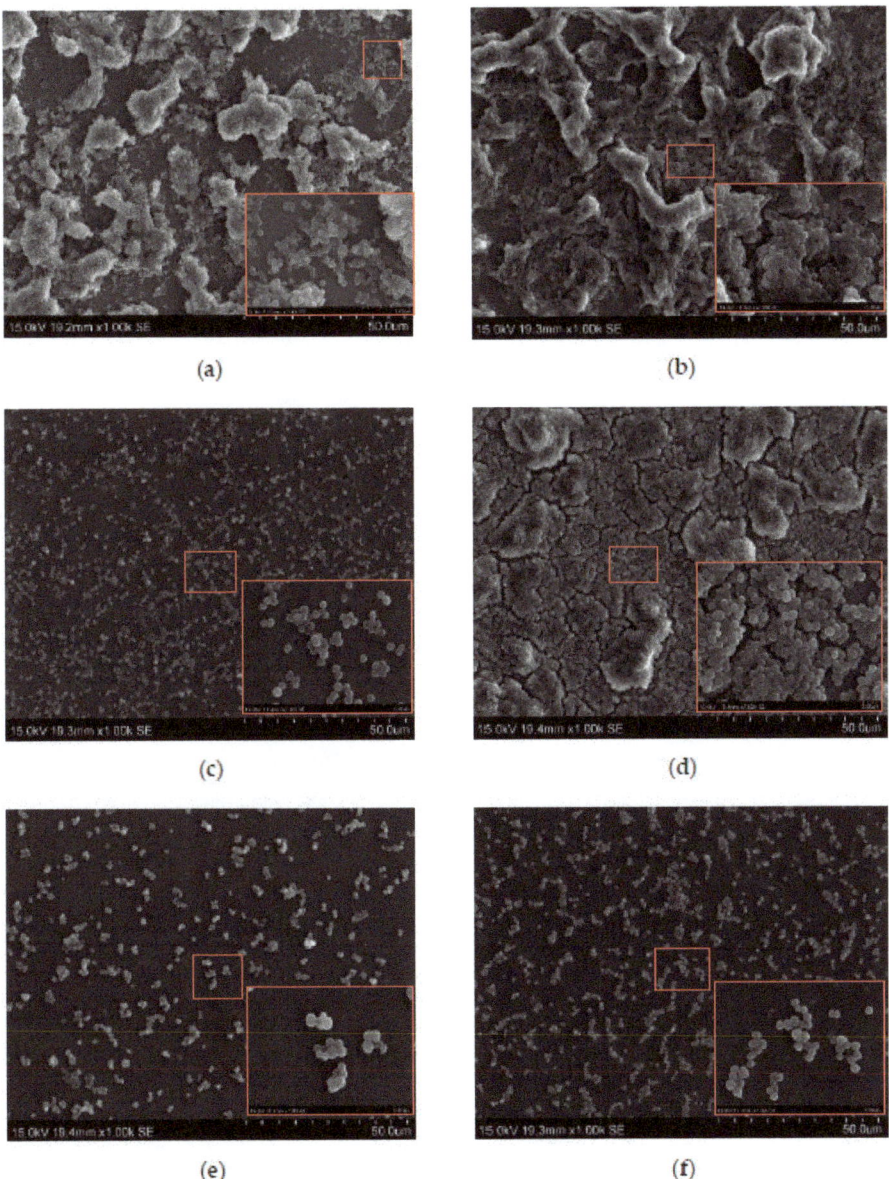

Figure 4. SEM images showing inhibitory activity of naringenin on biofilm formation of *S. aureus* cells. (**a,c,e**) for *S. aureus* cells grown at 25 °C with naringenin at 0, 20 and 40 mg/L, respectively; (**b,d,f**) for *S. aureus* cells grown at 37 °C with naringenin at 0, 20 and 40 mg/L, respectively.

3.4. Transcriptional Analysis of Biofilm-Related Genes in S. aureus Cells

The effect of naringenin on the level of expression of biofilm-related genes, including *cidA*, *icaA*, *dltB*, *agrA*, *sortaseA*, *sarA* and sigma factor *sigB* in *S. aureus* cells, were studied by RT-qPCR. Among the seven tested genes, *icaA*, *agrA*, *sarA* and *sigB* demonstrated significantly down-regulated ($p < 0.05$) gene expression when treated with naringenin, while *cidA* and *dltB* were up-regulated. Specifically, *icaA*, *agrA*, *sarA* and *sigB* were significantly down-regulated ($p < 0.05$) by 0.47-, 0.49, 0.58- and 0.63-fold for *S. aureus* grown at 25 °C in a culture medium with 20 mg/L naringenin (Figure 5a), and further decreased by 0.22-, 0.10-, 0.11- and 0.44-fold after the concentration of naringenin was increased to 40 mg/L, respectively. Under the same conditions, the expression of *cidA* and *dltB* were mildly up-regulated by 1.29- and 2.08-fold when *S. aureus* cells were exposed to naringenin at 40 mg/L. The expression of genes, including *cidA*, *dltB*, *icaA*, *agrA*, *sarA* and *sigB*, exhibited a similar trend for *S. aureus* cells cultivated at 37 °C (Figure 5b).

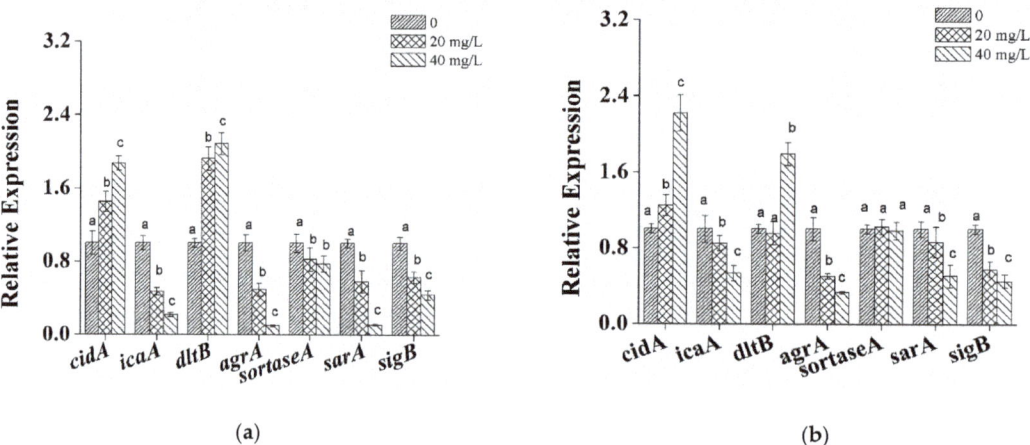

Figure 5. Effect of naringenin on the expression levels of biofilm-related genes in *S. aureus* at 25 °C (**a**) and 37 °C (**b**), where 16S rRNA was used as a reference gene. Data are presented as means ± standard deviations. Relative expression values are processed as mean ± SD and a–c indicate significant differences between different columns ($p < 0.05$).

By contrast, the levels of *sortaseA* expression in *S. aureus* cells grown at 25 °C were down-regulated in the presence of naringenin at concentrations of 20 and 40 mg/L, exhibiting 0.82- and 0.78-fold decreases ($p < 0.05$), respectively. At 37 °C, naringenin had no obvious effect on the transcription level of the studied *sortaseA* genes.

4. Discussion

In this work, no significant inhibitory effect of naringenin on S. aureus biofilm formation at 25 and 37 °C was found. However, the growth rate of S. aureus at 25 °C was significantly lower than that at 37 °C, without the final biomass being affected. Thus, different incubation times, i.e., 48 and 12 h, were used for 25 and 37 °C, respectively, to compensate for variations in the time required to reach the stationary phase. At subminimum inhibitory concentrations (sub-MICs) values ranging from 5 to 60 mg/L, naringenin dramatically inhibited *S. aureus* biofilm formation, with the biofilm formation further decreasing as the concentrations of naringenin with increased. Compared to an incubation temperature of 37 °C, the presence of naringenin resulted in a more significant effect on *S. aureus* biofilm formation for cultivation at 25 °C. There are various incubation temperatures that affect the growth of *S. aureus* and a series of changes caused by differences in subsequent growth.

Cell surface hydrophobicity is an important physical-chemical property of bacteria that facilitates their attachment to surfaces. Previously, it was reported that bacterial adherence of oral streptococci to the tooth surface was significantly suppressed by the reduction in cell surface hydrophobicity after treatment with constituents in cranberry juice and tea extract polyphenols [23,24]. Our results indicated that the cell surface hydrophobicity of *S. aureus* is reduced by treatment with naringenin (Figure 2a).

EPS is also important for biofilm production, forming multiple layers that help pathogens to adhere to surfaces and maintain biofilm architecture and acting as a protective barrier to prevent the entry of antibacterial agents into bacterial cells [25,26]. Therefore, the substantial decrease in EPS production by *S. aureus* after treatment with naringenin (Figure 2b) is consistent with this scenario. Since EPS production and surface hydrophobicity are important factors for biofilm formation, the decreases in these two properties following treatment with naringenin are the likely reasons that naringenin decreases biofilm formation in *S. aureus* cells.

Naringenin is not an antibiotic and, thus, the results obtained from CLSM and SEM are inconsistent with some previous studies that have reported that low concentrations of β-lactam and aminoglycoside antibiotics often facilitate biofilm formation by bacteria [27,28]. However, the inhibition of biofilm formation by naringenin is in agreement with the behavior of other flavonoids, including morin, rutin, quercetin and phloretin [25,29–31]. Notably, naringenin has a better effect on inhibiting the formation biofilms of *S. aureus* at the incubation temperature at 25 than at 37 °C.

To elucidate the underlying molecular mechanism for the inhibition of *S. aureus* biofilm by naringenin, we investigated the expression of some biofilm-related genes using RT-qPCR, including *dltB*, *sarA*, *sortaseA*, *agrA*, *icaA*, *cidA* and *sigB*. The *dltB* gene is responsible for D-alanylation of teichoic acids and the translocation of D-alanine through the cell membrane. It has been previously reported that *dltB* deficiency results in a higher negative net charge on the bacterial cell surface and defects in the initial binding of bacteria to a surface in the process of biofilm formation [32]. The expression of *cidA* was reported to be associated with extracellular DNA release, which is essential in the formation of *S. aureus* biofilm [33]. Our results revealed that the *dltB* and *cidA* genes are mildly up-regulated in *S. aureus* cells exposed to naringenin at 25 and 37 °C. Thus, it can be inferred that the suppression of biofilm formation by naringenin is probably not through *dltB* and *cidA* in *S. aureus*.

By contrast, the *icaA*, *agrA* and *sarA* genes are down-regulated in the presence of naringenin. The *ica* operons encode enzymes involved in the biosynthesis of polysaccharide intercellular adhesion [34]. Since it is well-known that the ability of *S. aureus* to form biofilm is dependent on polysaccharide intercellular adhesion, down-regulation of *icaA* likely decreases the production of polysaccharide intercellular adhesion, leading to reduction of biofilm formation. Thus, the decreased expression of *icaA* by naringenin might lead to a reduction in the biosynthesis of polysaccharide intercellular adhesion, which hinders the attachment of *S. aureus* cells to solid surfaces and subsequent biofilm formation. Both the *agr* (accessory gene regulator) and *sar* (staphylococcal accessory regulator) operons are two regulatory elements that control the production of virulence factors, as well mediate *S. aureus* biofilm formation by regulating the quorum sensing and polysaccharide intercellular adhesion production [35]. Thus, it can also be inferred that the down-regulated gene expression of *agrA* and *sarA* negatively affects biofilm formation by *S. aureus*. These data are in agreement with previous studies showing that *agr* down regulation has a negative impact on biofilm development by *S. aureus* [33,36].

Additionally, the expression levels of *sigB* were reduced for *S. aureus* cells exposed to naringenin. *sigB* is an alternative sigma factor modulating various stress responses in several Gram-positive bacteria, including *S. aureus* via a large regulon [37]. Overexpression of these genes can confer more resistance to heat, oxidative and antibiotic stresses. In summary, down-regulating the above genes may be the mechanism of action by which naringenin inhibits biofilm formation.

In some Gram-positive bacteria, including staphylococci, enterococci and streptococci, the *sortaseA* gene is responsible for coding a membrane enzyme that plays an important role in grappling some surface-exposed proteins to the cell wall envelope [38]. Recent reports have shown that *sortaseA* upregulation significantly increases the levels of biofilm formation in staphylococcal strains [39]. Our results show a mild down-regulation of *sortaseA* in the presence of naringenin for *S. aureus* grown at 25 °C. By contrast, at 37 °C no obvious change in the transcription level of *sortaseA* was observed. Since adhesion to the surface is an essential step for biofilm formation, the larger decrease in biofilm formation observed in *S. aureus* cultivated at 25 °C may be attributed to the down-regulated expression of *sortaseA* in *S. aureus* cells exposed to naringenin.

5. Conclusions

The data from our investigation indicates the potential of naringenin as a natural agent to prevent biofilm formation of S. aureus and possibly reduce health risks related to biofilm-formation in the food industry. However, more studies are necessary to gain a better understanding whether of there is any anti-biofilm activity toward other food-borne pathogens in food industry, and the efficiency when considering stainless steel and plastic surface.

Author Contributions: Conceptualization and methodology, Q.-H.W. and R.W.; software, data curation and formal analysis, Q.-H.W. and S.-Q.Z.; writing—original draft preparation, Q.-H.W. and R.W.; writing—review and editing, S.-Q.Z.; visualization, B.-R.C.; supervision, X.-A.Z.; funding acquisition, X.-A.Z. All authors have read and agreed to the published version of the manuscript.

Funding: This work was supported by the S&T projects of Guangdong Province (2019B020212004) and the 111 Project (B17018).

Institutional Review Board Statement: Not applicable.

Informed Consent Statement: Not applicable.

Data Availability Statement: Not applicable.

Conflicts of Interest: The authors declare no conflict of interest.

References

1. Normanno, G.; La, S.G.; Dambrosio, A.; Quaglia, N.C.; Corrente, M.; Parisi, A.; Santagada, G.; Firinu, A.; Crisetti, E.; Celano, G.V. Occurrence, characterization and antimicrobial resistance of enterotoxigenic *Staphylococcus aureus* isolated from meat and dairy products. *Int. J. Food Microbiol.* **2007**, *115*, 290–296. [CrossRef]
2. Tango, C.N.; Akkermans, S.; Hussain, M.S.; Khan, I.; Van Impe, J.; Jin, Y.-G.; Oh, D.H. Modeling the effect of pH, water activity, and ethanol concentration on biofilm formation of *Staphylococcus aureus*. *Food Microbiol.* **2018**, *76*, 287–295. [CrossRef] [PubMed]
3. Lauková, A.; Pogány Simonová, M.; Focková, V.; Kološta, M.; Tomáška, M.; Dvorožňáková, E. Susceptibility to Bacteriocins in Biofilm-Forming, Variable Staphylococci Isolated from Local Slovak Ewes' Milk Lump Cheeses. *Foods* **2020**, *9*, 1335. [CrossRef]
4. Vasudevan, P.; Nair, M.; Annamalai, T.; Venkitanarayanan, K.S. Phenotypic and genotypic characterization of bovine mastitis isolates of *Staphylococcus aureus* for biofilm formation. *Vet. Microbiol.* **2003**, *92*, 179–185. [CrossRef]
5. Kroning, I.S.; Iglesias, M.A.; Sehn, C.P.; Gandra, T.K.V.; Mata, M.M.; da Silva, W.P. *Staphylococcus aureus* isolated from handmade sweets: Biofilm formation, enterotoxigenicity and antimicrobial resistance. *Food Microbiol.* **2016**, *58*, 105–111. [CrossRef]
6. Kruk, M.; Trząskowska, M. Analysis of Biofilm Formation on the Surface of Organic Mung Bean Seeds, Sprouts and in the Germination Environment. *Foods* **2021**, *10*, 542. [CrossRef] [PubMed]
7. Ciccio, P.D.; Vergara, A.; Festino, A.R.; Paludi, D.; Zanardi, E.; Ghidini, S.; Ianieri, A. Biofilm formation by *Staphylococcus aureus* on food contact surfaces: Relationship with temperature and cell surface hydrophobicity. *Food Control* **2015**, *50*, 930–936. [CrossRef]
8. Abdallah, M.; Chataigne, G.; Ferreira-Theret, P.; Benoliel, C.; Drider, D.; Dhulster, P.; Chihib, N.E. Effect of growth temperature, surface type and incubation time on the resistance of *Staphylococcus aureus* biofilms to disinfectants. *Appl. Microbiol. Biotechnol.* **2014**, *98*, 2597–2607. [CrossRef] [PubMed]
9. Mandalari, G.; Bennett, R.; Bisignano, G.; Trombetta, D.; Saija, A.; Faulds, C.; Gasson, M.; Narbad, A. Antimicrobial activity of flavonoids extracted from bergamot (*Citrus bergamia* Risso) peel, a byproduct of the essential oil industry. *J. Appl. Microbiol.* **2007**, *103*, 2056–2064. [CrossRef]
10. Wen, Q.-H.; Wang, L.-H.; Zeng, X.-A.; Niu, D.-B.; Wang, M.-S. Hydroxyl-related differences for three dietary flavonoids as inhibitors of human purine nucleoside phosphorylase. *Int. J. Biol. Macromol.* **2018**, *118*, 588–598. [CrossRef]
11. Erlund, I.; Meririnne, E.; Alfthan, G.; Aro, A. Plasma kinetics and urinary excretion of the flavanones naringenin and hesperetin in humans after ingestion of orange juice and grapefruit juice. *J. Nutr.* **2001**, *131*, 235–241. [CrossRef]
12. Hashimoto, T.; Ide, T. Activity and mRNA Levels of Enzymes Involved in Hepatic Fatty Acid Synthesis in Rats Fed Naringenin. *J. Agric. Food Chem.* **2015**, *63*, 9536–9542. [CrossRef]
13. Kanaze, F.I.; Kokkalou, E.; Georgarakis, M.; Niopas, I. A validated solid-phase extraction HPLC method for the simultaneous determination of the citrus flavanone aglycones hesperetin and naringenin in urine. *J. Pharm. Biomed. Anal.* **2004**, *36*, 175–181. [CrossRef]

14. Denny, S.; West, P.W.J.; Mathew, T.C. Antagonistic interactions between the flavonoids hesperetin and naringenin and β-lactam antibiotics against *Staphylococcus aureus*. *Br. J. Biomed. Sci.* **2008**, *65*, 145–147. [CrossRef]
15. Lee, K.-A.; Moon, S.H.; Kim, K.-T.; Mendonca, A.F.; Paik, H.-D. Antimicrobial effects of various flavonoids on Escherichia coli O157: H7 cell growth and lipopolysaccharide production. *Food Sci. Biotechnol.* **2010**, *19*, 257–261. [CrossRef]
16. Vikram, A.; Jayaprakasha, G.K.; Jesudhasan, P.; Pillai, S.; Patil, B. Suppression of bacterial cell–cell signalling, biofilm formation and type III secretion system by citrus flavonoids. *J. Appl. Microbiol.* **2010**, *109*, 515–527. [CrossRef] [PubMed]
17. Yue, J.; Yang, H.; Liu, S.; Song, F.; Guo, J.; Huang, C. Influence of naringenin on the biofilm formation of Streptococcus mutans. *J. Dent.* **2018**, *76*, 24–31. [CrossRef] [PubMed]
18. Wang, L.H.; Wang, M.S.; Zeng, X.A.; Xu, X.M.; Brennan, C.S. Membrane and genomic DNA dual-targeting of citrus flavonoid naringenin against *Staphylococcus aureus*. *Integr. Biol.* **2017**, *9*, 820–829. [CrossRef]
19. Lee, J.H.; Kim, Y.G.; Lee, K.; Kim, S.C.; Lee, J. Temperature-dependent control of *Staphylococcus aureus* biofilms and virulence by thermoresponsive oligo (N-vinylcaprolactam). *Biotechnol. Bioeng.* **2015**, *112*, 716–724. [CrossRef]
20. Hsu, L.C.; Fang, J.; Borcatasciuc, D.A.; Worobo, R.W.; Moraru, C.I. Effect of Micro- and Nanoscale Topography on the Adhesion of Bacterial Cells to Solid Surfaces. *Appl. Environ. Microbiol.* **2013**, *79*, 2703–2712. [CrossRef]
21. Dubois, M.; Gilles, K.A.; Hamilton, J.K.; Rebers, P.A.; Smith, F. Colorimetric Method for Determination of Sugars and Related Substances. *Anal. Chem.* **1956**, *28*, 350–356. [CrossRef]
22. Wang, L.H.; Zeng, X.A.; Wang, M.S.; Brennan, C.S.; Gong, D. Modification of membrane properties and fatty acids biosynthesis-related genes in Escherichia coli and *Staphylococcus aureus*: Implications for the antibacterial mechanism of naringenin. *Biochim. Biophys. Acta (BBA)-Biomembr.* **2018**, *1860*, 481–490. [CrossRef] [PubMed]
23. Matsumoto, M.; Minami, T.; Sasaki, H.; Sobue, S.; Hamada, S.; Ooshima, T. Inhibitory effects of oolong tea extract on caries–inducing properties of Mutans streptococci. *Caries Res.* **1999**, *33*, 441–445. [CrossRef] [PubMed]
24. Yamanaka, A.; Kimizuka, R.; Kato, T.; Okuda, K. Inhibitory effects of cranberry juice on attachment of oral streptococci and biofilm formation. *Oral Microbiol. Immunol.* **2004**, *19*, 150–154. [CrossRef]
25. Al-Shabib, N.A.; Husain, F.M.; Ahmad, I.; Khan, M.S.; Khan, R.A.; Khan, J.M. Rutin inhibits mono and multi-species biofilm formation by foodborne drug resistant Escherichia coli and *Staphylococcus aureus*. *Food Control* **2017**, *79*, 325–332. [CrossRef]
26. Xiang, H.; Cao, F.; Ming, D.; Zheng, Y.; Dong, X.; Zhong, X.; Mu, D.; Li, B.; Zhong, L.; Cao, J. Aloe-emodin inhibits *Staphylococcus aureus* biofilms and extracellular protein production at the initial adhesion stage of biofilm development. *Appl. Microbiol. Biotechnol.* **2017**, *101*, 6671–6681. [CrossRef]
27. Hoffman, L.R.; D'Argenio, D.A.; MacCoss, M.J.; Zhang, Z.; Jones, R.A.; Miller, S.I. Aminoglycoside antibiotics induce bacterial biofilm formation. *Nature* **2005**, *436*, 1171–1175. [CrossRef]
28. Kaplan, J.B.; Izano, E.A.; Gopal, P.; Karwacki, M.T.; Kim, S.; Bose, J.L.; Bayles, K.W.; Horswill, A.R. Low Levels of β-Lactam Antibiotics Induce Extracellular DNA Release and Biofilm Formation in *Staphylococcus aureus*. *Mbio* **2012**, *3*, e00198-12. [CrossRef] [PubMed]
29. Lee, J.H.; Regmi, S.C.; Kim, J.A.; Cho, M.H.; Yun, H.; Lee, C.S.; Lee, J. Apple Flavonoid Phloretin Inhibits Escherichia coli O157:H7 Biofilm Formation and Ameliorates Colon Inflammation in Rats. *Infect. Immun.* **2011**, *79*, 4819–4827. [CrossRef]
30. Gopu, V.; Meena, C.K.; Shetty, P.H. Quercetin influences quorum sensing in food borne bacteria: In-vitro and in-silico evidence. *PLoS ONE* **2015**, *10*, e0134684. [CrossRef]
31. Sivaranjani, M.; Gowrishankar, S.; Kamaladevi, A.; Pandian, S.K.; Balamurugan, K.; Ravi, A.V. Morin inhibits biofilm production and reduces the virulence of Listeria monocytogenes—An in vitro and in vivo approach. *Int. J. Food Microbiol.* **2016**, *237*, 73–82. [CrossRef] [PubMed]
32. Peschel, A.; Otto, M.; Jack, R.W.; Kalbacher, H.; Jung, G.; Götz, F. Inactivation of the dlt Operon in *Staphylococcus aureus* Confers Sensitivity to Defensins, Protegrins, and Other Antimicrobial Peptides. *J. Biol. Chem.* **1999**, *274*, 8405–8410. [CrossRef]
33. Yan, X.; Gu, S.; Shi, Y.; Cui, X.; Wen, S.; Ge, J. The effect of emodin on *Staphylococcus aureus* strains in planktonic form and biofilm formation in vitro. *Arch. Microbiol.* **2017**, *199*, 1267–1275. [CrossRef]
34. Ma, Y.; Xu, Y.; Yestrepsky, B.D.; Sorenson, R.J.; Chen, M.; Larsen, S.D.; Sun, H. Novel inhibitors of *Staphylococcus aureus* virulence gene expression and biofilm formation. *PLoS ONE* **2012**, *7*, e47255. [CrossRef]
35. Pratten, J.; Foster, S.J.; Chan, P.F.; Wilson, M.; Nair, S.P. *Staphylococcus aureus* accessory regulators: Expression within biofilms and effect on adhesion. *Microbes Infect.* **2001**, *3*, 633–637. [CrossRef]
36. Coelho, L.R.; Souza, R.R.; Ferreira, F.A.; Guimaraes, M.A.; Ferreira-Carvalho, B.T.; Figueiredo, A. agr RNAIII divergently regulates glucose-induced biofilm formation in clinical isolates of *Staphylococcus aureus*. *Microbiology* **2008**, *154*, 3480–3490. [CrossRef] [PubMed]
37. Valle, J.; Toledo-Arana, A.; Berasain, C.; Ghigo, J.M.; Amorena, B.; Penadés, J.R.; Lasa, I. SarA and not σB is essential for biofilm development by *Staphylococcus aureus*. *Mol. Microbiol.* **2003**, *48*, 1075–1087. [CrossRef] [PubMed]
38. Hu, P.; Huang, P.; Chen, M.W. Curcumin reduces Streptococcus mutans biofilm formation by inhibiting sortase A activity. *Arch. Oral Biol.* **2013**, *58*, 1343–1348. [CrossRef]
39. Cascioferro, S.; Totsika, M.; Schillaci, D. Sortase A: An ideal target for anti-virulence drug development. *Microb. Pathog.* **2014**, *77*, 105–112. [CrossRef]

Article

Component Identification of Phenolic Acids in Cell Suspension Cultures of *Saussurea involucrata* and Its Mechanism of Anti-Hepatoma Revealed by TMT Quantitative Proteomics

Junpeng Gao [1], Yi Wang [2], Bo Lyu [2,3], Jian Chen [2] and Guang Chen [1,*]

[1] College of Life Science, Jilin Agricultural University, Changchun 130118, China; joepgao@163.com
[2] College of Food Science and Engineering, Jilin Agricultural University, Changchun 130118, China; Wangyi284419@163.com (Y.W.); michael_lvbo@163.com (B.L.); cj15068528679@163.com (J.C.)
[3] College of Food Science, Northeast Agricultural University, Harbin 150030, China
* Correspondence: chg61@163.com

Citation: Gao, J.; Wang, Y.; Lyu, B.; Chen, J.; Chen, G. Component Identification of Phenolic Acids in Cell Suspension Cultures of *Saussurea involucrata* and Its Mechanism of Anti-Hepatoma Revealed by TMT Quantitative Proteomics. *Foods* **2021**, *10*, 2466. https://doi.org/10.3390/foods10102466

Academic Editor: Guowen Zhang

Received: 2 September 2021
Accepted: 12 October 2021
Published: 15 October 2021

Publisher's Note: MDPI stays neutral with regard to jurisdictional claims in published maps and institutional affiliations.

Copyright: © 2021 by the authors. Licensee MDPI, Basel, Switzerland. This article is an open access article distributed under the terms and conditions of the Creative Commons Attribution (CC BY) license (https://creativecommons.org/licenses/by/4.0/).

Abstract: *Saussurea involucrata* (*S. involucrata*) had been reported to have anti-hepatoma function. However, the mechanism is complex and unclear. To evaluate the anti-hepatoma mechanism of *S. involucrata* comprehensively and make a theoretical basis for the mechanical verification of later research, we carried out this work. In this study, the total phenolic acids from *S. involucrata* determined by a cell suspension culture (ESPI) was mainly composed of 4,5-dicaffeoylquinic acid, according to the LC-MS analysis. BALB/c nude female mice were injected with HepG2 cells to establish an animal model of liver tumor before being divided into a control group, a low-dose group, a middle-dose group, a high-dose group, and a DDP group. Subsequently, EPSI was used as the intervention drug for mice. Biochemical indicators and differences in protein expression determined by TMT quantitative proteomics were used to resolve the mechanism after the low- (100 mg/kg), middle- (200 mg/kg), and high-dose (400 mg/kg) interventions for 24 days. The results showed that EPSI can not only limit the growth of HepG2 cells in vitro, but also can inhibit liver tumors significantly with no toxicity at high doses in vivo. Proteomics analysis revealed that the upregulated differentially expressed proteins (DE proteins) in the high-dose group were over three times that in the control group. ESPI affected the pathways significantly associated with the protein metabolic process, metabolic process, catalytic activity, hydrolase activity, proteolysis, endopeptidase activity, serine-type endopeptidase activity, etc. The treatment group showed significant differences in the pathways associated with the renin-angiotensin system, hematopoietic cell lineage, etc. In conclusion, ESPI has a significant anti-hepatoma effect and the potential mechanism was revealed.

Keywords: *S. involucrata*; anti-hepatoma; traditional Chinese medicine; pharmacological mechanism; proteomics

1. Introduction

Saussurea involucrata (*S. involucrata*) is a rare and slow-growing herb growing in the Tianshan and Altay Mountains of Xinjiang Province, China, at altitudes over 2600 m. *S. involucrata* has the effect of promoting blood circulation, relaxing tendons, dispersing cold, and removing dampness [1]. It is mainly used for the treatment of coughs, rheumatoid arthritis, high-altitude response, and stomach pain [2].

According to previous studies, *S. involucrata* has many bioactive compounds, including lignans [3], flavonoids [3], coumarins [4], sesquiterpene lactones [5], steroids, and phenylpropanoids [6]. These compounds show a wide range of biological activity, including anti-inflammatory [7], anti-aging [8], antioxidant, anti-fatigue [8], and anti-tumor effects [2]. In addition, it has been shown that the methanol extracted from *S. involucrata* could inhibit the expression of cytokines stimulated by lipopolysaccharide (LPS), and the ethyl acetate extract could effectively inhibit phosphorylation and activation of the EGFR,

Akt, and STAT3 pathway in PC-3 cells [9]. Meanwhile, *S. involucrata* also showed strong anti-rheumatic activity in the clinical environment [10]. It can be seen that *S. involucrata* has an obvious inhibitory effect on some inflammation.

Due to the unique growth environment and the long growth cycle of *S. involucrata*, artificial cultivation is very difficult. In addition, with the market demand and overexploitation, the stock of many rare medicinal plants had decreased sharply. Using biotechnology to regulate plants' secondary metabolism, obtain useful secondary metabolites and clarify their biosynthetic pathway has become an important goal of researchers [11]. Through the identification of metabolites and enzymes in the process of biosynthesis, people can better understand the metabolic pathway of plant components, combined with genomic and proteomic analysis technologies [12]. Under artificial control, a large number of products can be obtained in a short time from cell suspension cultures of *S. involucrata*, which can ensure stable product quality and high-efficacy components, and realize large-scale industrial production, to solve the problem of a shortage of *S. involucrata*.

This study was based on the previous research results of our team. We prepared the total phenolic acids from *S. involucrata* through a cell suspension culture (EPSI). However, its mechanism of action was not clear enough to guide clinical applications [13]. In order to explore the potential anti-hepatoma mechanism of EPSI more comprehensively and make a theoretical basis for later verification, this study depended on tumor-bearing BALB/c mice model combined with proteomics technology to verify the mechanism of action of *S. involucrata* on human hepatoma, while realizing the modern utilization of *S. involucrata* resources and the accurate analysis of its drug theory.

2. Materials and Methods

2.1. Cell Suspension Culture of S. involucrata

A cell suspension culture of *S. involucrata* was provided by the Engineering Research Center of Bioreactor and Pharmaceutical Development, Ministry of Education (College of Life Science, Jilin Agricultural University).

2.2. Extraction of the Phenolic Acids in Cell Suspension Cultures of S. involucrata (EPSI)

Phenolic acids are soluble in organic solvents, so the total phenolic acids were extracted by the organic solvent extraction method [14]. The method is as follows: The cell suspension culture of *S. involucrata* was mixed 1:8 (m/v) with absolute alcohol, heated at 50 °C, and stirred for 15 min. After the supernatant was collected, 8× absolute alcohol was added to the precipitate and stirred for 2 h; the above steps were repeated twice. The collected supernatant was filtered and subjected to rotary evaporation at 60 °C. After the rotary distillation, the raw materials were collected with water, and the organic phase was extracted with ethyl acetate. The above steps were repeated 3 times. After the materials had been subjected to rotary evaporation at 60 °C, the concentrate (EPSI) was collected and freeze-dried at −80 °C for storage.

2.3. Component Identification by Liquid Chromatography-Mass Spectrometry (LC-MS)

The composition of EPSI was analyzed by LC-MS (Thermo, Ultimate 3000LC, Q Exactive HF) [15–17]. The conditions were as follows. Chromatographic column: C18 column (Zorbax Eclipse C18 [1.8 μm × 2.1 × 100 mm]); separation conditions: column temperature: 30 °C; current speed: 0.3 mL/min; Mobile Phase A: water + 0.1% formic acid (CAS: 64-18-6, Xiya Chemical Technology (Shandong) Co., Ltd., Linyi, China), Mobile Phase B: acetonitrile (CAS: 75-05-8, Merck KGaA, Darmstadt, Germany); injection volume: 2 μL; autosampler temperature: 4 °C. The process of gradient elution is shown in Table 1.

The positive mode was used for analysis: Heater temperature: 325 °C; sheath gas velocity: 45 arb; flow rate of the auxiliary gas: 15 arb; purge gas flow rate: 1 arb; electrospray voltage: 3.5 kV; capillary temperature: 330 °C; S-lens RF level: 55%; scanning mode: full scan (M/Z 100–1500) and data dependent secondary mass spectrometry (dd-ms2,

TopN = 10); resolution: 120,000 (primary MS) and 60,000 (secondary MS); collision mode: high-energy collision dissociation (HCD).

Table 1. The mobile phase gradient elution process .

Time (min)	Current Speed (µL/mL)	Gradient	B (%)
0–2	300	-	5
2–6	300	Linear	30
6–7	300	-	30
7–12	300	Linear	78
12–14	300	-	78
14–17	300	Linear	95
17–20	300	-	95
20–21	300	Linear	5
21–25	300	-	5

Compound Discoverer 3.1 Software was used for correction of the retention time, peak identification, and peak extraction. According to the secondary MS spectrum information, the substances were identified by using the Thermo mzCloud Online Database (https://www.mzcloud.org/, accessed on 23 August 2021).

2.4. Effect of EPSI on the Multiplication of HepG2 Cells In Vitro

Hepatic cellular carcinoma (HCC) cell lines HepG2 were purchased from Bluef Biotechnology Development Co. Ltd. (Shanghai, China) and cultured in Dulbecco's Modified Eagle Medium (DMEM) (Thermo Fisher Scientific, Waltham, MA, USA) supplemented with 10% fetal bovine serum (FBS) (Thermo Fisher Scientific, Waltham, MA, USA).

After resuscitation, cells at the logarithmic phase were taken for standby. We adjusted the cell concentration to 10^4 units/mL and added them to a 96-well culture plate at a concentration of 100 µL/well for culturing (37 °C, CO_2 concentration: 5%, 24 h). The old medium was discarded after culturing, and a blank group with 100 µL of 5% medium was added. For the experimental group, EPSI samples of different concentrations were added and then cultured at 37 °C and 5% CO_2 for 72 h.

CCK8 (CCK-8 Cell Proliferation and Cytotoxicity Assay Kit, Solarbio, Beijing, China) and MTT (MTT Cell Proliferation and Cytotoxicity Assay Kit, Solarbio, Beijing, China) were used to evaluate the inhibitory effect of EPSI on HepG2 cells in vitro. All operations were carried out according to the operating instructions.

2.5. Acute Toxic Test

A fixed-dose procedure (FDP) was used to verify the acute toxicity of EPSI [18,19]. The dosage was set at 2000 mg/kg. Six- to eight-week-old BALB/c nude female mice (Beijing Vital River Laboratory Animal Technology Co. Ltd., Beijing, China) were used as the test animal. Intraperitoneal injection (I.P.) was used as the mode of administration. It was considered that EPSI had no acute toxicity if the LD50 was >2000 mg/kg.

2.6. Establishment of the Animal Model

Six- to eight-week-old BALB/c nude female mice were purchased from Beijing Vital River Laboratory Animal Technology Co. Ltd. (Beijing, China). All mice were housed under controlled conditions in individual cages at 22 ± 3°C and 60–70% relative humidity with a 12 h dark/light cycle in a specific germ-free environment and were allowed free access to sterile food and water. After 1 week of acclimation, each animal was subcutaneously injected with 10^6 units of HepG2 cells. When there were measurable tumors with an equivalent volume, the experiment animals were randomly divided into five groups. The groups consisted of the control group (no gavage, free feeding, n = 8), the low-dose group (100 mg/kg EPSI gavage for 24 days, free feeding, n = 8), the middle-dose group (200 mg/kg EPSI gavage for 24 days, free feeding, n = 8), the high-dose group (400 mg/kg

EPSI gavage for 24 days, free feeding, n = 8) and the DDP (PtCl$_2$[NH$_3$]$_2$) group (10% DDP intraperitoneal injection for 10 days, free feeding, n = 8). All operations were performed in a sterile environment.

The weights and the tumor volumes of the experimental animals in each group were measured once every 2 days at night during the intervention. The experimental animals were sacrificed when there were significant differences in tumor volume. The tumors were removed and stored at −80 °C after eyeball blood collection. All animal experiments were approved by the Laboratory Animal Welfare and Ethics Committee of Jilin Agricultural University (No. 20190410005), following National Research Council Guidelines.

2.7. Hematoxylin and Eosin (H&E) Staining

The tumor tissues obtained as described in Part 2.4 were separated by about 0.2 mm^3, without being frozen, and were fixed in a 4% paraformaldehyde solution for 12 h at room temperature and embedded in paraffin. After cutting the paraffin into 5 μm sections, the slides were dewaxed, rehydrated, and stained with 1% H&E at room temperature as described previously.

2.8. Enzyme Linked Immunosorbent Assay (ELISA) analysis

Blood was drawn from the eyeball and centrifuged (3000× g, 10 min, 4 °C) to obtain the serum. Tumor necrosis factor-α (TNF-α) was measured according to an ELISA kit protocol (No. ML002095, Shanghai Enzyme-linked Biotechnology Co., Ltd., Shanghai, China).

2.9. Tandem Mass Tag (TMT) Quantitative Proteomics

The process of TMT quantitative proteomics was basically consistent with the research methods for animal tissues in other studies [20]. The process is briefly described as follows [21–23].

2.9.1. Total Protein Extraction

The sample was ground into powder at a low temperature and transferred to a tube (cooled by liquid nitrogen). A certain amount of a PASP lysis buffer (100 mM NH$_4$HCO$_3$, 8 M Urea, pH = 8), was added, mixed, and ultrasonicated for 5 min under an ice bath to full splitting. The product was centrifuged at 4 °C and 12,000× g for 15 min. Next, 10 mM DTT (DL-dithiothreitol) was added to the supernatant and reacted at 56 °C for 1 h. Sufficient IAM (iodoacetamide) was then added and reacted at room temperature without light for 1 h. After completing the above process, 4× the volume of acetone (precooled at 20 °C) was added, precipitated at −20 °C for 2 h, and centrifuged at 4 °C and 12,000× g for 15 min, and the precipitate was collected. After that, 1 mL of acetone (precooled at 20 °C) was added to resuspend the precipitate, then the mixture was centrifuged at 4 °C and 12,000× g for 15 min, the precipitate was collected, air-dried, and an appropriate amount of a protein solution dissolution buffer (8 M urea, 100 mM TEAB (triethylammonium bicarbonate); pH 8.5) was added to dissolve the protein precipitate.

2.9.2. Protein Quality Test

The Bradford protein quantitative kit was used to prepare a BSA (bovine serum albumin) standard protein solution according to the instructions, and the concentration gradient range was from 0 to 0.5 g/L. The BSA standard protein solution with different concentration gradients and the sample solution were taken to be tested at different dilution ratios and were added into a 96-well plate, the volume was made up to 20 μL, and each gradient was repeated 3 times. After that, 180 μL of a G250 staining solution was added immediately, and the mixture was placed at room temperature for 5 min, and the absorbance at 595 nm was measured. The standard curve was drawn with the absorbance of the standard protein solution, and we calculated the protein concentration of the sample to be tested. For gel electrophoresis, 20 μg of the protein sample was loaded to 12% SDS-

PAGE (Sodium dodecyl sulfate—Polyacrylamide gel electrophoresis). The electrophoresis conditions were 80 V for 20 min in concentrated gel, then 120 V for 90 min in separation gel. After electrophoresis, Coomassie brilliant blue R-250 staining was performed to decolorize the sample until the bands were clear.

2.9.3. TMT Labeling of Peptides

A protein sample was taken and a DB dissolution buffer (8 M urea, 100 mM TEAB; pH 8.5) was added to make up the volume to 100 µL. Trypsin and the 100 mM TEAB buffer were added, mixed well, and digested at 37 °C for 4 h, then trypsin and $CaCl_2$ were added overnight. Formic acid was added to adjust the pH to be less than 3, then the sample was mixed well and centrifuged at room temperature at 12,000× g for 5 min. The supernatant was passed through the C18 demineralizer column slowly, then washed 3 times with a washing buffer (0.1% formic acid, 3% acetonitrile). After that, an appropriate amount of an elution buffer (0.1% formic acid, 70% acetonitrile) was added to collect the filtrate and freeze-dried. The lyophilized sample was reconstituted by adding 100 µL of 0.1 M TEAB buffer and 41 µL of TMT labeling reagent dissolved in acetonitrile. The reaction was reversed and mixed at room temperature for 2 h. After that, 8% ammonia was used to stop the reaction, and the samples marked with equal volume were mixed, desalted, and lyophilized [24].

2.9.4. Separation of Fractions

The mobile phase composition was as follows: Mobile Phase A: 2% acetonitrile, with the pH adjusted to 10.0 using ammonium hydroxide; B: 98% acetonitrile. The lyophilized powder was dissolved with Solution A and centrifuged at 12,000× g at room temperature for 10 min. The L-3000 HPLC (high-performance liquid chromatography) system with a C18 column (Waters BEH C18, 4.6 × 250 mm, 5 µm; column temperature: 45 ° C) was used to test the sample. The specific elution gradient is shown in Table 1. One tube was collected every minute and divided into 10 fractions. After freeze-drying, 0.1% formic acid was added to dissolve each fraction. The details of the elution gradient are shown in Table 2. One tube was collected per minute and divided into 10 fractions. After freeze-drying, 0.1% formic acid (FA) was added to dissolve each fraction.

Table 2. Peptide fraction separation: liquid chromatography elution gradient table.

Time (min)	Flow Rate (mL/min)	Mobile Phase A (%)	Mobile Phase B (%)
0	1	97	3
10	1	95	5
30	1	80	20
48	1	60	40
50	1	50	50
53	1	30	70
54	1	0	100

2.9.5. LC-MS/MS Analysis

The mobile phase composition was as follows: Mobile Phase A: 0.1% formic acid; Mobile Phase B: 80% acetonitrile and 0.1% formic acid. Next, 1 µg of the supernatant of each fraction was injected and tested. The EASY-nLC 1200 UHPLC system (Thermo Fisher) was coupled with the C18 Nano-Trap column (4.5 cm × 75 µm, 3 µm) as a homemade analytical column and the C18 Nano-Trap column (15 cm × 150 µm, 1.9 µm) as a homemade analytical column. The linear elution gradient is shown in Table 3. A Q Exactive HF-X mass spectrometer (Thermo Fisher) was used with the ion source of Nanospray Flex (ESI) to analyze the samples. The conditions were as follows: spray voltage: 2.1 kV; ion transport capillary temperature: 320 °C, full scan range: m/z 350 to 1500; primary MS resolution: 60,000 (at m/z 200); automatic gain control (AGC) target value: 3×10^6; maximum ion injection time: 20 ms. The top 40 precursors with the highest abundance in the full scan were selected and fragmented by higher-energy collisional dissociation (HCD) and analyzed by MS/MS. The conditions were as follows: resolution: 30,000 (at m/z 200); AGC target value: 5×10^4; maximum ion injection time: 54 ms; normalized collision energy: 32%; intensity threshold: 1.2×10^5; dynamic exclusion parameter: 20 s.

Table 3. Liquid chromatography elution gradient table.

Time (min)	Flow Rate (nL/min)	Mobile Phase A (%)	Mobile Phase B (%)
0	600	94	6
2	600	85	15
48	600	60	40
50	600	50	50
51	600	45	55
60	600	0	100

2.9.6. Data Analysis

Identification and Quantitation of Protein

The resulting spectra from each run were searched separately against the homo_sapiens _uniprot_2021_3_9 (194,557 sequences) database by the search engine Proteome Discoverer 2.4 (PD 2.4, Thermo). The search parameters were set as follows: mass tolerance for the precursor ion was 10 ppm and the mass tolerance for production was 0.02 Da. Carbamidomethyl was specified as a fixed modification; oxidation of methionine (M), and TMT plex were specified as dynamic modifications. Acetylation, TMT plex, Met loss, and Met-loss + Acetyl were specified as N-terminal modifications in PD 2.4. A maximum of 2 missed cleavage sites was allowed.

The software package PD 2.4 was used to improve the quality of the results for further filtering. Peptide spectrum matches (PSMs) with a reliability of more than 99% were identified as PSMs, which had to contain 1 unique peptide (5 unique peptides) or more. The identified PSMs and proteins with a FDR no more than 1.0% were retained and tested. The protein quantitation results were statistically analyzed by T-tests. The proteins whose quantitation was significantly different between the high-dose group and the control group ($p < 0.05$ and $|\log_2 FC| > 0.25$ (FC > 1.2 or FC < 0.83) (fold change, FC)) were defined as differentially expressed proteins (DEP).

Functional Analysis of Protein and DEP

Gene Ontology (GO) and InterPro (IPR) functional analyses were conducted using the interproscan program against the non-redundant protein database (including Pfam, PRINTS, ProDom, SMART, ProSite, and PANTHER) [21], and the COG (Clusters of Orthologous Groups) and KEGG (Kyoto Encyclopedia of Genes and Genomes) databases were used to analyze the protein families and pathways. DEPs were used for volcanic map analysis, cluster heat map analysis, and GO, IPR, and KEGG enrichment analysis [22]. The probable protein–protein interactions were predicted using the STRING-db server (http://string.embl.de/, accessed on 23 August 2021) [23].

2.10. Statistical Analysis

All the data are presented as means ± standard deviation (SD). Statistical analysis was carried out by GraphPad Prism version 6 (GraphPad Software, Inc., San Diego, CA, USA). Difference comparisons were carried out using the T-test or one-way analysis of variance (ANOVA) using IBM SPSS 25.0 (SPSS Inc., Chicago, CA, USA); $p < 0.05$ was considered statistically significant.

3. Results and Discussion

3.1. Composition of the Extract of the Phenolic Acids in Cell Suspension Cultures from S. involucrata (EPSI)

The total ion current is shown in Figure 1. The information of the qualitative metabolite results (positive mode) based on Figure 1 are shown in Table 4. In EPSI, 13 substances with a content of more than 1% were detected, including 4,5-dicaffeoylquinic acid (29.000%), linolenyl alcohol (8.453%), 7-hydroxycoumarine (7.691%), chlorogenic acid (6.134%), metronidazole (3.523%), apigenin 7-(6″-crotonylglucoside) (2.814%), 9-oxo-10(E),12(E)-octadecadienoic acid (2.618%), cynaroside (2.382%), phthalic anhydride (1.956%), hexadecanamide (1.883%), 3,4,5-tricaffeoylquinic acid (1.773%), diisobutyl phthalate (1.139%), and triethyl phosphate (1.061%). Many of them were reported to have biological activity and even pharmacology.

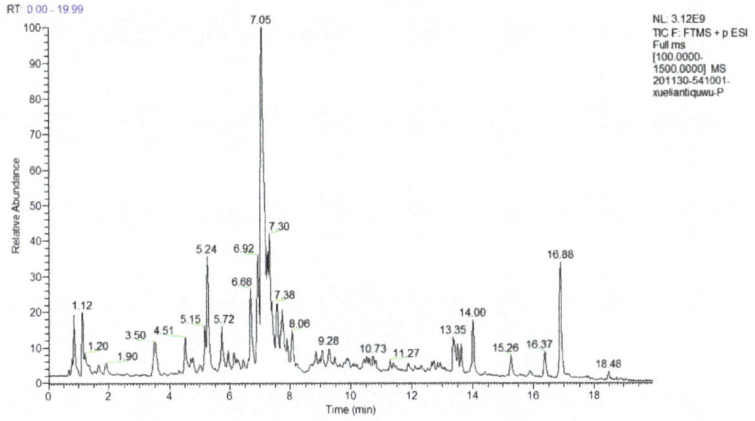

Figure 1. Total ion current (TIC) of EPSI.

Table 4. Composition of the extracts from cell suspension cultures of *S. involucrata* (content > 1%).

Name	Formula	CAS	Content (%)
4,5-Dicaffeoylquinic acid	$C_{25}H_{24}O_{12}$	14534-61-3	29.000
Linolenyl alcohol	$C_{18}H_{32}O$	506-44-5	8.453
7-Hydroxycoumarine	$C_9H_6O_3$	93-35-6	7.691
Chlorogenic acid	$C_{16}H_{18}O_9$	327-97-9	6.134
Metronidazole	$C_6H_9N_3O_3$	443-48-1	3.523
Apigenin 7- (6″-crotonylglucoside)	$C_{25}H_{24}O_{11}$	NA	2.814
9-Oxo-10(E),12(E)-octadecadienoic acid	$C_{18}H_{30}O_3$	54232-58-5	2.618
Cynaroside	$C_{21}H_{20}O_{11}$	5373-11-5	2.382
Phthalic anhydride	$C_8H_4O_3$	85-44-9	1.956
Hexadecanamide	$C_{16}H_{33}NO$	629-54-9	1.883
3,4,5-tricaffeoylquinic acid	$C_{34}H_{30}O_{15}$	86632-03-3	1.773
Diisobutyl phthalate	$C_{16}H_{22}O_4$	84-69-5	1.139
Triethyl phosphate	$C_6H_{15}O_4P$	78-40-0	1.061

4,5-Dicaffeoylquinic acid (isochlorogenic acid C), as the largest substance in the extract, is considered to have the function of promoting blood circulation in traditional Chinese medicine [2]. It has now been proven that it can improve the function of islet cells [25] and even inhibit cancer [26,27]. There are reasons to believe that it is the main anti-cancer ingredient of *S. involucrata*. As a lipid compound, linolenyl alcohol has been rarely reported as a functional compound; only a few reports showed that it has a certain antiviral activity [28,29]. 7-Hydroxycoumarine also occupied a certain ratio in the extract. 7-Hydroxycoumarin is one of the main chemical constituents of *Angelica dahurica* and can inhibit the growth of malignant tumors [30]. Meanwhile, its derivatives also have typical anti-inflammatory activities [31,32]. Chlorogenic acid is one of the main antibacterial and antiviral components of honeysuckle, which has good antioxidant and antibacterial properties [33,34]. 9-Oxo-10(E),12(E)-octadecadienoic acid is a potent PPARα agonist that decreases triglyceride accumulation [35], and can also induce the apoptosis of cancer cells [36].

Thus, after the identification of the components of EPSI, a variety of components with pharmacological effects were identified. These proved that EPSI may have potential anti-hepatoma ability.

3.2. Growth Inhibition of EPSI in HepG2 Cells

The CCK8 and MTT methods were used to determine EPSI's inhibition of HepG2 cell proliferation. As shown in Figure 2A (CCK8), the inhibitory ability of EPSI on the proliferation of HepG2 was at a low level (< 20%) when the concentration was 100–200 μg/mL. When the dose concentration was increased to 300 μg/mL, the proliferation inhibition of EPSI in HepG2 cells increased to about 50%, which was significantly different from that at a low dose ($p < 0.05$). When the dosage was over 400 μg/mL, the inhibitory ability of EPSI in HepG2 increased slightly, and the maximum rate was about 58%. This result showed that the IC50 value of EPSI was between 100–200 μg/mL for the CCK8 method. When the dosage was between 100–300 μg/mL, EPSI showed an obvious inhibition ability in HepG2 cells.

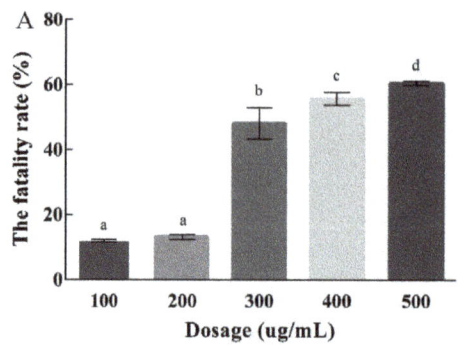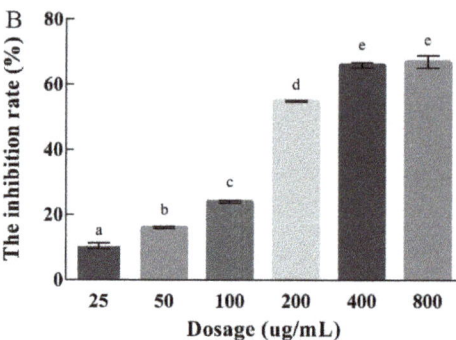

Figure 2. The growth inhibition of EPSI in HepG2 cells: (**A**) CCK8; (**B**) MTT. Different lowercase letters represent significant differences ($p < 0.05$).

In Figure 2B, we measured the growth inhibition of EPSI in HepG2 cells by the MTT method. When the dose concentration was 25–100 μg/mL, the proliferation of HepG2 cells was still at a high level and the maximum inhibition was less than 25%. When the dosage was 200 μg/mL, EPSI increased the inhibition of HepG2 cells significantly (55%, $p < 0.05$). When the dosage was 400 μg/mL or higher, the inhibition ability of HepG2 improved slightly. The maximum inhibition ability of EPSI in HepG2 cells was about 67%. These results showed that the IC50 value of EPSI was 100–200 μg/mL for the MTT method. When the dosage was 100–300 μg/mL, EPSI showed good inhibition in vitro.

3.3. Acute Toxicity Test

The fixed-dose procedure (FDP), a classic way to determine the acute toxicity, was used to verify the acute toxicity of EPSI. The acute toxicity test of EPSI in BALB/c mice showed a LD50 of >2 g/kg, indicating that the product was safe for the experiment of the next stage. This toxicity was lower than that of conventional drugs with anticancer activity [37–39].

3.4. Effect of EPSI on Tumors in BALB/c Nude Mice

The bodyweight changes of all experimental animals are shown in Figure 3A. Compared with the control group, the weight of the animals in the high-dose group showed a significant reduction ($p < 0.05$), which happened from Day 15. During the use of anticancer drugs, the situation of weight loss occurs [40,41]. The intersection of liver cancer, anticancer drugs, and weight change may be the thyroid gland [42,43], which is closely related to the proliferation of liver cancer cells and changes in bodyweight. Whether EPSI affects the thyroid needs further study. However, for cancer patients, keeping correct weight is necessary to maintain efficacy [44,45].

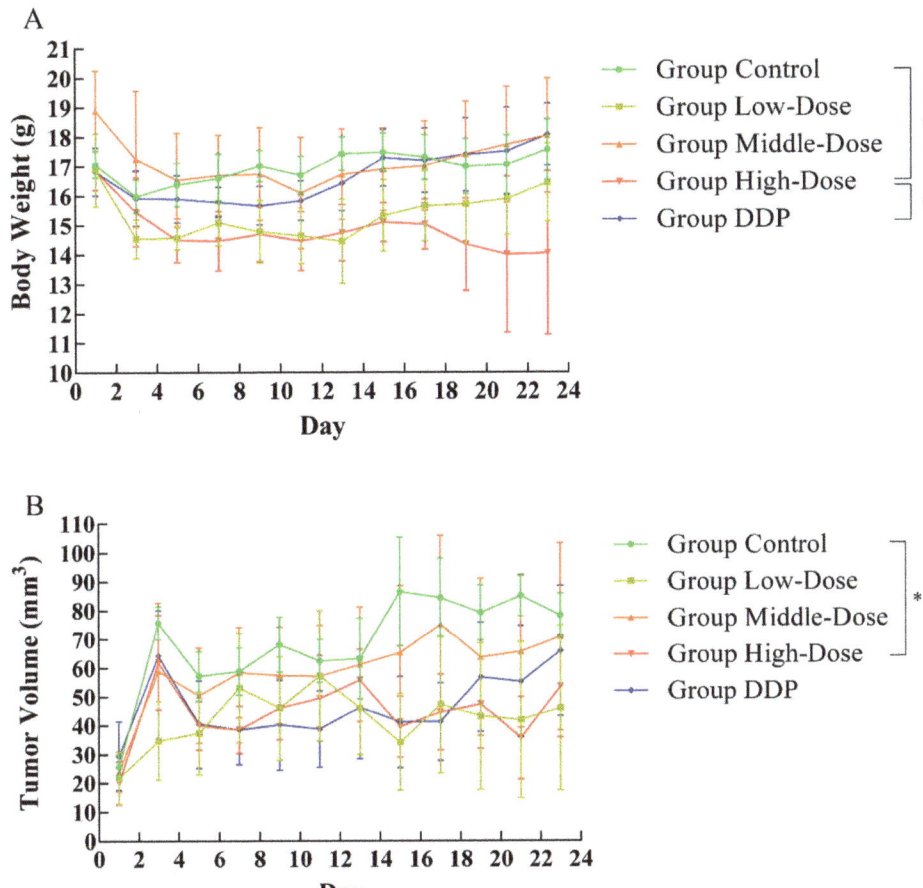

Figure 3. Bodyweight and tumor volume changes in the animal experiment during treatment. (**A**) Bodyweight changes and (**B**) tumor volume changes (n = 8; *: $p < 0.05$) at Day 23.

The changes in tumor volume are shown in Figure 3B. The state of the tumor at the sacrifice of experimental animals is shown in Figure 4. Because of the individual differences of experimental animals, the tumor volume was not very consistent. However, EPSI can certainly inhibit the growth of tumors, because the tumor volume of the treatment group (low/middle/high-dose groups) was lower than that of the control group (Figure 3B). A significant difference was shown between the high-dose group and the control group ($p < 0.05$). Meanwhile, the tumor volume of the medium-dose group and the high-dose group was significantly smaller than that of the DDP group ($p < 0.05$), showing good tumor inhibitory activity (Figure 3B). Observation of the tumor state after animal sacrificed also revealed the same situation (Figure 4): the tumor state of the treatment group was significantly different from that of other groups.

The results of H&E staining showed the state of the tumor tissue directly, and the results of the control group and the high-dose group are shown in Figure 4. Compared with the control group, the tumors in the high-dose group showed the significant reduction and destruction of tumor tissue. The tumor tissue in the control group was dense and uniform, while that of the high-dose group was disordered and less density, with even a lack of tissue. This result showed that EPSI had a significant inhibitory effect on hepatoma tumors, especially at high doses. Because EPSI passed the acute toxicity test, the high-dose group and the control group were selected for quantitative proteomic analysis.

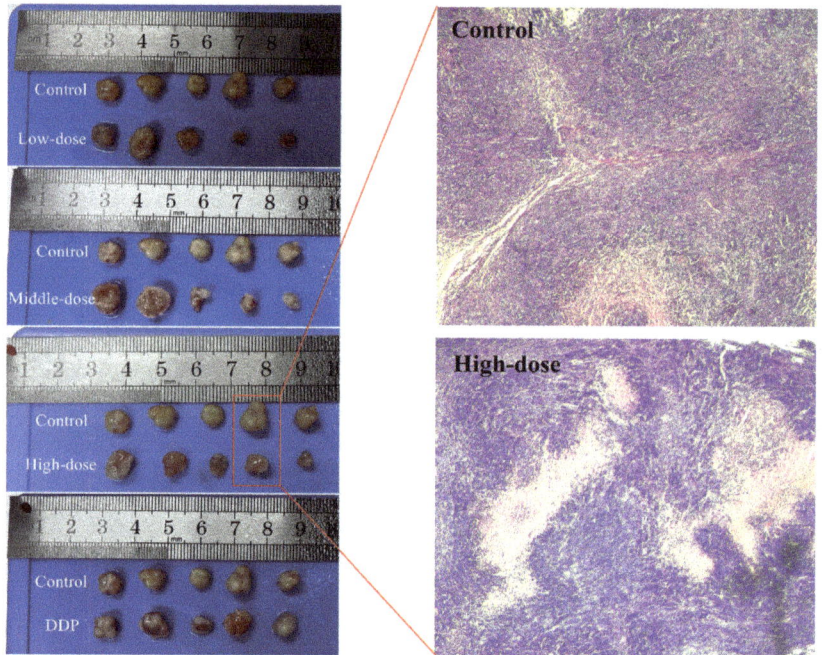

Figure 4. The state and hematoxylin and eosin (H&E) staining of tumors at the sacrifice of experimental animals (n = 8, representative samples are shown). Low dose: 100 mg/kg EPSI gavage for 24 days; middle dose: 200 mg/kg EPSI gavage for 24 days; high dose: 400 mg/kg EPSI gavage for 24 days; DDP: 10% DDP intraperitoneal injection for 10 days.

The effects of EPSI intake on TNF-α are shown in Figure 5. As a tumor necrosis factor, TNF-α is an important serological indicator used to measure the severity of a tumor. As shown in Figure 5, the serum TNF-α level in the control group was the lowest (116.11 pg/mL), which was basically the same as that in the DDP group (120.42 pg/mL). The level of TNF-α in the treatment group showed an obvious upward trend with an increase in the dosage. Although the low-dose group (124.82 pg/mL) increased slightly, no significant differences ($p > 0.05$) were seen compared with the control group. TNF-α in the middle-dose group (149.89 pg/mL) and the high-dose group (169.29 pg/mL) were significantly increased ($p < 0.05$) compared with the control group.

BALB/c mice are innately immune-deficient animals. The test animals have no thymus, resulting in T-cell defects and dysfunction. At this time, we can judge that the higher the TNF-α content, the stronger the immune response of the tested animals and the more obvious the improvement of their immunity. Therefore, these results mean that with the increase in the EPSI dosage, the limited immune system of the test animals was activated and produced a stronger immune response in the case of a lack of immune function.

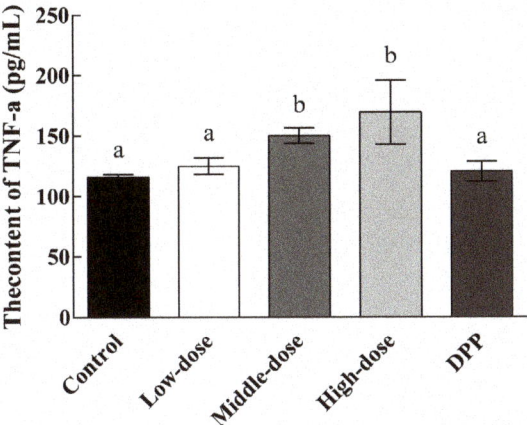

Figure 5. TNF-α content in the serum of experimental animals (n = 8). Low dose: 100 mg/kg EPSI gavage for 24 days; middle dose: 200 mg/kg EPSI gavage for 24 days; high dose: 400 mg/kg EPSI gavage for 24 days; DDP: 10% DDP intraperitoneal injection for 10 days. Different lowercase letters represent significant differences ($p < 0.05$).

3.5. Protein Expression Differences Induced by EPSI

The proteomic data of the tumors from the high-dose group of BALB/c mice are shown in Figure 6A. In total, 116 differentially expressed proteins (DE proteins) and 6787 proteins were identified in this experiment. On the whole, the number of upregulated DE proteins (99) was far more than that of downregulated DE proteins (17) between the control group and high-dose group after the intervention, and the upregulated DE protein amounts in the high-dose group were over 3 times higher than that in the control group when the fold change was 1.2×. The volcano map of the DE proteins with a 1.2× fold change is shown in Figure 6B. According to Figure 6B, the number of upregulated and downregulated proteins in the treatment group had unique proteins.

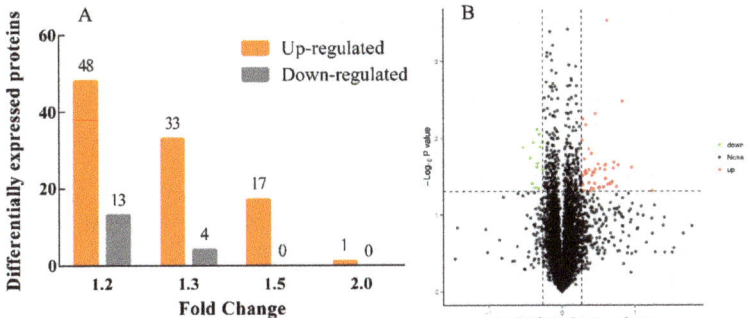

Figure 6. Distribution of proteins identified in tumor of BALB/c mice (n = 3). (**A**) Up-regulated and down-regulated of the DE proteins (Filtered with the threshold value of expression fold change, $p < 0.05$); (**B**) Volcano plot of the DE proteins. The X-axis was the fold change, and the Y-axis was the significant difference p value. Red dots represented the expression of up-regulated and green dots represented the down-regulated.

3.6. GO Enrichment of Differentially Quantified Proteins

Following the GO classifications for biological process (BP), cellular component (CC), and molecular function (MF), the enrichment results were examined to compare the functional correlations of DE proteins; the results are shown in Figure 7. In the upregulated proteins (Figure 7A), the proteins had functions in biological process and molecular function, but not cell component. These proteins are mainly concentrated (>20%) in protein metabolic processes (BP), metabolic processes (BP), catalytic activity (MF), and hydrolase activity (MF). In the downregulated proteins (Figure 7B), by comparison, all three classifications of proteins existed, and their functions showed a high concentration. Proteolysis (BP), endopeptidase activity (MF), and serine-type endopeptidase activity (MF) were the main roles of these proteins. These results suggest that the intake of EPSI associated with the anti-hepatoma effect may be achieved through specific mechanisms.

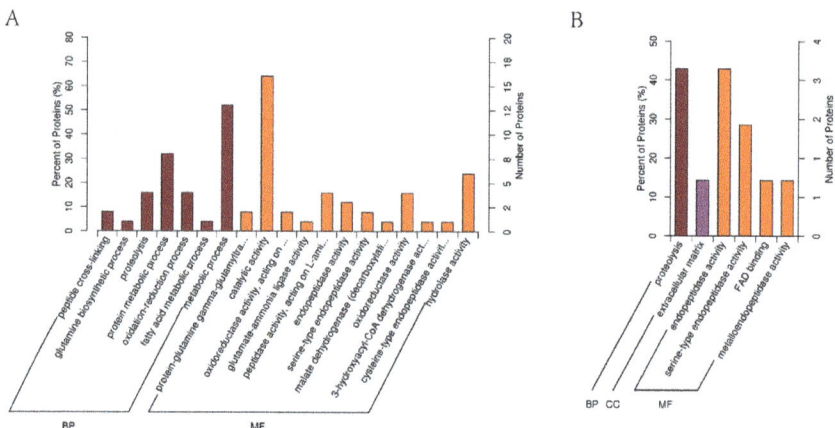

Figure 7. GO enrichment results. (CC: cell component; MF: molecular function; BP: biological process; n = 3). (**A**) Upregulated DE proteins ($p < 0.05$); (**B**) downregulated DE proteins ($p < 0.05$).

3.7. Specific Regulation Pathways for Inhibiting Liver Tumor Proliferation by EPSI

To further explore the related pathway of EPSI for anti-hepatoma effects and inhibition of liver tumors, KEGG pathway enrichment analysis was carried out (Figure 8). In the upregulated proteins (Figure 8A), the highest degree of protein enrichment was in caffeine metabolism, while most DE proteins with higher reliability appeared in the renin-angiotensin system. In the downregulated proteins (Figure 8B), two kinds appeared in neuroactive ligand-receptor interaction and hematopoietic cell lineage. The results showed that the inhibition of liver tumors by EPSI was closely related to the above pathways.

In the detailed KEGG pathway enrichment results (Table 5), we found 12 typical enrichment pathways in the process of inhibiting liver tumors by EPSI ($p < 0.05$). The names of the enriched KEGG pathways and the enriched proteins with significant changes are also listed in Table 5, while a description of the proteins is shown in Table 6.

Structural domain enrichment can identify the domain entries that are statistically significantly enriched. This function or positioning may be the cause of the differences. From the statistics of the enrichment results, a bubble chart of the structural domain is shown in Figure 9. As for the previous expression, peptidase S1/S6, chymotrypsin/Hap, peptidase cysteine/serine, and trypsin-like should be considered as the typical structural domains.

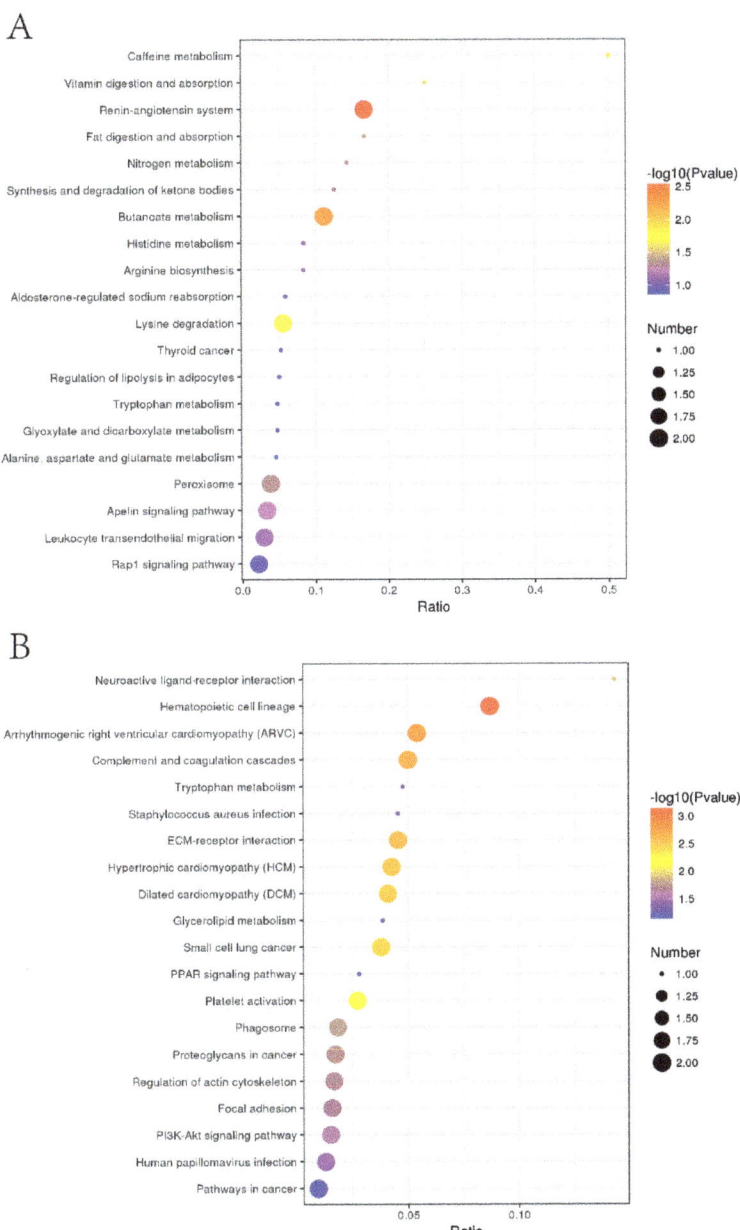

Figure 8. KEGG enrichment results. (**A**) Upregulated; (**B**) downregulated (n = 3). The X-axis is the ratio of the number of DE proteins in the corresponding pathway to the number of total proteins identified in the pathway. The greater the value, the higher the enrichment degree of the DE proteins in the pathway. The colors of the points represent the *p*-value of the hypergeometric test. The redder the color, the smaller the value, indicating that the reliability of the test is greater and more statistically significant. The size of the dot represents the number of DE proteins in the corresponding pathway. The larger the dot, the more DE proteins in the pathway.

Table 5. KEGG enrichment results (top 12, $p < 0.05$).

ID	Title	x	y	n	N	Prot ID
04614	Renin-angiotensin system	2	12	27	3214	B2R941 P23946
00650	Butanoate metabolism	2	18	27	3214	Q02338 Q08426
00380	Tryptophan metabolism	2	21	27	3214	Q9BS61 Q08426
04640	Hematopoietic cell lineage	2	23	27	3214	P17301 E7ESP4
00232	Caffeine metabolism	1	2	27	3214	P47989
04611	Platelet activation	3	74	27	3214	P17301 E7ESP4 O15264
04977	Vitamin digestion and absorption	1	4	27	3214	E1B4S8
00310	Lysine degradation	2	36	27	3214	Q9NVH6 Q08426
03320	PPAR signaling pathway	2	36	27	3214	A0A140VKG0 Q08426
05412	Arrhythmogenic right ventricular cardiomyopathy (ARVC)	2	37	27	3214	P17301 E7ESP4
04610	Complement and coagulation cascades	2	40	27	3214	A0A0F7G8J1 Q19UG4
04975	Fat digestion and absorption	1	6	27	3214	E1B4S8

ID: the ID of the enriched KEGG pathway; Title: the name of enriched KEGG pathway; X: the number of DE proteins associated with this pathway; Y: the number of background (all) proteins associated with the pathway; n: the number of DE proteins annotated by KEGG; N: the number of background (all) proteins annotated by KEGG; Prot ID: the enriched protein list.

Table 6. Description of the enriched proteins.

Prot ID	Description
B2R941	cDNA, FLJ94198, highly similar to Homo sapiens carboxypeptidase A3 (mast cell)
P23946	Chymase
Q02338	D-beta-hydroxybutyrate dehydrogenase, mitochondrial
Q08426	Peroxisomal bifunctional enzyme
Q9BS61	Kynurenine 3-monooxygenase
Q08426	Peroxisomal bifunctional enzyme
P17301	Integrin alpha-2
E7ESP4	Integrin alpha-2
P47989	Xanthine dehydrogenase/oxidase
O15264	Mitogen-activated protein kinase 13
E1B4S8	Apolipoprotein B (Fragment)
Q9NVH6	Trimethyllysine dioxygenase
Q08426	Peroxisomal bifunctional enzyme
A0A0F7G8J1	Plasminogen
Q19UG4	Christmas factor (Fragment)

Prot ID: the enriched protein list; description: the function of the protein described in the database.

3.8. Interaction Analysis of Differentially Expressed Proteins

According to the KEGG enrichment results (Table 5), we found that a few DE proteins appeared in multiple pathways, such as Q08426, E7ESP4, etc. These results showed that there should be a close relationship between the DE proteins, and between the pathways. The protein interaction network is shown in Figure 10. There are three main associated relationships in the network: the first is centered on P17301, the downregulation of which caused upregulation of other three proteins (E1B4S8, P12830, Q8TDR6) and downregulation of the other four (Q19UG4, A0A0F7G8J1, Q597H1, and Q99542). The second line established the relationships among Q6B051, B2R941, and P23946. In addition, there was also an association between Q08426 and P48163. The above results established the relationship between hematopoietic cell lineage, vitamin digestion and absorption, ARVC, complement and coagulation cascades, and fat digestion and absorption.

However, we were also aware that the situation that the metabolic pathways and related DE proteins summarized above may not be all related to EPSI. Because EPSI is not soluble in water, a certain amount of ethanol must be used as the solvent of EPSI during the feeding of experimental animals. Therefore, the above differences might have the effect of ethanol intervention in the body. We should probably pay more attention to what other pathways the related DE proteins appeared in. Therefore, we summarized the relevant information (Table 7).

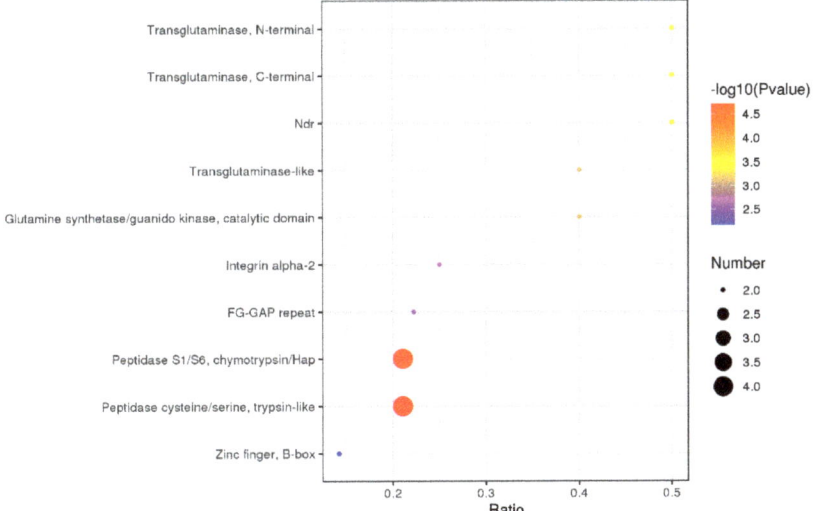

Figure 9. The enrichment results of the domain (n = 3). The X-axis is the ratio of the number of differential proteins in the corresponding domain to the number of total proteins identified in the domain. The greater the value, the higher the enrichment degree of DE proteins in the domain. The colors of the points represent the *p*-value of the hypergeometric test. The redder the color, the smaller the value, indicating that the reliability of the test is greater and more statistically significant. The size of the dot represents the number of DE proteins in the corresponding domain. The larger the dot, the more DE proteins in the domain.

In this result, nine DE proteins and 54 related KEGG pathways were shown; meanwhile, the pathways that have been reported to be directly associated with cancer were marked [46–74]. Among these, some are directly related to liver cancer or tumors, such as platelet activation [75], the PI3K-Akt signaling pathway [76], and the PPAR signaling pathway [77], etc. We found that, besides liver cancer, these signaling pathways appeared in studies on cancers of the breast, colon, and thyroid frequently. There are reasons to believe that the inhibitory effect of EPSI on liver tumors may be the same as the inhibitory mechanism of these cancers. We will carry out a demonstration of the metabolic pathways based on the results of this study in follow-up studies.

Table 7. KEGG pathways associated with DE proteins which had relationships in the network.

Prot ID	KEGG Pathway
P17301	Hematopoietic cell lineage * Platelet activation * Arrhythmogenic right ventricular cardiomyopathy (ARCV) ECM-receptor interaction * Hypertrophic cardiomyopathy (HCM) Dilated cardiomyopathy (DCM) Small-cell lung cancer * Proteoglycans in cancer * Focal adhesion * Phagosome * Regulation of actin cytoskeleton * PI3K-Akt signaling pathway * Human papillomavirus infection * Pathways in cancer *
E1B4S8	Vitamin digestion and absorption Fat digestion and absorption Cholesterol metabolism *
P12830	Apelin signaling pathway * Thyroid cancer * Rap1 signaling pathway * Bladder cancer * Endometrial cancer * Melanoma * Cell adhesion molecules (CAMs) * Pathogenic Escherichia coli infection * Gastric cancer * Adherens junction * Bacterial invasion of epithelial cells Pathways in cancer * Hippo signaling pathway *
Q19UG4	Complement and coagulation cascades *
A0A0F7G8J1	Complement and coagulation cascades * Neuroactive ligand-receptor interaction * Staphylococcus aureus infection * Influenza A
B2R941	Renin-angiotensin system * Protein digestion and absorption * Pancreatic secretion *
P23946	Renin-angiotensin system *
Q08426	Butanoate metabolism Tryptophan metabolism * Lysine degradation * PPAR signaling pathway * Peroxisome * Beta-alanine metabolism * Propanoate metabolismCarbon metabolism * Metabolic pathways * Fatty acid degradation * Fatty acid metabolism * Valine, leucine, and isoleucine degradation *
P48163	Carbon metabolism * Pyruvate metabolism Metabolic pathways *

Prot ID: the enriched protein list; *: reported to be directly associated with cancer or tumors.

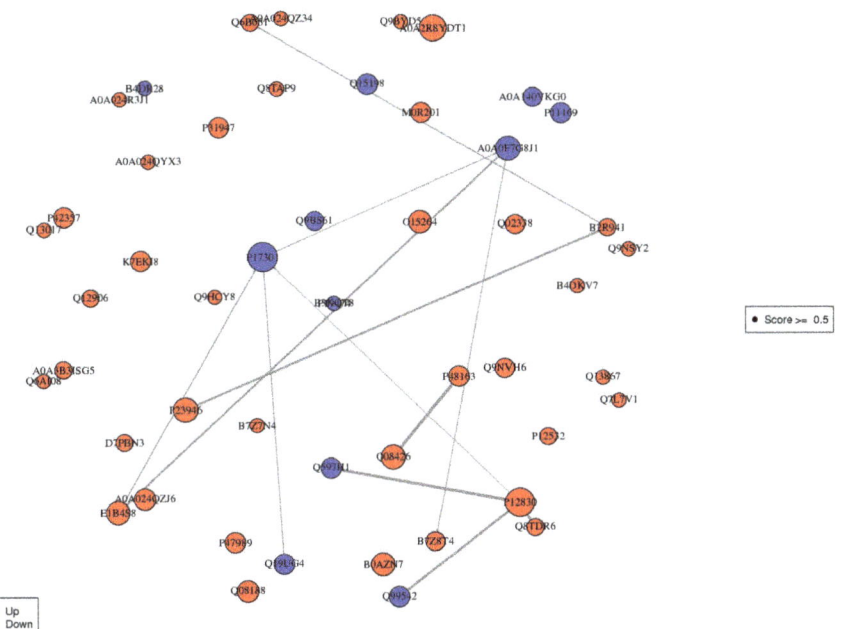

Figure 10. Protein interaction network. Each node in the interaction network represents a protein, and the node size represents the number of interacting proteins. The larger the node is, the more proteins interact with it. The color of the node indicates the expression level of the protein in the comparison pair: red represents significantly high expression of the protein and blue represents significantly low expression of the protein.

4. Conclusions

In summary, we identified the main components of EPSI by LC-MS and determined that it contained a variety of anti-inflammatory and anti-cancer components. On this foundation, we found that the EPSI showed good growth inhibition of HepG2 cells in vitro, while the high dose of EPSI had an obvious anti-hepatoma effect in vivo, and this effect was better than that of DDP without toxicity. To explore its anti-hepatoma mechanism, a TMT quantitative proteomic approach was used to examine the inhibition of liver tumors by EPSI. DE proteins and their relevance, and the related metabolic pathways were counted and displayed. The upregulated pathways, such as the renin-angiotensin system, and the downregulated pathways, such as hematopoietic cell lineage, and the changes in the PI3K-Akt signaling pathway and the PPAR signaling pathway may be the significant keys to the anti-hepatoma effects. Moreover, the key DE protein intervening in the pathways is probably a new target for the treatment of liver tumors. Moreover, our work shines a new light on the crucial inhibitive function of *S. involucrata* in the progression of liver disease.

Author Contributions: J.G.: Conceptualization, software, writing—original draft; Y.W.: visualization, writing—original draft; B.L.: methodology, data curation; J.C.: investigation, visualization; G.C.: supervision, funding acquisition, project administration. All authors have read and agreed to the published version of the manuscript.

Funding: This research received no external funding.

Institutional Review Board Statement: The study was conducted according to the guidelines of the Declaration of Helsinki, and approved by the Ethics Committee of Laboratory Animal Welfare and Ethics Committee of Jilin Agricultural University (protocol code No.20190410005, April 2019).

Informed Consent Statement: Not applicable.

Data Availability Statement: Raw data can be provided by the corresponding author on request.

Acknowledgments: The authors acknowledge Changchun WISH Technology Service Co., Ltd. for their continuous support.

Conflicts of Interest: The authors declare no conflict of interest.

References

1. Li, G.; Liu, F.; Zhao, R. Studies on pharmacological actions of *Saussurea involucrata* Kar et Kir ex Maxim (author's transl). *Yao Xue Xue Bao Acta Pharm. Sin.* **1980**, *15*, 368–370.
2. Chik, W.-I.; Zhu, L.; Fan, L.-L.; Yi, T.; Zhu, G.-Y.; Gou, X.-J.; Tang, Y.-N.; Xu, J.; Yeung, W.-P.; Zhao, Z.-Z.; et al. *Saussurea involucrata*: A review of the botany, phytochemistry and ethnopharmacology of a rare traditional herbal medicine. *J. Ethnopharmacol.* **2015**, *172*, 44–60. [CrossRef] [PubMed]
3. Li, Y.; Wang, C.; Guo, S.; Yang, J.; Xiao, P. Three guaianolides from *Saussurea involucrata* and their contents determination by HPLC. *J. Pharm. Biomed. Anal.* **2007**, *44*, 288–292. [CrossRef] [PubMed]
4. Chaudhary, N.; Aparoy, P. Deciphering the mechanism behind the varied binding activities of COXIBs through Molecular Dynamic Simulations, MM-PBSA binding energy calculations and per-residue energy decomposition studies. *J. Biomol. Struct. Dyn.* **2017**, *35*, 868–882. [CrossRef] [PubMed]
5. Yu, L.; Zhong-Jian, J. Guaianolides from *Saussurea involucrata*. *Phytochemistry* **1989**, *28*, 3395–3397. [CrossRef]
6. Seilgazy, M.; Li, J.; Aisa, H.A. Isolation of Steroidal Esters from Seeds of *Saussurea involucrata*. *Chem. Nat. Compd.* **2017**, *53*, 1196–1198. [CrossRef]
7. Ma, H.-P.; Fan, P.-C.; Jing, L.-L.; Yao, J.; He, X.-R.; Yang, Y.; Chen, K.-M.; Jia, Z.-P. Anti-hypoxic activity at simulated high altitude was isolated in petroleum ether extract of *Saussurea involucrata*. *J. Ethnopharmacol.* **2011**, *137*, 1510–1515. [CrossRef]
8. Zhai, K.-F.; Duan, H.; Xing, J.-G.; Huang, H. Study on the anti-inflammatory and analgesic effects of various parts from *Saussurea involucrate*. *Chin. J. Hosp. Pharm.* **2010**, *5*, 374–377.
9. Way, T.-D.; Lee, J.-C.; Kuo, D.-H.; Fan, L.-L.; Huang, C.-H.; Lin, H.-Y.; Shieh, P.-C.; Kuo, P.-T.; Liao, C.-F.; Liu, H.; et al. Inhibition of Epidermal Growth Factor Receptor Signaling by *Saussurea involucrata*, a Rare Traditional Chinese Medicinal Herb, in Human Hormone-Resistant Prostate Cancer PC-3 Cells. *J. Agric. Food Chem.* **2010**, *58*, 3356–3365. [CrossRef] [PubMed]
10. Gong, G.; Xie, F.; Zheng, Y.; Hu, W.; Qi, B.; He, H.; Dong, T.T.; Tsim, K.W. The effect of methanol extract from *Saussurea involucrata* in the lipopolysaccharide-stimulated inflammation in cultured RAW 264.7 cells. *J. Ethnopharmacol.* **2020**, *251*, 112532. [CrossRef] [PubMed]
11. Paek, K.-Y.; Murthy, H.N.; Zhong, J.-J. *Production of Biomass and Bioactive Compounds Using Bioreactor Technology*; Springer: Berlin/Heidelberg, Germany, 2014.
12. Wang, J.; Li, J.-L.; Li, J.; Li, J.-X.; Liu, S.-J.; Huang, L.-Q.; Gao, W.-Y. Production of active compounds in medicinal plants: From plant tissue culture to biosynthesis. *Chin. Herb. Med.* **2017**, *9*, 115–125. [CrossRef]
13. Rhiouani, H.; EL Hilaly, J.; Israili, Z.H.; Lyoussi, B. Acute and sub-chronic toxicity of an aqueous extract of the leaves of *Herniaria glabra* in rodents. *J. Ethnopharmacol.* **2008**, *118*, 378–386. [CrossRef]
14. Stalikas, C.D. Extraction, separation, and detection methods for phenolic acids and flavonoids. *J. Sep. Sci.* **2007**, *30*, 3268–3295. [CrossRef] [PubMed]
15. Chen, W.; Gong, L.; Guo, Z.; Wang, W.; Zhang, H.; Liu, X.; Yu, S.; Xiong, L.; Luo, J. A Novel Integrated Method for Large-Scale Detection, Identification, and Quantification of Widely Targeted Metabolites: Application in the Study of Rice Metabolomics. *Mol. Plant* **2013**, *6*, 1769–1780. [CrossRef] [PubMed]
16. Garcia, A.; Barbas, C. Gas Chromatography-Mass Spectrometry (GC-MS)-Based Metabolomics. In *Metabolic Profiling*; Springer: Berlin/Heidelberg, Germany, 2011; pp. 191–204.
17. Chen, Y.; Zhang, R.; Song, Y.; He, J.; Sun, J.; Bai, J.; An, Z.; Dong, L.; Zhan, Q.; Abliz, Z. RRLC-MS/MS-based metabonomics combined with in-depth analysis of metabolic correlation network: Finding potential biomarkers for breast cancer. *Analyst* **2009**, *134*, 2003–2011. [CrossRef]
18. Van den Heuvel, M.; Clark, D.; Fielder, R.; Koundakjian, P.; Oliver, G.; Pelling, D.; Tomlinson, N.; Walker, A. The international validation of a fixed-dose procedure as an alternative to the classical LD50 test. *Food Chem. Toxicol.* **1990**, *28*, 469–482. [CrossRef]
19. Stallard, N.; Whitehead, A. Reducing animal numbers in the fixed-dose procedure. *Hum. Exp. Toxicol.* **1995**, *14*, 315–323. [CrossRef] [PubMed]
20. Wen, Y.; Jin, L.; Zhang, D.; Zhang, L.; Xie, C.; Guo, D.; Wang, Y.; Wang, L.; Zhu, M.; Tong, J.; et al. Quantitative proteomic analysis of scleras in guinea pig exposed to wavelength defocus. *J. Proteom.* **2021**, *243*, 104248. [CrossRef]
21. Kachuk, C.; Stephen, K.; Doucette, A. A Comparison of sodium dodecyl sulfate depletion techniques for proteome analysis by mass spectrometry. *J. Chromatogr. A* **2015**, *1418*, 158–166. [CrossRef]
22. Wiśniewski, J.R.; Zougman, A.; Nagaraj, N.; Mann, M. Universal sample preparation method for proteome analysis. *Nat. Methods* **2009**, *6*, 359–362. [CrossRef] [PubMed]
23. Gillette, M.A.; Satpathy, S.; Cao, S.; Dhanasekaran, S.M.; Vasaikar, S.V.; Krug, K.; Petralia, F.; Li, Y.; Liang, W.-W.; Reva, B.; et al. Proteogenomic Characterization Reveals Therapeutic Vulnerabilities in Lung Adenocarcinoma. *Cell* **2020**, *182*, 200–225. [CrossRef] [PubMed]

24. Zhang, H.; Liu, T.; Zhang, Z.; Payne, S.H.; Zhang, B.; McDermott, J.E.; Zhou, J.-Y.; Petyuk, V.A.; Chen, L.; Ray, D.; et al. Integrated Proteogenomic Characterization of Human High-Grade Serous Ovarian Cancer. *Cell* **2016**, *166*, 755–765. [CrossRef]
25. Yin, X.-L.; Xu, B.-Q.; Zhang, Y.-Q. Gynura divaricata rich in 3, 5−/4, 5-dicaffeoylquinic acid and chlorogenic acid reduces islet cell apoptosis and improves pancreatic function in type 2 diabetic mice. *Nutr. Metab.* **2018**, *15*, 73. [CrossRef] [PubMed]
26. Lodise, O.; Patil, K.; Karshenboym, I.; Prombo, S.; Chukwueke, C.; Pai, S.B. Inhibition of Prostate Cancer Cells by 4,5-Dicaffeoylquinic Acid through Cell Cycle Arrest. *Prostate Cancer* **2019**, *2019*, 4520645. [CrossRef] [PubMed]
27. Byambaragchaa, M.; de la Cruz, J.; Yang, S.H.; Hwang, S.-G. Anti-metastatic potential of ethanol extract of *Saussurea involucrata* against hepatic cancer in vitro. *Asian Pac. J. Cancer Prev.* **2013**, *14*, 5397–5402. [CrossRef] [PubMed]
28. Crout, R.; Gilbertson, J.R.; Gilbertson, J.D.; Platt, D.; Langkamp, H.; Connamacher, R. Effect of linolenyl alcohol on the in-vitro growth of the oral bacterium *Streptococcus mutans*. *Arch. Oral Biol.* **1982**, *27*, 1033–1037. [CrossRef]
29. Sola, A.; Vilas, P.; Rodriguez, S.; Gancedo, A.G.; Gil-Fernández, C. Inactivation and inhibition of African swine fever virus by monoolein, monolinolein, and γ-linolenyl alcohol. *Arch. Virol.* **1986**, *88*, 285–292. [CrossRef]
30. Marshall, M.E.; Kervin, K.; Benefield, C.; Umerani, A.; Albainy-Jenei, S.; Zhao, Q.; Khazaeli, M.B. Growth-inhibitory effects of coumarin (1,2-benzopyrone) and 7-hydroxycoumarin on human malignant cell lines in vitro. *J. Cancer Res. Clin. Oncol.* **1994**, *120*, S3–S10. [CrossRef] [PubMed]
31. Timonen, J.M.; Nieminen, R.M.; Sareila, O.; Goulas, A.; Moilanen, L.J.; Haukka, M.; Vainiotalo, P.; Moilanen, E.; Aulaskari, P.H. Synthesis and anti-inflammatory effects of a series of novel 7-hydroxycoumarin derivatives. *Eur. J. Med. Chem.* **2011**, *46*, 3845–3850. [CrossRef]
32. Kostova, I.P.; Manolov, I.I.; Nicolova, I.N.; Danchev, N.D. New metal complexes of 4-methyl-7-hydroxycoumarin sodium salt and their pharmacological activity. *Il Farm.* **2001**, *56*, 707–713. [CrossRef]
33. Sato, Y.; Itagaki, S.; Kurokawa, T.; Ogura, J.; Kobayashi, M.; Hirano, T.; Sugawara, M.; Iseki, K. In vitro and in vivo antioxidant properties of chlorogenic acid and caffeic acid. *Int. J. Pharm.* **2011**, *403*, 136–138. [CrossRef]
34. Lou, Z.; Wang, H.; Zhu, S.; Ma, C.; Wang, Z. Antibacterial Activity and Mechanism of Action of Chlorogenic Acid. *J. Food Sci.* **2011**, *76*, M398–M403. [CrossRef]
35. Kim, Y.I.; Hirai, S.; Takahashi, H.; Goto, T.; Ohyane, C.; Tsugane, T.; Konishi, C.; Fujii, T.; Inai, S.; Iijima, Y.; et al. 9-oxo-10(E),12(E)-octadecadienoic acid derived from tomato is a potent PPAR α agonist to decrease triglyceride accumulation in mouse primary hepatocytes. *Mol. Nutr. Food Res.* **2011**, *55*, 585–593. [CrossRef] [PubMed]
36. Zhao, B.; Tomoda, Y.; Mizukami, H.; Makino, T. 9-Oxo-(10E, 12E)-octadecadienoic acid, a cytotoxic fatty acid ketodiene isolated from eggplant calyx, induces apoptosis in human ovarian cancer (HRA) cells. *J. Nat. Med.* **2015**, *69*, 296–302. [CrossRef] [PubMed]
37. Xargay-Torrent, S.; López-Guerra, M.; Saborit-Villarroya, I.; Rosich, L.; Navarro, A.; Villamor, N.; Aymerich, M.; Perez-Galan, P.; Roue, G.; Campo, E.; et al. The Multi-Kinase Inhibitor Sorafenib Blocks Migration, BCR Survival Signals, Protein Translation and Stroma-Mediated Bortezomib Resistance in Mantle Cell Lymphoma. *Blood* **2012**, *120*, 1647. [CrossRef]
38. Li, H.-T.; Zhu, X. Quinoline-based compounds with potential activity against drugresistant cancers. *Curr. Top. Med. Chem.* **2021**, *21*, 426–437. [CrossRef]
39. Mirmalek, S.A.; Jangholi, E.; Jafari, M.; Yadollah-Damavandi, S.; Javidi, M.A.; Parsa, Y.; Parsa, T.; Salimi-Tabatabaee, S.A.; Kolagar, H.G.; Jalil, S.K.; et al. Comparison of in Vitro Cytotoxicity and Apoptogenic Activity of Magnesium Chloride and Cisplatin as Conventional Chemotherapeutic Agents in the MCF-7 Cell Line. *Asian Pac. J. Cancer Prev.* **2016**, *17*, 131–134. [CrossRef]
40. Spindler, S.R. Rapid and reversible induction of the longevity, anticancer and genomic effects of caloric restriction. *Mech. Ageing Dev.* **2005**, *126*, 960–966. [CrossRef] [PubMed]
41. Rephaeli, A.; Waks-Yona, S.; Nudelman, A.; Tarasenko, I.; Tarasenko, N.; Phillips, D.; Cutts, S.; Kessler-Icekson, G. Anticancer prodrugs of butyric acid and formaldehyde protect against doxorubicin-induced cardiotoxicity. *Br. J. Cancer* **2007**, *96*, 1667–1674. [CrossRef] [PubMed]
42. Gionfra, F.; De Vito, P.; Pallottini, V.; Lin, H.-Y.; Davis, P.J.; Pedersen, J.Z.; Incerpi, S. The Role of Thyroid Hormones in Hepatocyte Proliferation and Liver Cancer. *Front. Endocrinol.* **2019**, *10*, 532. [CrossRef] [PubMed]
43. Kowalik, M.A.; Columbano, A.; Perra, A. Thyroid Hormones, Thyromimetics and Their Metabolites in the Treatment of Liver Disease. *Front. Endocrinol.* **2018**, *9*, 382. [CrossRef] [PubMed]
44. Baker, S.D.; Grochow, L.B.; Donehower, R.C. Should Anticancer Drug Doses Be Adjusted in the Obese Patient? *J. Natl. Cancer Inst.* **1995**, *87*, 333–334. [CrossRef]
45. Felici, A.; Verweij, J.; Sparreboom, A. Dosing strategies for anticancer drugs: The good, the bad and body-surface area. *Eur. J. Cancer* **2002**, *38*, 1677–1684. [CrossRef]
46. Ueno, N.; Rizzo, J.; Demirer, T.; Cheng, Y.; Hegenbart, U.; Zhang, M.; Bregni, M.; Carella, A.; Blaise, D.; Bashey, A. Allogeneic hematopoietic cell transplantation for metastatic breast cancer. *Bone Marrow Transplant.* **2008**, *41*, 537–545. [CrossRef]
47. Osada, J.; Rusak, M.; Kamocki, Z.; Dabrowska, M.I.; Kedra, B. Platelet activation in patients with advanced gastric cancer. *Neoplasma* **2010**, *57*, 145–150. [CrossRef] [PubMed]
48. Bao, Y.; Wang, L.; Shi, L.; Yun, F.; Liu, X.; Chen, Y.; Chen, C.; Ren, Y.; Jia, Y. Transcriptome profiling revealed multiple genes and ECM-receptor interaction pathways that may be associated with breast cancer. *Cell. Mol. Biol. Lett.* **2019**, *24*, 38. [CrossRef]
49. Van Meerbeeck, J.P.; Fennell, D.A.; De Ruysscher, D.K. Small-cell lung cancer. *Lancet* **2011**, *378*, 1741–1755. [CrossRef]
50. Iozzo, R.V.; Sanderson, R.D. Proteoglycans in cancer biology, tumour microenvironment and angiogenesis. *J. Cell. Mol. Med.* **2011**, *15*, 1013–1031. [CrossRef] [PubMed]

51. Zhao, J.; Guan, J.-L. Signal transduction by focal adhesion kinase in cancer. *Cancer Metastasis Rev.* **2009**, *28*, 35–49. [CrossRef]
52. Chen, W.Y.; Wu, F.; You, Z.Y.; Zhang, Z.M.; Guo, Y.L.; Zhong, L.X. Analyzing the differentially expressed genes and pathway cross-talk in aggressive breast cancer. *J. Obstet. Gynaecol. Res.* **2015**, *41*, 132–140. [CrossRef]
53. Imai-Sumida, M.; Chiyomaru, T.; Majid, S.; Saini, S.; Nip, H.; Dahiya, R.; Tanaka, Y.; Yamamura, S. Silibinin suppresses bladder cancer through down-regulation of actin cytoskeleton and PI3K/Akt signaling pathways. *Oncotarget* **2017**, *8*, 92032–92042. [CrossRef] [PubMed]
54. Grm, H.S.; Bergant, M.; Banks, L. Human papillomavirus infection, cancer & therapy. *Indian J. Med. Res.* **2009**, *130*, 277. [PubMed]
55. Ding, X.; Zhang, W.; Li, S.; Yang, H. The role of cholesterol metabolism in cancer. *Am. J. Cancer Res.* **2019**, *9*, 219–227. [PubMed]
56. Kälin, R.E.; Kretz, M.P.; Meyer, A.M.; Kispert, A.; Heppner, F.L.; Brändli, A.W. Paracrine and autocrine mechanisms of apelin signaling govern embryonic and tumor angiogenesis. *Dev. Biol.* **2007**, *305*, 599–614. [CrossRef] [PubMed]
57. Bailey, C.L.; Kelly, P.; Casey, P.J. Activation of Rap1 Promotes Prostate Cancer Metastasis. *Cancer Res.* **2009**, *69*, 4962–4968. [CrossRef] [PubMed]
58. Jain, S.; Jagtap, V.; Pise, N. Computer Aided Melanoma Skin Cancer Detection Using Image Processing. *Procedia Comput. Sci.* **2015**, *48*, 735–740. [CrossRef]
59. Makrilia, N.; Kollias, A.; Manolopoulos, L.; Syrigos, K. Cell adhesion molecules: Role and clinical significance in cancer. *Cancer Investig.* **2009**, *27*, 1023–1037. [CrossRef] [PubMed]
60. Veziant, J.; Gagnière, J.; Jouberton, E.; Bonnin, V.; Sauvanet, P.; Pezet, D.; Barnich, N.; Miot-Noirault, E.; Bonnet, M. Association of colorectal cancer with pathogenic Escherichia coli: Focus on mechanisms using optical imaging. *World J. Clin. Oncol.* **2016**, *7*, 293. [CrossRef]
61. Bajenova, O.; Chaika, N.; Tolkunova, E.; Davydov-Sinitsyn, A.; Gapon, S.; Thomas, P.; O'Brien, S. Carcinoembryonic antigen promotes colorectal cancer progression by targeting adherens junction complexes. *Exp. Cell Res.* **2014**, *324*, 115–123. [CrossRef]
62. Pan, D. The Hippo Signaling Pathway in Development and Cancer. *Dev. Cell* **2010**, *19*, 491–505. [CrossRef]
63. Liu, Y.; Xiong, S.; Liu, S.; Chen, J.; Yang, H.; Liu, G.; Li, G. Analysis of gene expression in bladder cancer: Possible involvement of mitosis and complement and coagulation cascades signaling pathway. *J. Comput. Biol.* **2020**, *27*, 987–998. [CrossRef] [PubMed]
64. Huan, J.; Wang, L.; Xing, L.; Qin, X.; Feng, L.; Pan, X.; Zhu, L. Insights into significant pathways and gene interaction networks underlying breast cancer cell line MCF-7 treated with 17β-estradiol (E2). *Gene* **2014**, *533*, 346–355. [CrossRef]
65. Raad, I.; Narro, J.; Khan, A.; Tarrand, J.; Vartivarian, S.; Bodey, G. Serious complications of vascular catheter-related *Staphylococcus aureus* bacteremia in cancer patients. *Eur. J. Clin. Microbiol. Infect. Dis.* **1992**, *11*, 675–682. [CrossRef]
66. Pinter, M.; Jain, R.K. Targeting the renin-angiotensin system to improve cancer treatment: Implications for immunotherapy. *Sci. Transl. Med.* **2017**, *9*, eaan5616. [CrossRef]
67. Reber, H.A.; Johnson, F.E.; Montgomery, C.; Carl, W.R. Pancreatic secretion in hamsters with pancreatic cancer. *Surgery* **1977**, *82*, 34–41.
68. Platten, M.; Nollen, E.A.A.; Röhrig, U.F.; Fallarino, F.; Opitz, C.A. Tryptophan metabolism as a common therapeutic target in cancer, neurodegeneration and beyond. *Nat. Rev. Drug Discov.* **2019**, *18*, 379–401. [CrossRef]
69. Li, X.; Zhang, C.; Zhao, T.; Su, Z.; Li, M.; Hu, J.; Wen, J.; Shen, J.; Wang, C.; Pan, J.; et al. Lysine-222 succinylation reduces lysosomal degradation of lactate dehydrogenase a and is increased in gastric cancer. *J. Exp. Clin. Cancer Res.* **2020**, *39*, 1–17. [CrossRef] [PubMed]
70. Budczies, J.; Brockmöller, S.F.; Müller, B.M.; Barupal, D.K.; Richter-Ehrenstein, C.; Kleine-Tebbe, A.; Griffin, J.L.; Orešič, M.; Dietel, M.; Denkert, C.; et al. Comparative metabolomics of estrogen receptor positive and estrogen receptor negative breast cancer: Alterations in glutamine and beta-alanine metabolism. *J. Proteom.* **2013**, *94*, 279–288. [CrossRef]
71. Newman, A.C.; Maddocks, O.D.K. One-carbon metabolism in cancer. *Br. J. Cancer* **2017**, *116*, 1499–1504. [CrossRef] [PubMed]
72. Mikalayeva, V.; Ceslevičienė, I.; Sarapinienė, I.; Žvikas, V.; Skeberdis, V.A.; Jakštas, V.; Bordel, S. Fatty acid synthesis and degradation interplay to regulate the oxidative stress in cancer cells. *Int. J. Mol. Sci.* **2019**, *20*, 1348. [CrossRef]
73. Low, E.N.D.; Mokhtar, N.M.; Wong, Z.; Ali, R.A.R. Colonic Mucosal Transcriptomic Changes in Patients with Long-Duration Ulcerative Colitis Revealed Colitis-Associated Cancer Pathways. *J. Crohn's Coliti* **2019**, *13*, 755–763. [CrossRef] [PubMed]
74. Conde, V.R.; Oliveira, P.F.; Nunes, A.R.; Rocha, C.S.; Ramalhosa, E.; Pereira, J.A.; Alves, M.G.; Silva, B.M. The progression from a lower to a higher invasive stage of bladder cancer is associated with severe alterations in glucose and pyruvate metabolism. *Exp. Cell Res.* **2015**, *335*, 91–98. [CrossRef] [PubMed]
75. Malehmir, M.; Pfister, D.; Gallage, S.; Szydlowska, M.; Inverso, D.; Kotsiliti, E.; Leone, V.; Peiseler, M.; Surewaard, B.G.J.; Rath, D.; et al. Platelet GPIbα is a mediator and potential interventional target for NASH and subsequent liver cancer. *Nat. Med.* **2019**, *25*, 641–655. [CrossRef] [PubMed]
76. Zheng, L.; Gong, W.; Liang, P.; Huang, X.; You, N.; Han, K.Q.; Li, Y.M.; Li, J. Effects of AFP-activated PI3K/Akt signaling pathway on cell proliferation of liver cancer. *Tumor Biol.* **2014**, *35*, 4095–4099. [CrossRef]
77. Ferrara-Romeo, I.; Martínez, P.; Blasco, M.A. Mice lacking RAP1 show early onset and higher rates of DEN-induced hepatocellular carcinomas in female mice. *PLoS ONE* **2018**, *13*, e0204909. [CrossRef] [PubMed]

Article

Metabolism of Phenolics of *Tetrastigma hemsleyanum* Roots under In Vitro Digestion and Colonic Fermentation as Well as Their In Vivo Antioxidant Activity in Rats

Yong Sun [1,2], Fanghua Guo [1], Xin Peng [3], Kejun Cheng [4], Lu Xiao [2], Hua Zhang [5], Hongyan Li [2,*], Li Jiang [5] and Zeyuan Deng [1,*]

[1] State Key Laboratory of Food Science and Technology, Nanchang University, Nanchang 330047, China; yongsun@ncu.edu.cn (Y.S.); gfh1376234247@outlook.com (F.G.)
[2] College of Food Science and Technology, Nanchang University, Nanchang 330047, China; xl2691915147@163.com
[3] Ningbo Research Institute, Zhejiang University, Ningbo 315100, China; pengx@nit.zju.edu.cn
[4] Chemical Biology Center, Lishui Institute of Agriculture and Forestry Sciences, Lishui 323000, China; chengkejun@gmail.com
[5] College of Pharmacy, Jiangxi University of Traditional Chinese Medicine, Nanchang 330004, China; sunnymay_z@hotmail.com (H.Z.); viviface@yeah.net (L.J.)
* Correspondence: lihongyan@ncu.edu.cn (H.L.); zeyuandeng@hotmail.com (Z.D.)

Abstract: *Tetrastigma hemsleyanum* Diels et Gilg is a herbaceous perennial species distributed mainly in southern China. The *Tetrastigma hemsleyanum* root (THR) has been prevalently consumed as a functional tea or dietary supplement. In vitro digestion models, including colonic fermentation, were built to evaluate the release and stability of THR phenolics with the method of HPLC-QqQ-MS/MS and UPLC-Qtof-MS/MS. From the oral cavity, the contents of total phenolic and flavonoid began to degrade. Quercetin-3-rutinoside, quercetin-3-glucoside, kaempferol-3-rutinoside, and kaempferol-3-glucoside were metabolized as major components and they were absorbed in the form of glycosides for hepatic metabolism. On the other hand, the total antioxidant capacity (T-AOC), superoxide dismutase (SOD), glutathione peroxidase (GSH-Px) activity, and glutathione (GSH) content were significantly increased, while malondialdehyde (MDA) content was decreased in plasma and tissues of rats treated with THR extract in the oxidative stress model. These results indicated that the THR extract is a good antioxidant substance and has good bioavailability, which can effectively prevent some chronic diseases caused by oxidative stress. It also provides a basis for the effectiveness of THR as a traditional functional food.

Keywords: *Radix Tetrastigma*; colonic fermentation; phenolics; antioxidant capacity; bioavailability

1. Introduction

Tetrastigma hemsleyanum Diels et Gilg, a herbaceous perennial species, is distributed mainly in southern China. The *Tetrastigma hemsleyanum* root (THR, called "Sanyeqing" in Chinese) is one of the functional foods commonly used in China. As an edible plant, THR has been prevalently consumed as a dietary supplement, nutrient, or health tea for its health benefits. Meanwhile, the specific bioactive properties of THR such as immunoregulatory, antioxidant, anti-inflammatory, antipyretic, analgesic, and antiproliferactive capacity have been widely reported. It has also been applied to treat high fever, infantile febrile convulsion, pneumonia, asthma, hepatitis, rheumatism, menstrual disorders, sore throat, and scrofula [1,2]. In our previous publication, abundant flavonoids and phenolic acids were found in THR, which may be responsible for their bioactivity above. Moreover, the phenolics extract from THR showed strong antioxidant activities in several in vitro assay systems (DPPH, FRAP, and ABTS) [3]. Although the phenolics extract from THR has strong in vitro antioxidant capacity, few studies have reported its in vivo antioxidant effects.

The overproduction of reactive oxygen/nitrogen species (ROS/RNS), e.g., superoxide (O_2^-), hydrogen peroxide (H_2O_2), hydroxyl radical (OH), singlet oxygen (O_2), nitric oxide (NO·), peroxynitrite ($ONOO^-$), and nitrogen dioxide (NO_2), leads to an imbalance between the pro-oxidant and the antioxidant, which is the main cause of oxidative stress [4,5]. Under normal physiological conditions, these active substances are not necessarily a threat to the body. However, when they are unable to be cleared by the body to a certain degree, the risk of diseases increases, e.g., inflammation, cancer, type 2 diabetes mellitus (T2DM), and atherosclerosis [5,6]. To combat the oxidative stress, a defense system was established, which refers to various antioxidant enzymes, including superoxide dismutase (SOD), gluthatione peroxidase (GSH-Px) and catalase (CAT) [7]. On the other hand, oxidative stress induces lipid peroxidation [8], and malondialdehyde (MDA), which is a peroxidation product of polyunsaturated fatty acids, was detected as the index of lipid peroxidation [9]. There is lack of information on how these phenolics extracts from THR affect the in vivo antioxidant enzymes and MDA.

Dietary polyphenols have received increasing attention in recent years, due to their potential antioxidant properties [10]. Generally, the in vivo bioactivity of phenolics (including antioxidant activity) is significantly distinct from the in vitro data; therefore, the phenolic content in food does not reflect its absorption and metabolism in the human body [11]. For example, the first pass effect of xenobiotics, as well as low phenolic absorbability, leads to an extremely low phenolic concentration in systematic circulation [12]. Meanwhile, the structure modification during metabolism in the body has a considerable influence on the bioactivity as well. In contrast, the liberation of phenolics from matrices during digestion and colonic fermentation potentially promote the concentration of phenolics in plasma. In parallel, the structure alteration yielded by pH, enzymes, and gut microbials also contribute to the bioavailability and bioactivity of phenolics [13]. Mosele et al. [14] found that intestinal microbes were beneficial for polyphenol metabolism when they studied pomegranate products (juice, pulp, and peel extract) with in vitro gastrointestinal digestion and colonic fermentation models. However, the release rate and stability of THR phenolics during the digestion process as well as the specific routine of microbial metabolism during colonic fermentation are still unknown.

In this paper, in vitro simulated gastrointestinal digestion and colonic fermentation models were built to evaluate the stability and catabolism of *Tetrastigma hemsleyanum* phenolics combined with HPLC-QqQ-MS/MS and UPLC-Qtof-MS/MS methods, which have been used to identify metabolites from animals or humans. Furthermore, after oral administration of THR extract to rats, the regulating effects of these phenolics and/or their metabolites on the antioxidant enzyme activities, the total non-enzymatic antioxidant capacity, and lipid peroxidation in plasma and different tissues of rats were also investigated.

2. Materials and Methods

2.1. Materials and Reagents

Fresh *Tetrastigma hemsleyanum* roots were purchased from Jiangxi Shangrao Red Sun Agricultural Development Co., Ltd. (Nanchang, China). Pepsin, bile salts, and pancreatin were obtained from Aladdin Co. Ltd. (Shanghai, China). Human salivary a-amylase, DPPH, and DMSO were purchased from Sigma-Aldrich Co. (St. Louis, MO, USA). Methanol, acetonitrile, formic acid, and other solvents and chemical agents were obtained from Merck (Darmstadt, Germany). Water was purified in-house by a Milli-Q system (Bedford, MA, USA). The superoxide dismutase (SOD), glutathione peroxidase (GSH-Px), total antioxidant capacity (T-AOC), reduced glutathione (GSH), and malondialdehyde (MDA) kits were purchased from Nanjing Jiancheng Bioengineering Company (Nanjing, China).

2.2. Extract Preparation

THR samples were dried until constant weight by an electro-thermostatic blast oven (Senxin, Shanghai, China) and ground into fine powder (through a 200 mesh sieve). The powder (1.0 kg) was extracted with 20 L 80% methanol (v/v) for 45 min at 50 °C by ultrasonication (Shumei Instrument Factory, Kunshan, China). The suspension was centrifuged at 3000× g for 10 min (Thermo Scientific, Waltham, MA, USA), and was evaporated at 50 °C using a rotary vacuum evaporator (Eyela N-100, Miyagi, Japan). About 20% of its original volume remained, which was thoroughly freeze-dried (SIM International Group Co. Ltd., San Francisco, CA, USA). The dried THR extract was stored at −80 °C before use.

2.3. In Vitro Gastrointestinal Digestion

The in vitro digestion model was slightly adapted from Minekus et al. [15] and McClements et al. [16]. Three stages including in vitro oral, gastric, and intestinal digestion were carried out to revivify the physical and chemical effects of THR extract in the human body. All digestion processes were placed into a thermostatic water-bath shaker (37 °C, 400 rev/min), as well as away from light. Gastric and intestinal stages were air-free under nitrogen.

2.3.1. Oral Digestion

The simulated salivary fluid (SSF) consisted of human salivary a-amylase (75 U/mL), $CaCl_2 \cdot 2H_2O$ (0.75 mM) and HBSS solution (10×) to 1× in the final stage. The SSF mixture was pre-warmed for 2 min at 30 °C. The freeze-dried THR extract was weighed (1 g) and mixed with 10 mL SSF in 50 mL polyethylene tube to perform the digestion for 5 min.

2.3.2. Gastric Digestion

After oral digestion, 6 mL simulated gastric fluid (SGF) was added to the 10 mL extract-SSF mixture, and shaken for 30 min in the 37 °C water bath. To be specific, the SGF was phosphate buffer, which was adjusted to pH 2 by hydrochloride acid (1 M), and contained porcine pepsin (2000 U/mL) and $CaCl_2 \cdot 2H_2O$ (0.15 mM). Prior to seal-capping, most air was blown off by nitrogen.

2.3.3. Intestinal Digestion

After gastric digestion, the gastric samples-chime was neutralized to pH 5.8 by dicarbonate (2 M), then mixed with simulated intestinal fluid (SIF), which consisted of pancreatin, bile salts, and $CaCl_2 \cdot 2H_2O$ at a final concentration of 100 U/mL, 0.01 mM, and 0.3 mM, respectively. Then, the mixed digestive juice was adjusted to pH 8 with NaOH for digestion for 8 h. PMSF was added to terminate the reaction with a final concentration of 0.174 mg/mL.

The digestive juices of each stage were freeze-dried, including oral digestive products (THR-O), gastric digestion products (THR-G), and intestinal digestive products (THR-I). Then, all dried samples were re-dissolved with 80% methanol, which were purified by Agilent C_{18} solid phase extraction column and stored at −80 °C before analysis. All samples were taken 3 times in parallel.

2.4. In Vitro Colonic Fermentation

The in vitro fermentation colonic model was based on Maccaferri and Pereira-Caro with slight modification [17,18].

2.4.1. Preparation of Anaerobic Medium

A total of 7.5 g bouillon culture-medium was dissolved in 250 mL of distilled water and the mixture was filtered. The filtrate was sterilized at 120 °C for 20 min, and 1.5 mg hemin and 0.25 μL vitamin K_1 were added when the temperature of filtrate dropped to room temperature.

2.4.2. Preparation of Human Intestinal Bacterial Suspension

Three healthy young female volunteers (24–40 years old, who had not taken any antibiotics or fungi within 3 months before sampling and had no intestinal or metabolic diseases) were selected to collect their fresh feces. Then, the feces were mixed with anaerobic medium under anaerobic condition to make the final concentration of the fecal suspension 5%, which was filtered with gauze to obtain the intestinal bacteria solution. The operation of stools and filtration was restricted in 10 min.

2.4.3. Intestinal Flora Fermentation

Under anaerobic condition, the THR extract after in vitro gastrointestinal digestion was added to the intestinal bacteria solution at a final concentration of 0.01 g/mL, which was placed into temperature-controlled anaerobic incubator (10% CO_2, 10% H_2, and 80% N_2, 37 °C) for 48 h. Samples were taken at 0, 12, 24, and 48 h and freeze-dried. All dried samples were re-dissolved with 80% methanol, which were purified by Agilent C_{18} solid phase extraction column and stored at -80 °C before analysis. All samples were taken 3 times in parallel at all time points.

2.5. Qualitive and Quantitative Analysis of Phenolic Compounds

2.5.1. Qualitive Analysis by LC-QTOF-MS/MS

LC conditions: UPLC system (Agilent 1290 infinity series) was applied to Chromatographic separation, including a degasser, a binary pump (Bin Pump SL), an injector (HiP-ALS), a column oven (TCC SL), and a DAD detector (Agilent Technologies, Santa Clara, CA, USA). Agilent Zorbax EclipseXDB-C_{18} columns (4.6 mm × 250 mm, 5 μm) and Agilent Eclipse XDB-C18 guard columns (4.6 mm × 12.5 mm, 5 μm) were used. The mobile phase consisted of A (water containing 0.2% formic acid, v/v) and B (acetonitrile containing 0.2% formic acid, v/v). The gradient elution program was as follows: 0–10 min, 14–18% B; 10–30 min, 18–20% B; 30–35 min, 20–18% B; and 35–40 min, 18–14% B. The injection volume was 10 μL, the column temperature was 40 °C, and the flow rate was 0.5 mL/min. The DAD was set between 280 and 320 nm to obtain a real-time chromatogram with absorption peaks in the range of 190–400 nm.

MS conditions: Quadrupole time-of-flight precision mass spectrometer (Agilent 6538) equipped with an electrospray ionization source (ESI) was applied in negative mode. The full mass spectral data were obtained at m/z 50–1000 mass range. The best mass spectrometry parameters were as follows: capillary voltage 3.5 kV, injection cone voltage 35 V, desolvation gas flow (N_2) velocity 900 L/h, injection cone flow (N_2) velocity 50 L/h, desolvation temperature 325 °C, ion source temperature 150 °C; inlet pressure was set to 10 psi in CID mode with high purity argon gas as collision gas; the CID collision energy of polyphenol monomer was set to 5, 10, 15, 20, 25, and 30 eV, and the collision energy was set to 5, 10, 15, 20, 25, 30, 40, and 50 eV; and the cracking voltage was 120 V.

2.5.2. Quantitative Analysis by LC-QqQ-MS/MS

LC conditions: UPLC system (Agilent 1290 infinity series) was applied to Chromatographic separation, including a degasser, a binary pump (Bin Pump SL), a injector (HiP-ALS), a column oven (TCC SL), and a DAD detector (Agilent Technologies, Santa Clara, CA, USA). Agilent Zorbax EclipseXDB-C_{18} columns (4.6 mm × 250 mm, 5 μm) and Agilent Eclipse XDB-C18 guard columns (4.6 mm × 12.5 mm, 5 μm) were used. The mobile phase consisted of A (water containing 0.1% formic acid, v/v) and B (methanol containing formic acid, v/v). The gradient elution program was as follows: 0–5 min, 15–20% B; 5–15 min, 20–35% B; 15–25 min, 35–40% B; 25–38 min, 40–45% B; 38–45 min, 45–60% B; 45–55 min, 60–75% B; 55–65 min, 75–100% B; and 65–70 min, 100–15% B. The injection volume was 10 μL, the column temperature was 40 °C, and the flow rate was 1 mL/min. The DAD was set between 280 and 320 nm to obtain a real-time chromatogram with absorption peaks in the range of 190–400 nm.

MS conditions: Triple series quadrupole mass spectrometer (Agilent 6430 QqQ) equipped with an electrospray ionization source (ESI) was applied in negative + MRM mode. The full mass spectral data was obtained at m/z 50–1000 mass range. The best mass spectrometry parameters were as follows: capillary voltage 4 kV, injection cone voltage

35 V, desolvation gas flow (N_2) velocity 900 L/h, injection cone gas flow (N_2) velocity 50 L/h, desolvation temperature 300 °C, and ion source temperature 150 °C; inlet pressure was set to 15 psi with high purity argon gas as collision gas. Quantitative ion pair, collision energy, and cracking voltage were individually optimized for each monomer.

2.6. In Vivo Antioxidant Activities

2.6.1. Animals

Sprague–Dawley (SD) rats (male, Batch No. SCXK 2015-0003), weighing approximately 200–220 g, were purchased from Shanghai Laboratory Animal Center, Chinese Academy of Sciences (Shanghai, China). They were acclimatized in an environmentally controlled breeding room for 7 days prior to treatments. The experimental protocols were approved by the Animal Ethics Committee of Nanchang University (No. 20150829).

2.6.2. Experimental Design

Forty SD rats were randomly divided into five groups (n = 8) as follows:

Group I, normal control (saline); Group II, model control (D-galactose (D-gal) solution, 200 mg/kg BW); Group III, positive control (Vitamin C (VC) solution, 200 mg/kg BW); Group IV, low dosage (THR extract, 200 mg/kg BW); and Group V, high dosage (THR extract, 1000 mg/kg BW).

Rats in all groups were fasted for 12 h before the treatments but with access to deionized water [7]. The doses were also chosen based on the common intakes of THR in China and there was no toxic effect on the rats in our pre-experiments (data not shown). Rats in Group I were intraperitoneally injected with physiological saline, and rats in Groups II–V were intraperitoneally injected with D-gal solution (200 mg/kg BW per day). In addition, rats in Group III, IV, and V were intragastrically administered with VC, low-, and high-dose THR extract solutions at 200, 200, and 1000 mg/kg BW per day, respectively. This process lasted 20 days.

Blood samples were collected and centrifuged at $4000\times g$ for 10 min. The organs (liver, heart, and kidney) of the rats were immediately dissected out, cleaned, and weighed after plasma collected. Then, they were homogenized in 10 mM Tris-HCl buffer (pH 7.4) and centrifuged at $3000\times g$ for 15 min. The supernatants were analyzed for the antioxidant activities. All the above procedures were conducted at 4 °C.

2.6.3. Antioxidant Assays

For the SOD, GSH-Px, T-AOC, and GSH and MDA levels, the 96-well microplates (BD Falcon, Franklin Lakes, NJ, USA) with a multi-well plate reader (Thermo Scientific varioskan flash, Waltham, MA, USA) were used according to the method reported by Li et al. [4]. Results of the enzyme activities and T-AOC capacity were expressed as units per milliliter (U/mL) in plasma or per milligram of protein (U/mg prot) in tissues.

GSH can react with 5,5'-dithio-bis-(2-nitrobenzoic acid), which was measured at 412 nm and the results were expressed as milligram per milliliter (mg/mL) in plasma or milligram per gram of protein (mg/g prot) in tissues. MDA is a product of lipid peroxide degradation that condenses with thiobarbituric acid (TBA) to form a red product with strong absorption at 532 nm [19]. The results were expressed as nanomoles per milliliter (nmol/mL) in plasma or nanomoles per milligram of protein (nmol/mg prot) in tissues [7].

2.7. Statistical Analysis

Results were expressed as means ± SD. All data were analyzed by the SPSS statistical software, version 19.0 (SPSS Inc., Chicago, IL, USA), which was carried out using one-way analysis of variance (ANOVA) followed by Duncan's multiple range tests to estimate statistical significance at the level of $p < 0.05$. Furthermore, the LC-MS data were acquired and analyzed by MassHunter Acquisition B.03.01 and Qualitative Analysis B.03.01. The MassBank (http://www.massbank.jp, accessed on 25 April 2017), ChemSpider (http://www.chemspider.com, accessed on 6 May 2017), and Phenol-Explorer (www.phenol-explorer.eu, accessed on 2 May 2017) databases were used to analyze the MS^n data.

3. Results and Discussion

3.1. Change in Phenolic Profiles of THR Extract during In Vitro Digestion

The changes of total phenolic and flavonoid content of THR extract during the in vitro simulated digestion were explored. In vitro digestion model has been widely used to study changes in food composition [20]. As shown in Table 1, the total phenolic and flavonoid contents of the THR extract decreased obviously after oral, stomach, and intestinal digestions. After oral digestion, the total phenolic content decreased slightly from 225.38 to 214.13 mg GAE/g DW, and the total flavonoid decreased from 124.95 to 109.47 mg CAE/g DW. This result was different from the previous reports, which showed that short-term oral digestion and α-amylase had little effect on the polyphenols, and that oral digestion can be omitted [11,14]. However, Nada Bahloul et al. [21] reported that polyphenols, the aqueous extracts from Tunisian diplotaxis, were digested in large quantities in the oral digestion compared to stomach and intestinal digestions. The discrepant results may be due to the different structures of the phenolic substances extracted by various substances. Obviously, after gastric digestion, the total phenolic content reduced significantly from 214.13 to 104.61 mg GAE/g DW and total flavonoid decreased from 109.47 to 83.64 mg CAE/g DW, which indicated that many components were unstable in the stomach environment, such as the ring cleavage of anthocyanins [22]. However, Mosele et al. [11] studied the stability and metabolism of *Arbutus unedo* bioactive compounds and found that polyphenols were stable in the stomach, which might be protected by a certain amount of pectin formed in the gel. According to our previous study [3], rutin and isoquercitrin were the main components in THR extract, which had a high rate of metastasis when passed through the small intestine wall [23,24]. Interestingly, the polyphenol and flavonoid in the THR extract appeared to be more stable in the intestine, especially flavonoids. After intestinal digestion, total phenolic content decreased from 104.61 to 97.53 mg GAE/g DW and total flavonoid content decreased slightly from 83.64 to 81.98 mg CAE/g DW. It was suggested that the phenolic components of the THR extract were degraded to some extent during the digestion of the gastrointestinal tract.

Table 1. Total phenolic content and total flavonoid content of THR extract during in vitro gastrointestinal digestion and colonic fermentation by human microflora [A].

Sample	Total Phenolic Content (mg GAE/g DW) [B]	Total Flavonoid Content (mg CAE/g DW) [C]
Raw	225.38 ± 2.62 [h]	124.95 ± 3.31 [g]
In vitro gastrointestinal digestion		
Oral digestion	214.13 ± 2.34 [g]	109.47 ± 1.78 [f]
Gastric digestion	104.61 ± 1.51 [f]	83.64 ± 2.14 [e]
Intestinal digestion	97.53 ± 2.47 [e]	81.98 ± 1.37 [e]
In vitro colonic fermentation by human microflora		
0 h	86.93 ± 2.09 [d]	72.44 ± 1.56 [d]
12 h	78.27 ± 1.88 [c]	61.25 ± 1.08 [c]
24 h	65.42 ± 3.53 [b]	49.84 ± 1.67 [b]
48 h	33.85 ± 2.05 [a]	17.56 ± 1.29 [a]

[A] Results are expressed as mean ± standard deviation of three replicates. Values followed by different letters (a–h) within the same column are significantly different ($p < 0.05$). [B] Phenolic contents are expressed as mg gallic acid equivalents (GAE) in 1 g of dry weight of extracts ± standard deviation. [C] Flavonoid contents are expressed as mg catechin equivalents (CAE) in 1 g of dry weight of extracts ± standard deviation.

Gastrointestinal digested samples were analyzed by LC-QTOF-MS/MS and LC-QqQ-MS/MS. There are four distinct peaks in Figure 1. According to our previous report [3], quercetin-3-rutinoside (rutin, peak 1), quercetin-3-glucoside (isoquercitrin, peak 2), kaempferol-3-rutinoside (peak 3), and kaempferol-3-glucoside (peak 4) were identified. As shown in Table 2, the changes in the contents of four substances mainly occurred in the gastric digestion compared to the oral and intestinal digestions. Kaempferol-3-rutinoside was the most abundant of the four substances, as shown in Table 2, reaching 53.48 mg kaempferol/g DW. From the mouth to the intestine, the content of rutin reduced by about half, but it was more stable in the gastrointestinal tract [25]. This might have been caused by two factors. Firstly, rutin may interact with other substances (such as proteins or polysaccharides) to form a complex, which could not be detected in the present analytical method. Secondly, there may be significant differences in the in vitro digestion parameters and representation methods, such as filtration, centrifugation, membrane dialysis, etc. [25]. On the other hand, isoquercitrin can be degraded into quercetin due to digestive enzymes in the stomach [26] or degraded into aglycon and quercetin in the gut, which were absorbed or entered the colon degraded by microorganisms [27]. In addition, it has been reported that isoquercitrin can be absorbed directly into the body [23,24]. Kaempferol glycosides are also unstable under oral and gastrointestinal conditions. It has been reported that kaempferol glycosides can be hydrolyzed to aglycone kaempferol in oral and intestine digestion by glycosidases, which will bind to starch in the digestive tract [28]. During gastrointestinal digestion, part of kaempferol glycosides were degraded or absorbed, and another part entered the lower digestive tract to continue to metabolize. It has been reported that about 48% of dietary polyphenols were bioaccessible in the small intestine and 42% in the large intestine, while the rest were not accessible [29]. Because of the limitation of chromatographic conditions, the corresponding metabolites were not found.

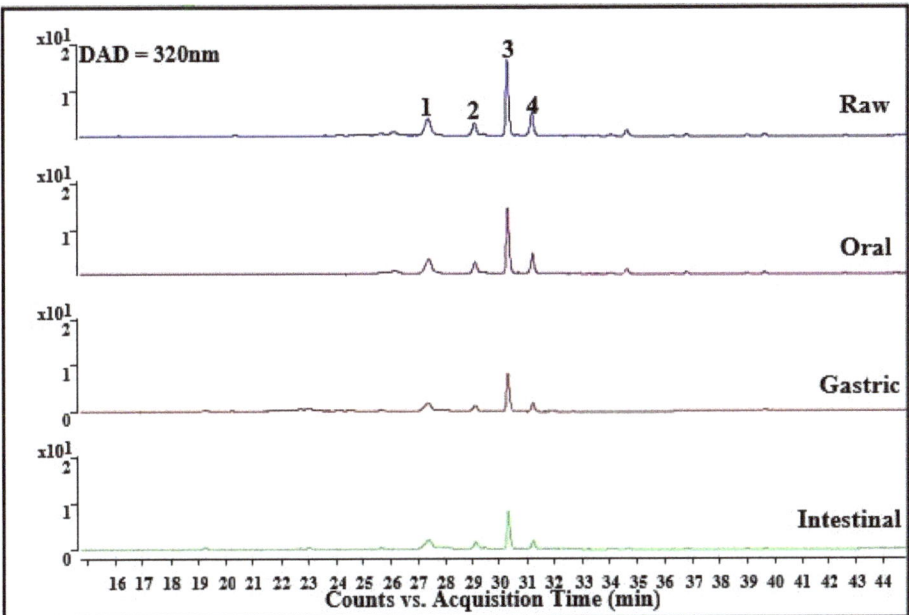

Figure 1. The change of the major phenolics in THR extract during the in vitro gastrointestinal digestion.

Table 2. Identification and quantification of the major phenolics in THR extract during the in vitro gastrointestinal digestion and colonic fermentation by human microflora [A].

No.	Compounds [B]	t_R (min)	Parent/Daughter Ions (m/z)	Raw	Gastrointestinal Digestion			Colonic Fermentation by Human Microflora			
					Oral	Gastric	Intestinal	0 h	12 h	24 h	48 h
1	Quercetin-3-rutinoside	27.41	609/301	16.86 ± 1.19 [h]	14.09 ± 1.25 [g]	8.77 ± 0.86 [f]	8.04 ± 0.45 [d]	8.12 ± 0.79 [e]	6.70 ± 0.26 [c]	3.37 ± 0.03 [b]	3.19 ± 0.40 [a]
2	Quercetin-3-glucoside	29.18	463/301	12.55 ± 1.03 [h]	11.24 ± 1.31 [g]	7.18 ± 1.12 [f]	6.63 ± 0.58 [e]	6.57 ± 0.58 [d]	4.92 ± 0.68 [c]	2.48 ± 0.11 [b]	2.42 ± 0.21 [a]
3	Kaempferol-3-rutinoside	30.39	593/285	53.48 ± 3.47 [h]	49.13 ± 2.76 [g]	27.94 ± 1.34 [f]	27.88 ± 1.07 [e]	24.5 ± 2.01 [d]	21.98 ± 1.91 [c]	21.46 ± 1.31 [b]	7.67 ± 0.55 [a]
4	Kaempferol-3-glucoside	31.26	447/285	19.97 ± 1.57 [h]	16.40 ± 1.44 [g]	9.54 ± 0.95 [e]	9.73 ± 1.04 [f]	9.38 ± 0.82 [d]	7.60 ± 0.64 [c]	4.54 ± 0.08 [b]	2.88 ± 0.09 [a]

[A] Results are expressed as mean ± standard deviation of three replicates. Values followed by different letters (a–h) within the same line are significantly different ($p < 0.05$). [B] Peaks 1 and 2 were expressed as quercetin equivalents per g dry weight extract (mg quercetin/g DW). Peaks 3 and 4 were expressed as kaempferol equivalents per g dry weight extract (mg kaempferol/g DW).

The polyphenol content in food does not represent the amount of metabolic absorption in the body. During the digestion process, various components are continuously exposed to physical (temperature), chemical (pH), and biological (enzyme) conditions [11], which will affect the bioavailability and biological activity of potential food bioactive compounds [13]. On the other hand, food matrices should be considered, which affect the stability of polyphenols during gastric and intestinal digestion and the proportion of phenolic compounds that reach the colon or are absorbed [14]. The food we consume every day contains a lot of polyphenols, but, in fact, only a small part can be absorbed. Demethylation and deglycosylation are the main ways polyphenols are metabolized, which can improve their effectiveness of [30]. Some in vitro studies showed that the phenolic compounds can be transported through the intestinal epithelial cells in the form of glycosides via sugar transporters [22]. After absorption, they were hydrolyzed to aglycons by β-glucosidase. Moreover, aglycones can also be formed in the lumen through the action of membrane-bound lactase phlorizin hydrolase (LPH), which was absorbed passively through the epithelium, as compared with conjugation in the ileal epithelium or the liver. Hepatic metabolites (methylated, sulfated, or glucuronidated conjugates) were returned via the enterohepatic circulation (in bile) to the gut lumen [22,31].

3.2. Change in Phenolic Profiles of THR Extract during In Vitro Colonic Fermentation

The gastrointestinal digested samples were further used to carry out colonic fermentation because intestinal microbes are an important part of our entire digestive ecosystem. It is important to study the effects of microorganisms on the digestion and absorption of phenols. As shown in Table 1, one-third of the total phenolics and half of the flavonoids, which were not digested in the gastrointestinal tract, entered the colon. However, this result was not the same as Mosele et al. [14], who reported that most of the phenolics entered into the colon. When polyphenols reached the colon, they were absorbed intactly through the epithelium or metabolized by the colonic microbiotas [22]. Analyzed with LC-QTOF-MS/MS, four substances were found as shown in Figure 2, including quercetin-3-rutinoside (rutin, peak 1), quercetin-3-glucoside (isoquercitrin, peak 2), kaempferol-3-rutinoside (peak 3), and kaempferol-3-glucoside (peak 4). Some small peaks were also shown in Figure 2, which could not be detected with the existing chromatographic conditions due to the low content. After 48 h of fermentation, the total phenolic content reduced from 86.93 to 33.85 mg GAE/g DW and the total flavonoid content reduced from 72.44 to 17.56 mg CAE/g DW, indicating that colonic microorganisms can decompose these polyphenols. As Mosele et al. [14] reported, tannic acid and ellagic acid were metabolized into urolithins by microorganisms, which were considered to be a potentially biologically active microbial metabolite.

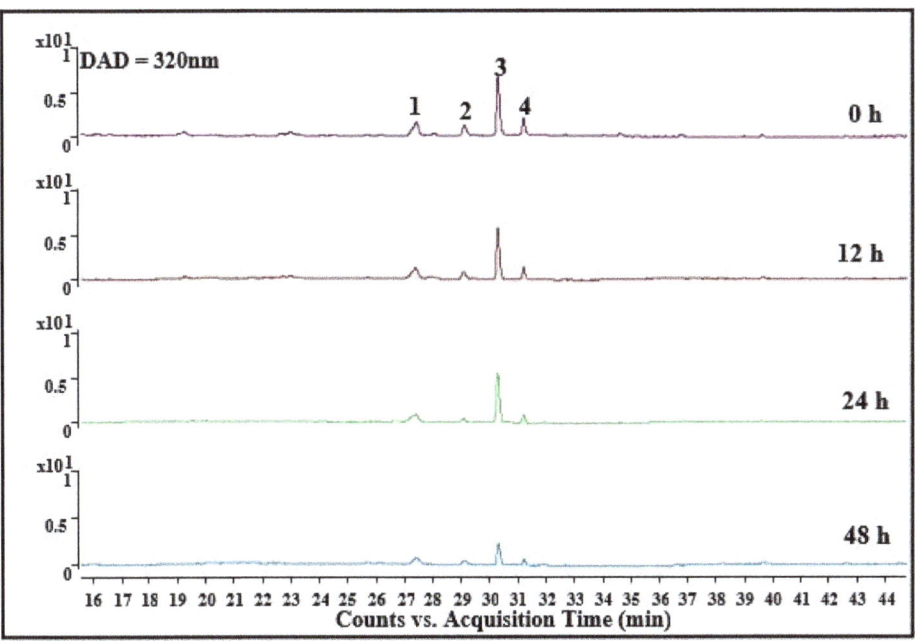

Figure 2. The change of the major phenolics in THR extract during colonic fermentation by human microflora.

As shown in Table 2, it was obvious that the contents of the four substances dropped to a lower level after 48 h of microbial metabolism. After 24 h of fermentation, the contents of quercetin-3-rutinoside and quercetin-3-glucoside remained basically unchanged. However, kaempferol-3-rutinoside and kaempferol-3-glucoside were degraded throughout the fermentation process. It has been proven that rutin was degraded into quercetin and further converted into 3-hydroxylphenylacetic acid under the action of microorganisms in the colon [20]. In addition, rutin can be converted to isoquercitrin by microorganisms under anaerobic condition [32], and isoquercitrin was decomposed into quercetin and glucose by microorganisms through O-deglycosylation [33]. Kaempferol glycosides were reported to be degraded to kaempferol and corresponding sugar ligands through deglycosylation. Kaempferol was further degraded to form 3-(4-hydroxyphenyl) propionic acid through ring fission under the action of microorganisms, which was further degraded to form 3-phenylpropionic acid and phenylacetic acid. 3-Phenylpropionic acid can also be converted to phenylacetic acid [34]. Many substances, which are digested difficultly in the gastrointestinal tract, can be degraded under the action of colonic microorganisms. Chlorogenic acid is considered to be a substance that is difficult to be absorbed in the gastrointestinal tract. It has even been reported to be resistant to intestinal enzymes [35] but can be absorbed after being broken down into small molecules by microorganisms [36]. Chlorogenic acid was one of the main components of THR extract in our previous identification [3]. It can be degraded into caffeic acid and quinic acid by microorganisms, and caffeic acid can be converted into ferulic acid by deacetylation [20]. 3-(3-Hydroxyphenyl) propionic acid, a decomposition product of caffeic acid, was metabolized through the transport of Caco-2 cells in liver [37]. In the metabolic absorption of the colon, phenolic compounds entering the colon were mainly composed of unabsorbed glycosides and conjugates that pass through the ileal and hepatic metabolic cycles [22,31]. Flavonoids that entered the colon were metabolized into simple phenolic acid under the action of microorganisms with the form of fission, ring-opening, and other degradation [22], which entered the circulation via transporters or by passive diffusion [37].

3.3. In Vivo Antioxidant Activities of THR

The MDA and GSH contents and the antioxidant capacity of SOD, GSH-Px, and T-AOC enzymes in plasma, liver, heart, and kidney samples of rats under different treatment regimens were tested to evaluate the antioxidant activity of THR extract. The effects of THR extract were also compared with the positive control of VC. Results are shown in Figure 3.

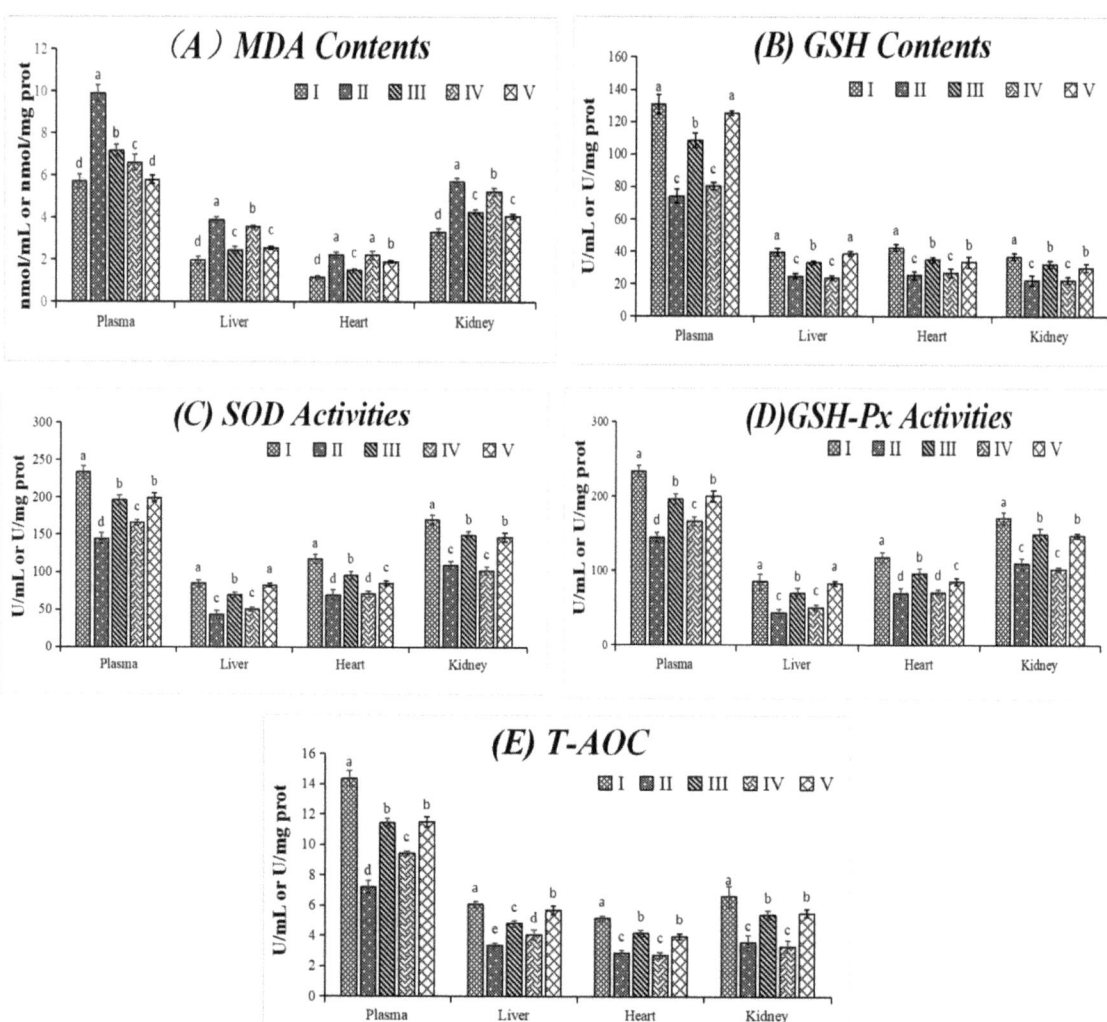

Figure 3. The changes of (**A**) MDA contents; (**B**) GSH contents; (**C**) SOD activities; (**D**) GSH-Px activities; and (**E**) T-AOC in different groups. Data were presented as mean ± SD (n = 8); values not sharing a common superscript letter denote significant difference (p < 0.05). Rats were divided into 5 groups as follows: I: normal control group (saline); II: model control group (D-gal solution, 200 mg/kg BW); III: positive control group (VC solution, 200 mg/kg BW), IV: low dosage group (THR extract, 200 mg/kg BW), and V: high dosage group (THR extract, 1000 mg/kg BW). Please note that the data of the I, II, and III groups were the same as we used in our previous paper (*Journal of Functional Foods* 2017, 30, 179–193), experiments for which were carried out simultaneously with the current paper.

3.3.1. Effect on MDA and GSH Contents

MDA and GSH contents can reflect the level of oxidative stress in the body. As shown in Figure 3A,B, compared with the normal control group (group I), the MDA content increased significantly and GSH content reduced in plasma and various tissues in the model control group (group II), which showed that the D-gal model was established successfully. As shown in Figure 3A, MDA contents were reduced significantly, or even lower than the positive control group in plasma, in both low- and high-dosage groups. It indicated that THR extract can effectively reduce the lipid peroxidation. As shown in Figure 3B, compared with the model control group (group I), the GSH contents remained basically unchanged in the low dosage group (group IV), but increased significantly in the high-dosage group (group V). It was indicated that the effect of the THR extract was dose-dependent. A high-dose (1000 mg/kg BW) THR extract had a significant effect on the reduction of MDA and the increase of GSH, which was similar to or stronger than the positive control (group III, 200 mg VC/kg BW). Interestingly, the contents of MDA and GSH were the highest in plasma compared to other tissues. THR extract showed significant beneficial effects, especially at higher doses, that ultimately contributed to reducing oxidative stress and protected rats from oxidative damage in different tissues and organs.

3.3.2. Effects on SOD, GSH-Px, and T-AOC Activities

SOD and GSH-Px are important antioxidant enzymes in the body. As shown in Figure 3C–E, the SOD, GSH-Px, and T-AOC activities were reduced significantly by D-gal solution (group II) compared with the blank control group (group I), which showed that the D-gal model was established successfully. From group IV and group V, it was clear that there was a dose-dependent effect on SOD, GSH-Px and T-AOC activities. Low-dose THR extract (group IV) showed significant effects on SOD, GSH-Px, and T-AOC activities in plasma, but no significant effects were found in other tissues except the T-AOC in the liver. However, the SOD, GSH-Px, and T-AOC activities were improved obviously in plasma and other tissues treated with high-dose THR extract. The activity of SOD and GSH-Px in the plasma and kidney were higher than those in the liver and heart (Figure 3C,D). However, the T-AOC activity was higher in the plasma than in other tissues (Figure 3). High-dose THR extract (1000 mg/kg BW) showed significantly stronger T-AOC activity than VC (Figure 3E). In general, the use of THR extract can increase the antioxidant capacity of rats and reduce the damage caused by D-gal.

3.3.3. The Relationship between Metabolites and In Vivo Antioxidant Activities

It was shown that gastrointestinal digestion and colonic fermentation had significant effects on the antioxidant activity of polyphenol. The structure of polyphenols was changed in these degradation processes, leading to the change of antioxidant activity [20]. Luzardo-Ocampo et al. [24] found that the antioxidant activity of digested corn (*Zea mays* L.) and common beans (*Phaseolus vulgaris* L.) were higher than their methanol extracts, but Bao et al. [38] found that the antioxidant activity of flavonoids from tartary buckwheat rice decreased after in vitro digestion. Food matrix showed a significant impact on digestion; generally, free polyphenols were extracted but there are many bound polyphenols in food that may be released during digestion. Of course, the antioxidant activity changed due to the polyphenol structural modification when the extracts were used for in vitro digestion. It has been reported that the antioxidant activity of the parent phenols was reduced under microbial action, but the overall antioxidant capacity did not diminish due to the accumulation of metabolites that might produce the same antioxidant capacity as the parents [20]. However, the antioxidant activity of metabolites was higher than that of their parents in some cases, such as dihydroferulic acid, a metabolite of chlorogenic acid [20]. Although in vitro antioxidant experiments cannot reflect in vivo antioxidants, they can be used as a reference.

It has been reported that bioactive components in diets act in two ways: by directly scavenging free radicals and/or by activating the transcription of cytoprotective enzymes involved in the detoxification of xenobiotics [22]. On one hand, polyphenols and their metabolites play an antioxidant role directly, which can be absorbed directly or after being metabolized in gastrointestinal digestion and colon fermentation. On the other hand, polyphenols and some simple phenolic acids as their metabolites can activate the Keap1/Nrf2/ARE (Kel-Ch ECH associating protein 1/nf-e2-related factor 2/Antioxidant The Response Elements) pathway, and these active substances can modify the key cysteine residues on Keap1, which enables binding to NRF-2 and migration of the activated complex into the nucleus, where it activates genes with antioxidant elements, coding antioxidant enzymes such as superoxide dismutase, catalase, glutathione synthetase, etc. [22]. Rats treated with THR extract showed positive effects on MDA, GSH, SOD, GSH-Px, and T-AOC (Figure 3). The dose-dependence indicated that polyphenols in THR have good bioavailability, and the dietary intake of foods with higher polyphenols has potential benefits for oxidative stress. Therefore, THR as a traditional functional ingredient with good biological value has great potential.

4. Conclusions

The stability and catabolism of THR were evaluated under in vitro simulated gastrointestinal digestion and colonic fermentation models. Total phenolic and flavonoid content were degraded during each digestion process. THR extract showed higher gastrointestinal digestibility and less reaching the colon; in other words, it may be absorbed in the gastrointestinal tract and have high bioavailability. Quercetin-3-rutinoside, quercetin-3-glucoside, kaempferol-3-rutinoside, and kaempferol-3-glucoside were the main substances of THR, which were degraded at various stages, while their in vivo metabolites [3] could help explain the favorable changes in several antioxidant biomarkers (SOD, GSH-Px, T-AOC) and the lipid peroxidation product (MDA) in rats treated with THR extract. This also proved that THR is reasonable as a traditional functional food and is effective to treat chronic diseases caused by oxidative stress.

Author Contributions: Conceptualization, H.L. and Z.D.; methodology, Y.S.; investigation, Y.S., F.G., X.P. and K.C.; data curation, Y.S., L.X., L.J. and H.Z.; writing—original draft preparation, Y.S. and F.G.; writing—review and editing, H.L., L.J. and Z.D.; supervision, H.L. and Z.D.; funding acquisition, Y.S. All authors have read and agreed to the published version of the manuscript.

Funding: This research was funded by the National Natural Science Foundation of China (No. 82060781) and China Postdoctoral Science Foundation (No. 2020M671975).

Institutional Review Board Statement: The study was conducted according to the guidelines of the Declaration of Helsinki, and approved by the Animal Ethics Committee of Nanchang University (No. 20150829).

Informed Consent Statement: Informed consent was obtained from all subjects involved in the study.

Conflicts of Interest: The authors declare no conflict of interest.

References

1. Huang, Z.; Mao, Q.Q.; Wei, J.P. Evaluation of anti-inflammatory, analgesic and antipyretic actions for the extracts from *Radix Tetrastigmae*. *Chin. New Drugs J.* **2005**, *14*, 861.
2. Sun, Y.; Li, H.; Hu, J.; Li, J.; Fan, Y.W.; Liu, X.R.; Deng, Z.Y. Qualitative and quantitative analysis of phenolics in *Tetrastigma hemsleyanum* and their antioxidant and antiproliferative activities. *J. Agric. Food Chem.* **2013**, *61*, 10507–10515. [CrossRef]
3. Sun, Y.; Qin, Y.; Li, H.; Peng, H.; Chen, H.; Xie, H.R.; Deng, Z.Y. Rapid characterization of chemical constituents in Radix Tetrastigma, a functional herbal mixture, before and after metabolism and their antioxidant/antiproliferative activities. *J. Funct. Foods* **2015**, *18*, 300–318. [CrossRef]
4. Li, X.; Zhang, Y.; Yuan, Y.; Sun, Y.; Qin, Y.; Deng, Z.; Li, H. Protective effects of selenium, vitamin E, and purple carrot anthocyanins on D-galactose-induced oxidative damage in blood, liver, heart and kidney rats. *Biol. Trace Elem. Res.* **2016**, *173*, 433–442. [CrossRef]

15. Ito, F.; Sono, Y.; Ito, T. Measurement and clinical significance of lipid peroxidation as a biomarker of oxidative stress: Oxidative stress in diabetes, atherosclerosis, and chronic inflammation. *Antioxidants* **2019**, *8*, 72. [CrossRef] [PubMed]
16. Apak, R.; Ozyurek, M.; Guclu, K.; Capanoglu, E. Antioxidant activity/capacity measurement. 3. Reactive oxygen and nitrogen species (ROS/RNS) scavenging assays, oxidative stress biomarkers, and chromatographic/chemometric assays. *J. Agric. Food Chem.* **2016**, *64*, 1046–1070. [CrossRef]
17. Sun, Y.; Tsao, R.; Chen, F.; Li, H.; Wang, J.; Peng, H.; Zhang, K.; Deng, Z. The phytochemical composition, metabolites, bioavailability and in vivo antioxidant activity of *Tetrastigma hemsleyanum* leaves in rats. *J. Funct. Foods* **2017**, *30*, 179–193. [CrossRef]
18. Samarghandian, S.; Borji, A.; Delkhosh, M.B.; Samini, F. Safranal treatment improves hyperglycemia, hyperlipidemia and oxidative stress in streptozotocin-induced diabetic rats. *J. Pharm. Pharm. Sci.* **2013**, *16*, 352–362. [CrossRef] [PubMed]
19. Gupta, D. Methods for determination of antioxidant capacity: A review. *Int. J. Pharm. Sci. Res.* **2015**, *6*, 546–566.
20. Gowd, V.; Bao, T.; Wang, L.; Huang, Y.; Chen, S.; Zheng, X.; Cui, S.L.; Chen, W. Antioxidant and antidiabetic activity of blackberry after gastrointestinal digestion and human gut microbiota fermentation. *Food Chem.* **2018**, *269*, 618–627. [CrossRef]
21. Mosele, J.I.; Macià, A.; Romero, M.P.; Motilva, M.J. Stability and metabolism of Arbutus unedo bioactive compounds (phenolics and antioxidants) under in vitro digestion and colonic fermentation. *Food Chem.* **2016**, *201*, 120–130. [CrossRef]
22. Wu, B.; Kulkarni, K.; Basu, S.; Zhang, S.; Hu, M. First-Pass Metabolism via UDP-Glucuronosyltransferase: A Barrier to Oral Bioavailability of Phenolics. *J. Pharm. Sci.* **2011**, *100*, 3655–3681. [CrossRef] [PubMed]
23. Hur, S.J.; Lim, B.O.; Decker, E.A.; McClements, D.J. In vitro human digestion models for food applications. *Food Chem.* **2011**, *125*, 1–12. [CrossRef]
24. Mosele, J.I.; Macià, A.; Romero, M.P.; Motilva, M.J.; Rubiò, L. Application of in vitro gastrointestinal digestion and colonic fermentation models to pomegranate products (juice, pulp and peel extract) to study the stability and catabolism of phenolic compounds. *J. Funct. Foods* **2015**, *14*, 529–540. [CrossRef]
25. Minekus, M.; Alminger, M.; Alvito, P.; Ballance, S.; Bohn, T.; Bourlieu, C.; Carriere, F.; Boutrou, R.; Corredig, M.; Dupont, D. A standardised static in vitro digestion method suitable for food—An international consensus. *Food Funct.* **2014**, *5*, 1113–1124. [CrossRef]
26. McClements, D.J.; Li, Y. Review of in vitro digestion models for rapid screening of emulsion-based systems. *Food Funct.* **2010**, *1*, 32–59. [CrossRef]
27. Maccaferri, S.; Klinder, A.; Cacciatore, S.; Chitarrari, R.; Honda, H.; Luchinat, C.; Bertini, I.; Carnevali, P.; Gibson, G.R.; Brigidi, P. In vitro fermentation of potential prebiotic flours from natural sources: Impact on the human colonic microbiota and metabolome. *Mol. Nutr. Food Res.* **2012**, *56*, 1342–1352. [CrossRef]
28. Pereira-Caro, G.; Moreno Rojas, J.M.; Brindani, N.; Del Rio, D.; Lean, M.; Hara, Y.; Crozier, A. Bioavailability of Black Tea Theaflavins: Absorption, Metabolism and Colonic Catabolism. *J. Agric. Food Chem.* **2017**, *65*, 5365–5374. [CrossRef] [PubMed]
29. Ni, S.; Qian, D.; Duan, J.A.; Guo, J.; Shang, E.X.; Shu, Y.; Xue, C. UPLC–QTOF/MS-based screening and identification of the constituents and their metabolites in rat plasma and urine after oral administration of Glechoma longituba extract. *J. Chromatogr. B* **2010**, *878*, 2741–2750. [CrossRef]
30. Ekbatan, S.S.; Sleno, L.; Sabally, K.; Khairallah, J.; Azadi, B.; Rodes, L.; Prakash, S.J.; Donnelly, D.; Kubow, S. Biotransformation of polyphenols in a dynamic multistage gastrointestinal model. *Food Chem.* **2016**, *204*, 453–462. [CrossRef] [PubMed]
31. Bahloul, N.; Bellili, S.; Aazza, S.; Chérif, A.; Faleiro, M.L.; Antunes, M.D.; Miguel, M.G.; Mnif, W. Aqueous extracts from tunisian diplotaxis: Phenol content, antioxidant and anti-acetylcholinesterase activities, and impact of exposure to simulated gastrointestinal fluids. *Antioxidants* **2016**, *5*, 12. [CrossRef] [PubMed]
32. Correa-Betanzo, J.; Allen-Vercoe, E.; McDonald, J.; Schroeter, K.; Corredig, M.; Paliyath, G. Stability and biological activity of wild blueberry (*Vaccinium angustifolium*) polyphenols during simulated in vitro gastrointestinal digestion. *Food Chem.* **2014**, *165*, 522–531. [CrossRef]
33. Crespy, V.; Morand, C.; Manach, C.; Besson, C.; Demigne, C.; Remesy, C. Part of quercetin absorbed in the small intestine is conjugated and further secreted in the intestinal lumen. *Am. J. Physiol.-Gastrointest. Liver Physiol.* **1999**, *277*, 120–126. [CrossRef] [PubMed]
34. Luzardo-Ocampo, I.; Campos-Vega, R.; Gaytán-Martínez, M.; Preciado-Ortiz, R.; Mendoza, S.; Loarca-Piña, G. Bioaccessibility and antioxidant activity of free phenolic compounds and oligosaccharides from corn (*Zea mays* L.) and common bean (*Phaseolus vulgaris* L.) chips during in vitro gastrointestinal digestion and simulated colonic fermentation. *Food Res. Int.* **2017**, *100*, 304–311. [CrossRef]
35. Gayoso, L.; Claerbout, A.S.; Calvo, M.I.; Cavero, R.Y.; Astiasarán, I.; Ansorena, D. Bioaccessibility of rutin, caffeic acid and rosmarinic acid: Influence of the in vitro gastrointestinal digestion models. *J. Funct. Foods* **2016**, *26*, 428–438. [CrossRef]
36. Attri, S.; Sharma, K.; Raigond, P.; Goel, G. Colonic fermentation of polyphenolics from Sea buckthorn (*Hippophae rhamnoides*) berries: Assessment of effects on microbial diversity by Principal Component Analysis. *Food Res. Int.* **2018**, *105*, 324–332. [CrossRef]
37. Chang, Q.; Zuo, Z.; Chow, M.S.; Ho, W.K. Difference in absorption of the two structurally similar flavonoid glycosides, hyperoside and isoquercitrin, in rats. *Eur. J. Pharm. Biopharm.* **2005**, *59*, 549–555. [CrossRef] [PubMed]

28. Takahama, U.; Hirota, S. Effects of starch on nitrous acid-induced oxidation of kaempferol and inhibition of α-amylase-catalysed digestion of starch by kaempferol under conditions simulating the stomach and the intestine. *Food Chem.* **2013**, *141*, 313–319. [CrossRef]
29. Gião, M.S.; Gomes, S.; Madureira, A.R.; Faria, A.; Pestana, D.; Calhau, C.; Pintado, M.E.; Azevedo, I.; Malcata, F.X. Effect of in vitro digestion upon the antioxidant capacity of aqueous extracts of *Agrimonia eupatoria*, *Rubus idaeus*, *Salvia* sp. and *Satureja montana*. *Food Chem.* **2012**, *131*, 761–767. [CrossRef]
30. McGhie, T.K.; Walton, M.C. The bioavailability and absorption of anthocyanins: Towards a better understanding. *Mol. Nutr. Food Res.* **2007**, *51*, 702–713. [CrossRef]
31. Scalbert, A.; Williamson, G. Dietary intake and bioavailability of polyphenols. *J. Nutr.* **2000**, *130*, 2073S–2085S. [CrossRef] [PubMed]
32. Shin, N.R.; Moon, J.S.; Shin, S.Y.; Li, L.; Lee, Y.B.; Kim, T.J.; Han, N.S. Isolation and characterization of human intestinal *Enterococcus avium* EFEL009 converting rutin to quercetin. *Lett. Appl. Microbiol.* **2016**, *62*, 68–74. [CrossRef] [PubMed]
33. Yuan, M.; Shi, D.Z.; Wang, T.Y.; Zheng, S.Q.; Liu, L.J.; Sun, Z.X.; Wang, T.Y.; Ding, Y. Transformation of trollioside and isoquercetin by human intestinal flora in vitro. *Chin. J. Nat. Med.* **2016**, *14*, 220–226. [CrossRef]
34. Vollmer, M.; Esders, S.; Farquharson, F.M.; Neugart, S.; Duncan, S.H.; Schreiner, M.; Louis, P.; Maul, R.; Rohn, S. Mutual interaction of phenolic compounds and microbiota: Metabolism of complex phenolic apigenin-c-and kaempferol-o-derivatives by human fecal samples. *J. Agric. Food Chem.* **2018**, *66*, 485–497. [CrossRef] [PubMed]
35. Lafay, S.; Morand, C.; Manach, C.; Besson, C.; Scalbert, A. Absorption and metabolism of caffeic acid and chlorogenic acid in the small intestine of rats. *Br. J. Nutr.* **2006**, *96*, 39–46. [CrossRef] [PubMed]
36. Ludwig, I.A.; de Peña, M.; Cid, C.; Crozier, A. Catabolism of coffee chlorogenic acids by human colonic microbiota. *Biofactors* **2013**, *39*, 623–632. [CrossRef]
37. Sadeghi Ekbatan, S.; Iskandar, M.M.; Sleno, L.; Sabally, K.; Khairallah, J.; Prakash, S.; Kubow, S. Absorption and metabolism of phenolics from digests of polyphenol-rich potato extracts using the Caco-2/HepG2 co-culture system. *Foods* **2018**, *7*, 8. [CrossRef] [PubMed]
38. Bao, T.; Wang, Y.; Li, Y.T.; Gowd, V.; Niu, X.H.; Yang, H.Y.; Chen, L.S.; Chen, W.; Sun, C.D. Antioxidant and antidiabetic properties of tartary buckwheat rice flavonoids after in vitro digestion. *J. Zhejiang Univ. Sci. B* **2016**, *17*, 941–951. [CrossRef]

Article

4-Hydroxyderricin Promotes Apoptosis and Cell Cycle Arrest through Regulating PI3K/AKT/mTOR Pathway in Hepatocellular Cells

Xiang Gao [1,†], Yuhuan Jiang [1,†], Qi Xu [1], Feng Liu [2,3], Xuening Pang [1], Mingji Wang [3], Qun Li [3,4] and Zichao Li [1,3,*]

1. Institute of Biomedical Engineering, College of Life Sciences, Qingdao University, Qingdao 266071, China; gaoxiang@qdu.edu.cn (X.G.); 2018025210@qdu.edu.cn (Y.J.); xq@qdu.edu.cn (Q.X.); 2018025216@qdu.edu.cn (X.P.)
2. Department of Horticultural Technology, Ningbo City College of Vocational Technology, Ningbo 315199, China; coy24@163.com
3. Joint Institute of *Angelica keiskei* Health Industry Technology, Qingdao University, Qingdao 266071, China; 13589228558@139.com (M.W.); qunli@qdu.edu.cn (Q.L.)
4. College of Chemistry and Chemical Engineering, Qingdao University, Qingdao 266071, China
* Correspondence: zichaoli@qdu.edu.cn
† These authors contributed equally to this work as co-first authors.

Abstract: 4-hydroxyderricin (4-HD), as a natural flavonoid compound derived from *Angelica keiskei*, has largely unknown inhibition and mechanisms on liver cancer. Herein, we investigated the inhibitory effects of 4-HD on hepatocellular carcinoma (HCC) cells and clarified the potential mechanisms by exploring apoptosis and cell cycle arrest mediated via the PI3K/AKT/mTOR signaling pathway. Our results show that 4-HD treatment dramatically decreased the survival rate and activities of HepG2 and Huh7 cells. The protein expressions of apoptosis-related genes significantly increased, while those related to the cell cycle were decreased by 4-HD. 4-HD also down-regulated PI3K, p-PI3K, p-AKT, and p-mTOR protein expression. Moreover, PI3K inhibitor (LY294002) enhanced the promoting effect of 4-HD on apoptosis and cell cycle arrest in HCC cells. Consequently, we demonstrate that 4-HD can suppress the proliferation of HCC cells by promoting the PI3K/AKT/mTOR signaling pathway mediated apoptosis and cell cycle arrest.

Keywords: *Angelica keiskei*; chalcone; anti-tumor; mechanism; apoptosis; cell cycle

1. Introduction

Hepatocellular carcinoma (HCC) is the second most common cause of cancer-related death and is currently the sixth most diagnosed cancer worldwide. The global incidence varies with geographic location, but is generally higher in East Asia and Africa [1,2]. To date, chemotherapy is still the most frequently used and effective way to cure HCC. Currently, numerous first-line treatment drugs, such as sorafenib, donafenib, anlotinib, and lenvatinib have been utilized for the clinical treatment of HCC [3]. Although side effects of targeted chemotherapeutic agents are rare, there have been several cases of acquired skin perforation [4]. More seriously, HCC patients usually develop tolerance to chemotherapy drugs due to the intrinsic resistance or the acquired resistance [5]. Hence, developing notable anti-HCC ingredients derived from natural products is of great significance, and among which, plant flavonoids have been continuously attractive to researchers globally so far [6]. Studies have also shown that flavonoids exhibit good tolerance and efficacy against a variety of tumors, including liver cancer [7].

Nowadays, the regulation of apoptosis and cell cycle has attracted much attention for the treatment of cancers, such as HCC. Apoptosis, also called programmed cell death, plays a prominent role in a diversity of physiological and pathological processes [8]. Existing studies have found that apoptosis pathways include mitochondrial pathways, endoplasmic reticulum pathways, and death receptor pathways [9]. Mitochondrial apoptosis pathways are the most common, where the abnormal expression of Bcl-2 family proteins, cytochrome c, and caspases may occur [10]. Moreover, apoptosis is recognized as one of the valid strategies for tumor suppression, which involves a variety of morphological changes and cell signal transduction pathways [11]. The cell cycle is closely related to DNA replication and cell proliferation, including four phases of G0/G1, S, G2, and M. Cell cycle activities are usually regulated by cyclins (CCNs), cycle-dependent kinases (CDKs), and cyclin-dependent kinase inhibitors (CKIs) to maintain the activities of normal cell [12]. Abnormal expressions of these cell cycle factors can lead to uncontrolled cell proliferation and promote the occurrence of carcinogenesis [13]. It is well documented that more than 90% of human cancers are related to the accelerated G1 phase due to the abnormal expression of CCNs, CDKs, and CKIs [1,14]. Increasing evidence suggests that abnormal activation of the phosphatidylinositol-3-kinase (PI3K), AKT, and mammalian target of rapamycin (mTOR) pathway is a frequent event in numerous malignant tumors, including prostate cancer [15], gastrointestinal cancer [16], breast cancer [17], non-small cell lung cancer [18], acute myeloid leukemia [19], and liver cancer [20]. Recently, the activation of the PI3K/AKT/mTOR signaling pathway was found to be closely related to the regulation of apoptosis and cell cycle arrest in human endometrial cancer cells [21] and HCC [22]. A new report also showed that the natural flavonoid pectolinarigenin could induce cell apoptosis and G2/M phase cell cycle arrest of HCC by regulating the PI3K/AKT/mTOR pathway [23].

Angelica keiskei Koidzumi (*A. keiskei*), a traditionally healthy vegetable which is originally planted in pacific-coast islands in Japan, has been reported to contain varieties of bioactive compounds, especially chalcones [24]. It was officially recognized by the National Health Commission of China as a source of new food ingredients in 2019 after its introduction in the early 1990s, and its in-depth development and industrialization have been rapidly heating domestically ever since [25–28]. 4-hydroxyderricin (4-HD), as one of the most abundant chalcone in *A. keiskei*, exists in all parts of the plant. It has been proved to exhibit antibacterial [29], anti-inflammatory [24], antidiabetic [30,31], hypotensive [32], lipid regulation [32–34], and prevention of muscle atrophy [35] and loss [36], making it a valuable food-source active compound, which shows the promising potentiality of application in formulation or preparation for nutraceutical and functional foods. Specifically, in the research field of anti-cancer effects, it is reported to show anti-osteosarcoma effect by inhibiting the activation and differentiation of M2 macrophages [37]. Moreover, another study demonstrates that 4-HD can suppress melanomatogenesis by targeting BRAF and PI3K [38]. Currently, the literature relating the inhibitory effect and mechanisms of 4-HD on liver cancer is limited. Furthermore, no studies have reported the regulating roles of 4-HD on apoptosis and cell cycle arrest in HCC. Herein, we aimed to study the inhibitory effect of 4-HD on HCC and clarify the potential mechanisms by exploring PI3K/AKT/mTOR signaling pathway mediated apoptosis and cell cycle arrest.

2. Materials and Methods

2.1. Materials and Cell Culture

HepG2 and Huh7 of human HCC cell lines were supplied by the Cell Bank of the Chinese Academy of Sciences (Shanghai, China) [39]. The DMEM medium (Invitrogen, Carlsbad, CA, USA) supplemented with 10% FBS (Invitrogen, Carlsbad, CA, USA) and 1% penicillin/streptomycin (Invitrogen, Carlsbad, CA, USA) were employed for cell culture at 37 °C in a humidified incubator (5% CO_2).

2.2. Cell Counting Kit-8 (CCK-8) Assay

A single-cell suspension was firstly prepared based on digestion of the HepG2 and Huh7 cells in the logarithmic growth phase with trypsin. It was then counted, seeded into a 96-well plate at a density of 5×10^3 cells/mL, and placed in an incubator for culture. After the cells adhered to the wall, they were treated with 4-HD or 4-HD+LY294002 for

48 h, protected from light, and 10 µL of CCK-8 (Solarbio, Beijing, China) solution was added to each well without air bubbles. Then the cells were incubated for 30 min. Lastly, a microplate reader (Thermo Fisher Scientific, Waltham, MA, USA) was employed to read the absorbance, which was detected at 450 nm [40].

2.3. Wound Healing Assay

HepG2 and Huh7 cells were collected in the logarithmic growth phase and spread evenly in a six-well plate. When the cell growth density reached 95%, a 10 µL pipette tip was used to streak the line gently with a straight edge. After slowly washing with PBS, 1% serum-containing medium was added to the 6-well plate. Subsequently, 4-HD or a mixture of 4-HD and LY294002 was added for 48 h. An inverted phase-contrast microscope (Olympus, Tokyo, Japan) was used to observe cell migration.

2.4. Transwell Assay

The serum-free medium and Matrigel were diluted at a ratio of 1:8, added vertically to the upper chamber of the Transwell chamber (Corning, NY, USA), and placed in the incubator overnight. 2×10^5 cells/mL were added to the upper chamber of the small chamber, and 4-HD, LY294002 and DMEM medium containing 20% bovine fetal serum were added to the lower chamber of the small chamber. 4% paraformaldehyde was poured into the 24-well plate, then the lower chamber was placed on a 24-well plate. Crystal violet was added to the 24-well plate after the cells were fixed for 30 min. After staining, the 24-well plate was placed under a microscope (Olympus, Tokyo, Japan) for observation.

2.5. Cell-Cycle Analysis

The cells were firstly cultured for 48 h. Afterwards, the cell cycle was analyzed via the cell cycle analysis kit (Beyotime, Shanghai, China). Briefly, the collected cells were fixed with cold ethanol (70%) and kept at 4 °C overnight. DNA was stained with propidium iodide (PI, 0.05 mg/mL) and RNase (2.0 mg/mL). Then, the pretreated cells were placed on the FACScan flow cytometer (BD Biosciences, San Jose, CA, USA) for cell cycle analysis. The cell percentages in G0/G1, S and G2/M phases were calculated by the Cell Lab Quanta SC software (BD Biosciences, San Jose, CA, USA).

2.6. Apoptosis Analysis

An Annexin V-FITC apoptosis detection kit (Sigma-Aldrich, USA) was utilized for the evaluation of the effect of 4-HD treatment on cell apoptosis via flow cytometry. HepG2 and Huh7 cells were prepared into a single cell suspension. Then it was seeded in a 6-well plate (2×10^5 cells/mL) and treated with 4-HD or LY294002 for 48 h. Subsequently, the cells were digested, centrifuged, mixed with Annexin V-FITC, and then the solution of PI staining was added. Then the cells were incubated for 15 min and kept from light at room temperature. Finally, the cell apoptosis was detected by flow cytometry.

2.7. TUNEL Assay

TUNEL assays were implemented according to the kit manufacturer's protocol (Abcam, USA). The HCC cells were incubated in a 48-well plate for 48 h, and then fixed by 4% paraformaldehyde. After that, 0.1% TritonX-100 was added to the 48-well plate. 3% H_2O_2 was used to block the cells, then Tunel reaction buffer was added, and the cells were incubated in the 37 °C incubator in the dark for 60 min. A confocal fluorescence microscope (Nikon, Tokyo, Japan) was employed to observe the cell morphology.

2.8. Immunofluorescence Staining

Cells were seeded on clean glass slides and treated with 4-HD or 4-HD+LY294002 for 48 h. 4% paraformaldehyde was used to fix the cells and TritonX-100 was added in 0.3% PBS to permeabilize the cells at room temperature. The following primary antibodies were used: anti-p-AKT (Abcam, Cambridge, UK). PBS contains 5% BSA was used to dilute the antibodies, and after washing, Cy3-conjugated goat anti-mouse IgG (Beyotime, Shanghai, China) was utilized to incubate the cells at room temperature for 1 h. Lastly, it was determined by a fluorescence microscope (Nikon, Tokyo, Japan).

2.9. Western Blotting Analysis

Western blot was performed via the standard methods. HepG2 and Huh7 cells were placed in a 6-well plate with a density of 5×10^5 cells and incubated at 37 °C for 24 h. The total protein of each treatment group was extracted with lysis buffer containing 1% phenylmethanesulfonyl fluoride (PMSF). After separated, the proteins were then transferred to PVDF membranes (Solarbio, Beijing, China). The following primary antibodies were used for Western blot analysis: PI3K p85 (ABclonal A4992), Bax (ABclonal A19684), Bcl-2 (ABclonal A0208), cleaved caspase-3 (WL02348), CDK1/CDC2 (WL02373), cyclin B1 (WL01760), GAPDH (ab8245), AKT (#4685S), p-AKT (#4060S), p-PI3K p85 (#4228), m-TOR (#2983), p-mTOR (#5536), caspase-3 (#9662), caspase-9 (#9508S), cytochrome c (#11940T), PARP (#9532S), cyclin D1 (#2978P), CDK4 (#12790S), and CDK6 (#13331S). The antibodies and all secondary antibodies were supplied by CST (Danvers, MA, USA) and the membranes were visualized by BM Chemiluminescence Western Blotting Kit (Sigma-Aldrich, Schnelldoff, Germany). All the experiments were repeated three times.

2.10. Statistical Analysis

The results are presented as the standard error of mean ± mean, and one-way analysis of variance (ANOVA) was employed to evaluate the data. Duncan's multiple range test was performed to compare the differences among the groups via the SPSS software (version 22.0), and statistical significance was represented by $p < 0.05$ [41]. Each experiment was repeated at least three times.

3. Results

3.1. Subsection

3.1.1. 4-HD Inhibited the Proliferation and Metastasis of HepG2 and Huh7

The CCK-8 assay was employed to explore the effects of 4-HD on the cell viability of human HCC cells HepG2 and Huh7. As shown in Figure 1A,B, different concentrations of 4-HD (0, 5, 10, 20, 30, 40, 50, 80, and 100 µM) were selected and incubated with HepG2 and Huh7 cells for 24 h and 48 h, respectively. When the concentration of 4-HD was above 40 µM, the cell viability of the two types of HCC cells was significantly decreased after 24 h incubation ($p < 0.05$ for all) (Figure 1A). When the incubation was prolonged to 48 h, the cell viability was remarkably decreased by 4-HD with a concentration higher than 20 µM ($p < 0.01$ for all) (Figure 1B). A notable dose-dependent manner was observed.

The wound-healing assay was implemented to observe the migration effect of 4-HD on HCC cells. As shown in Figure 1C,D, the wound healing of HepG2 and Huh7 cells was markedly reduced after co-incubation with 20 µM and 40 µM of 4-HD for 48 h ($p < 0.001$ for all), compared with the control group, suggesting that 4-HD can effectively inhibit the migration process of HepG2 and Huh7. Subsequently, the Transwell experiment was employed to record the number of HepG2 and Huh7 cells passing through the Transwell chamber after 48 h of exposure to different concentrations of 4-HD. As seen in Figure 1E,F, the number of lower chambers was decreased when the concentration of 4-HD was raised ($p < 0.001$ for all). And obviously, 4-HD inhibited the invasion ability of HepG2 and Huh7 in a dose-dependent manner. Combining the results of wound healing and Transwell experiments, it can be inferred that 4-HD can significantly restrain the migration and invasion of HepG2 and Huh7.

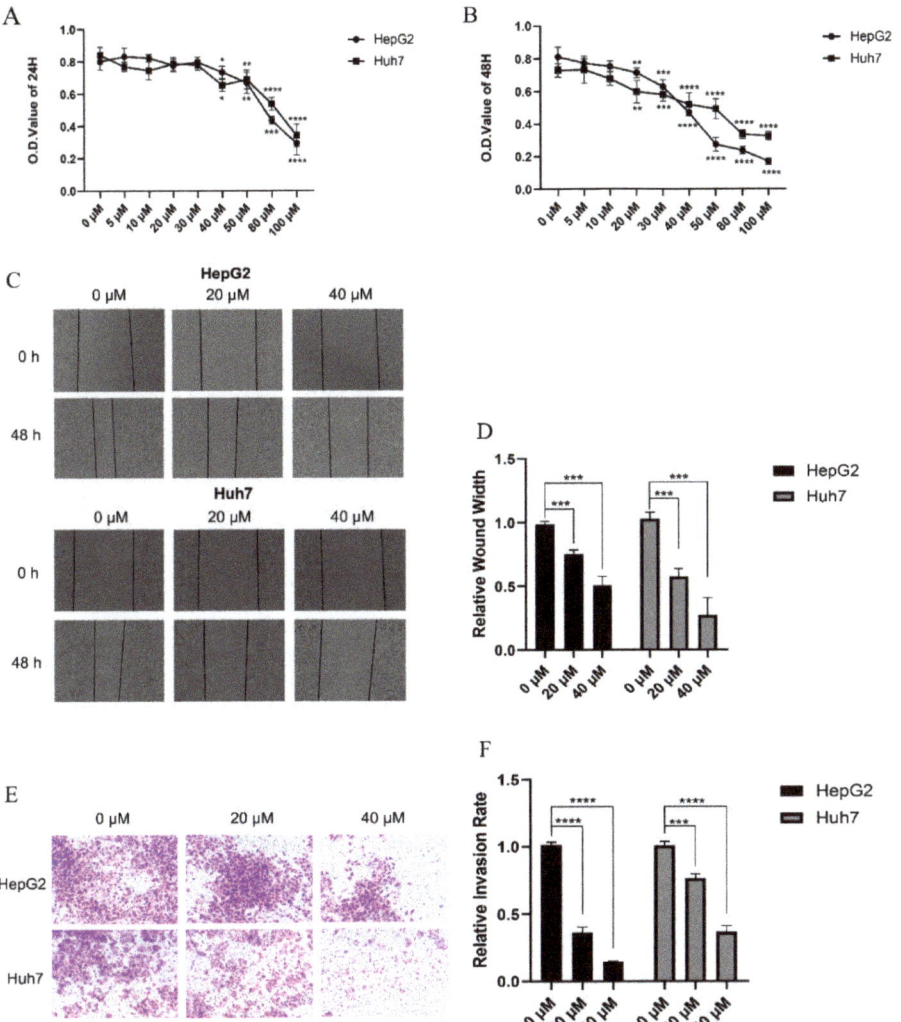

Figure 1. 4-HD inhibits the proliferation and metastasis of HCC cells. (**A**,**B**) Determination of the survival rate of HepG2 cells and Huh7 cells treated with 4-HD (0–100 μM) after 24 h or 48 h by CCK-8 assay; (**C**,**D**) Evaluation of effects for 4-HD on cell migration by wound healing assay; (**E**,**F**) Assessment of effects for 4HD-treated cell invasion by Transwell assay. * $p < 0.05$, ** $p < 0.01$, *** $p < 0.001$, **** $p < 0.0001$.

3.1.2. 4-HD Induced Apoptosis and Cell Cycle Arrest in HepG2 and Huh7 Cells

4-HD Induced Apoptosis in HepG2 and Huh7 Cells

To further evaluate the pro-apoptosis effects of 4-HD on HepG2 and Huh7 cells, TUNEL and flow cytometry were successively performed. As shown in Figure 2A,B, the number of positive cells in the 4-HD treatment group was increased in a dose-dependent manner ($p < 0.01$ for all), compared with the control group. It can be observed from Figure 2C,D that, as the concentrations of 4-HD went up, the rate of apoptosis exhibited an upward trend ($p < 0.05$ for all). Furthermore, Western blot experiments were employed to explore the mechanism of 4-HD-induced apoptosis of HepG2 and Huh7 cells. As seen in Figure 2E,F, the expressions of cytochrome c, cleaved caspase-3, cleaved caspase-9, cleaved

PARP, and Bax proteins were up-regulated as the concentrations of 4-HD rose, while the expressions of pro-caspase-3, pro-caspase-9, PARP, and Bcl-2 proteins were remarkably down-regulated ($p < 0.05$ for all). These results indicate that 4-HD can promote HepG2 and Huh7 cells apoptosis by activating the mitochondrial apoptosis pathway.

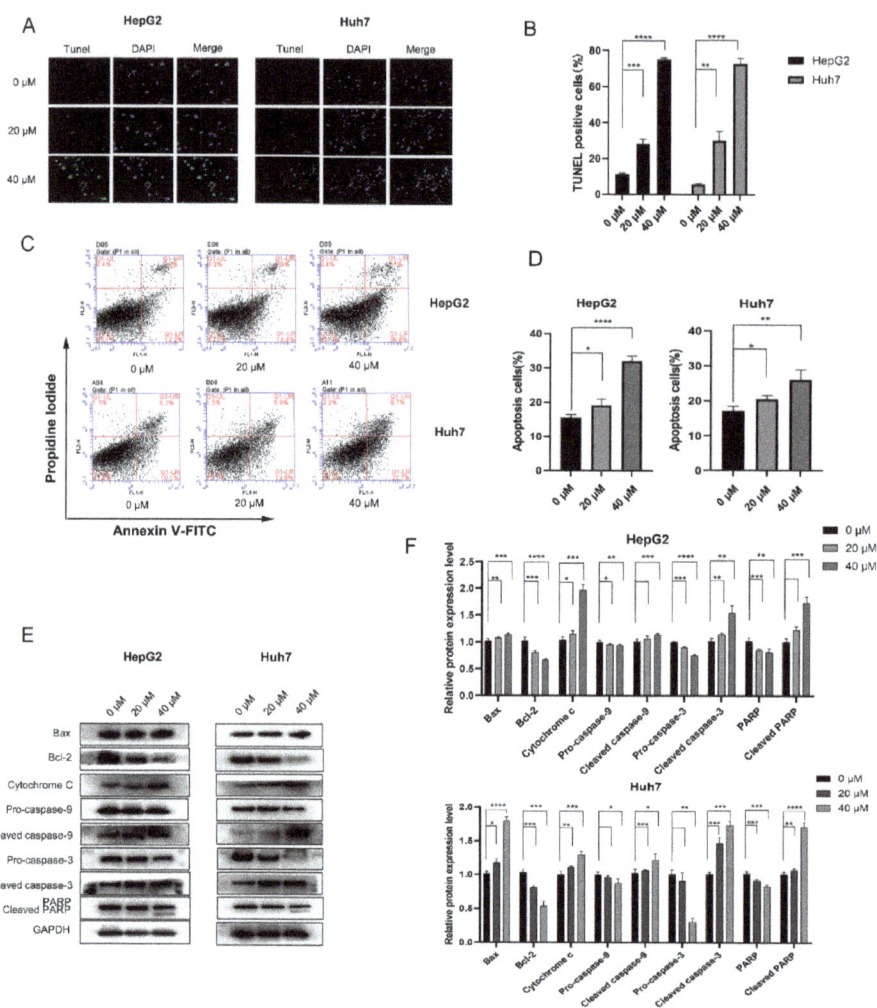

Figure 2. Effects of 4-HD on apoptosis of HepG2 and Huh7 cells. (**A,B**) TUNEL (green) and DAPI (blue) double-positive cells were elevated after treatment with various concentrations of 4-HD (magnification, 400×); (**C,D**) Cells were treated with 4-HD for 48 h, stained with annexin V-FITC/PI, and then analyzed by flow cytometry; (**E,F**) The effects of 4-HD on the expressions of Bax, Bcl-2, cytochrome c, caspase-9, caspase-3 and PARP proteins were evaluated via Western blot. Relative expressions of the proteins were normalized to GAPDH. * $p < 0.05$, ** $p < 0.01$, *** $p < 0.001$, **** $p < 0.0001$.

4-HD Induced Cell Cycle Arrest in HCC Cells

Cells can be divided into the G0/G1, S, and G2/M phases according to the DNA content and detected by flow cytometry. Herein, the effects of 4-HD on the cell cycle of HepG2 and Huh7 cells were investigated. After PI staining, the cells in different cell cycles were distinguished according to the fluorescence intensity. As shown in Figure 3A,B,E,F, the proportion of HepG2 cells in the G2/M phase was increased with the elevation of

4-HD concentrations. However, the proportion of Huh7 cells in the G0/G1 phase was surprisingly increased as well, suggesting that 4-HD can block HepG2 cells in the G2/M phase and Huh7 cells in the G0/G1 phase ($p < 0.01$ for all).

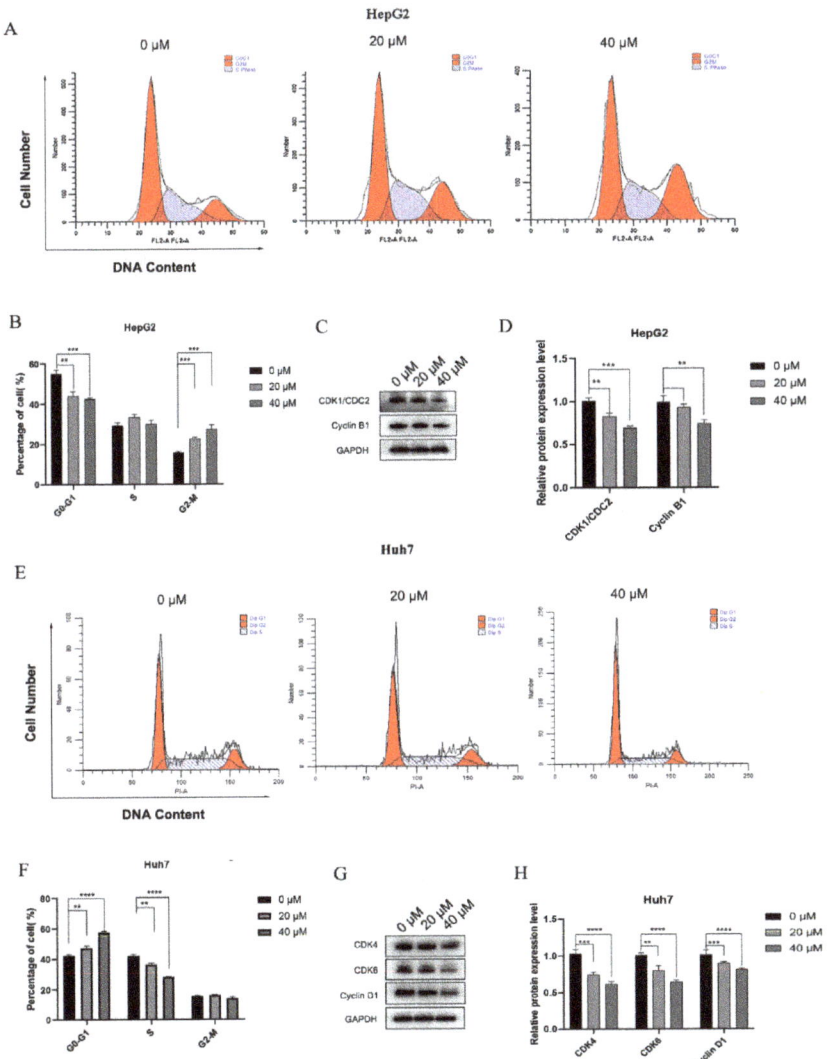

Figure 3. The effect of 4-HD on HCC cell cycle distribution. (**A,B**) Phase distribution of HepG2 treated with 4-HD for 48 h by flow cytometry analysis; (**C,D**) The effects of 4-HD on the expression of Cyclin B1 and CDK1/CDC2 in HepG2 cells were evaluated by Western blot; (**E,F**) Phase distribution of Huh7 treated with 4-HD for 48 h by flow cytometry analysis; (**G,H**) The effects of 4-HD on the expressions of Cyclin D1, CDK4 and CDK6 in Huh7 cells were evaluated by Western blot. Relative expressions of the proteins were normalized to GAPDH. ** $p < 0.01$, *** $p < 0.001$, **** $p < 0.0001$.

HepG2 and Huh7 cells were respectively incubated with 4-HD at concentrations of 20 µM and 40 µM for 48 h, and the proteins were extracted. As displayed in Figure 3C,D,G,H, the relative protein expressions of cyclin B1 and CDK1/CDC2 were determined. In HepG2 cells, the levels of cyclin B1 and CDK1/CDC2 were declined with the increasing concentrations of 4-HD, while in Huh7 cells, the expressions of cyclin D1, CDK4, and CDK6 were dramatically down-regulated. It can be inferred that 4-HD can induce HepG2 cells arrest at the G2/M phase and Huh7 cells arrest at the G0/G1 phase ($p < 0.01$ for all).

3.1.3. 4-HD Regulated the PI3K/AKT/mTOR Pathway

To investigate that 4-HD regulates the PI3K/AKT/mTOR pathway, the protein expressions of genes related to the pathway were detected. It can be observed from Figure 4A,B that, as the concentration of 4-HD raised, the expressions of PI3K, p-PI3K, p-AKT, and p-mTOR proteins were down-regulated, while the expressions of AKT and mTOR proteins were up-regulated ($p < 0.05$ for all). Moreover, the same results were obtained in the immunofluorescence experiment, as shown in Figure 4C, indicating that 4-HD can inhibit the phosphorylation of AKT.

In order to confirm whether 4-HD governed the PI3K/AKT/mTOR axis, PI3K inhibitor LY294002 was selected, Western blot analysis and immunofluorescence experiments were performed to detect related proteins. Figure 4D,E shows that the expressions of p-PI3K, p-AKT, and p-mTOR proteins in the 4-HD+LY294002 group were notably decreased compared with the 4-HD group ($p < 0.05$ for all). As seen in Figure 4F from immunofluorescence results, the expressions of p-AKT gradually were decreased, compared with the control group. What's more, the inhibitory effect of p-AKT was enhanced by LY294002+4HD group.

Then, aiming to validate whether the PI3K/AKT/mTOR signaling pathway was involved in 4-HD inhibiting the migration and invasion of HCC cells, LY294002 was incubated with HepG2 cells, followed by wound healing and Transwell experiments. It can be observed from Figure 4G,H that the wound healing width of 4-HD and LY294002 combined treatment group was markedly higher than that of LY294002 alone treatment group, indicating that LY294002 combined treatment can enhance the inhibition of cell migration ($p < 0.05$ for all). As noticed in Figure 4I,J, the number of HepG2 cells that invaded the lower chamber in the 4-HD and LY294002 combination group was eminently lower than that in the 4-HD group ($p < 0.01$ for all). It can be inferred from these results that 4-HD can inhibit the proliferation and metastasis of HCC cells by regulating the PI3K/AKT/mTOR pathway.

3.1.4. 4-HD Induced Apoptosis and Cycle Arrest of HCC Cells by Regulating the PI3K/AKT/mTOR Pathway

To further verify whether PI3K/Akt/mTOR signaling pathway is involved in the process of 4-HD inducing apoptosis and cell cycle arrest of hepatoma cells, as shown in Figure 5A,B, the proportion of apoptosis in the 4-HD + LY294002 treatment group was increased, compared with the control group ($p < 0.001$). As seen in Figure 5C,D, compared with the 4-HD group, the expressions of cytochrome c, cleaved caspase-3, cleaved caspase-9, cleaved PARP, and Bax proteins were remarkably elevated in the 4-HD + LY294002 group, accompanied by the decreasing expressions of pro-caspase-3, pro-caspase-9, PARP and Bcl-2 proteins ($p < 0.05$ for all).

The cell cycle distribution of HepG2 was assessed by flow cytometry. As exhibited in Figure 5E,F, compared with the 4-HD group, the cell proportion at the G0/G1 phase was declined, while the one at the G2/M phase was increased after combining with LY294002 ($p < 0.01$ for all). Furthermore, Western blot was performed to verify the expression of cyclin. As shown in Figure 5G,H, 4-HD+LY294002 treatment significantly down-regulated the expressions of CDK1/CDC2 and Cyclin B1 proteins ($p < 0.05$ for all). Combining the above experimental data, it can be concluded that 4-HD can induce apoptosis and cycle arrest of HCC cells through the PI3K/AKT/mTOR signaling axis.

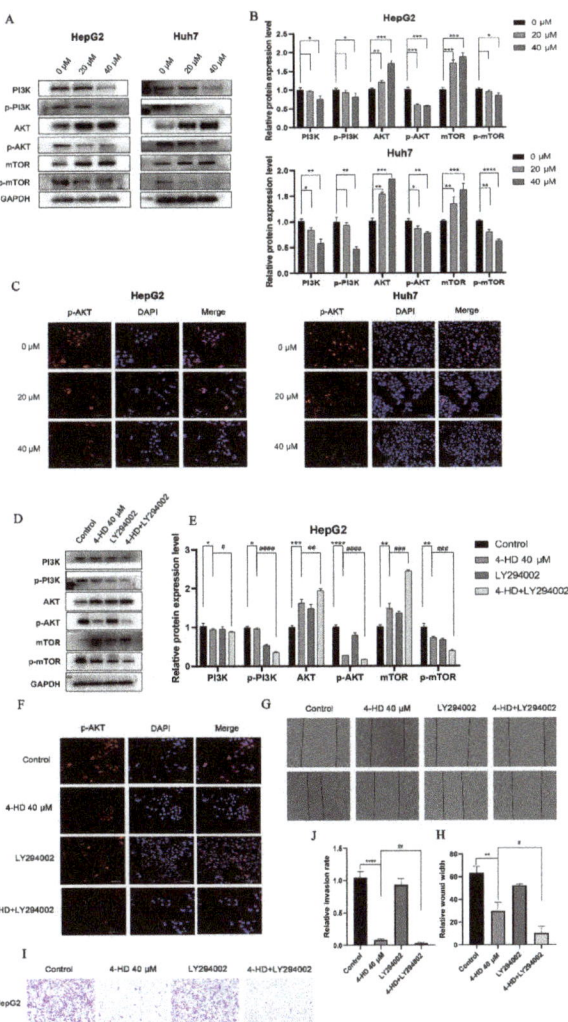

Figure 4. 4-HD inhibited the proliferation and metastasis of HCC cells by regulating the PI3K/AKT/mTOR signaling pathway. (**A,B**) Relative proteins expressions of PI3K/AKT/mTOR pathway in HepG2 cells and Huh7 cells treated with 0 μM, 20 μM, and 40 μM of 4-HD for 48 h; (**C**) Immunofluorescence was employed to quantify the expression of p-AKT protein in HepG2 cells (magnification: 400×); (**D,E**) Relative proteins expressions of PI3K/AKT/mTOR pathway in HepG2 cells were treated with PBS (control), 4-HD (40 μM 4-HD), LY294002 (10 μM LY294002) and 4-HD + LY294002 (40 μM 4-HD + 10 μM LY294002) for 48 h; (**F**) immunofluorescence was performed to quantify the expression of p-AKT protein in HepG2 cells treated with LY294002 (magnification: 400×); (**G,H**) wound healing assay was carried out to conduct the effect of LY294002 on the migration of HepG2 cells; (**I,J**) Transwell assay was implemented to detect the effect of LY294002 on the invasion of HepG2 cells. * $p < 0.05$, ** $p < 0.01$, *** $p < 0.001$, **** $p < 0.0001$ vs. control (0 μM); # $p < 0.05$, ## $p < 0.01$, ### $p < 0.001$, #### $p < 0.0001$ vs. 40 μM 4-HD.

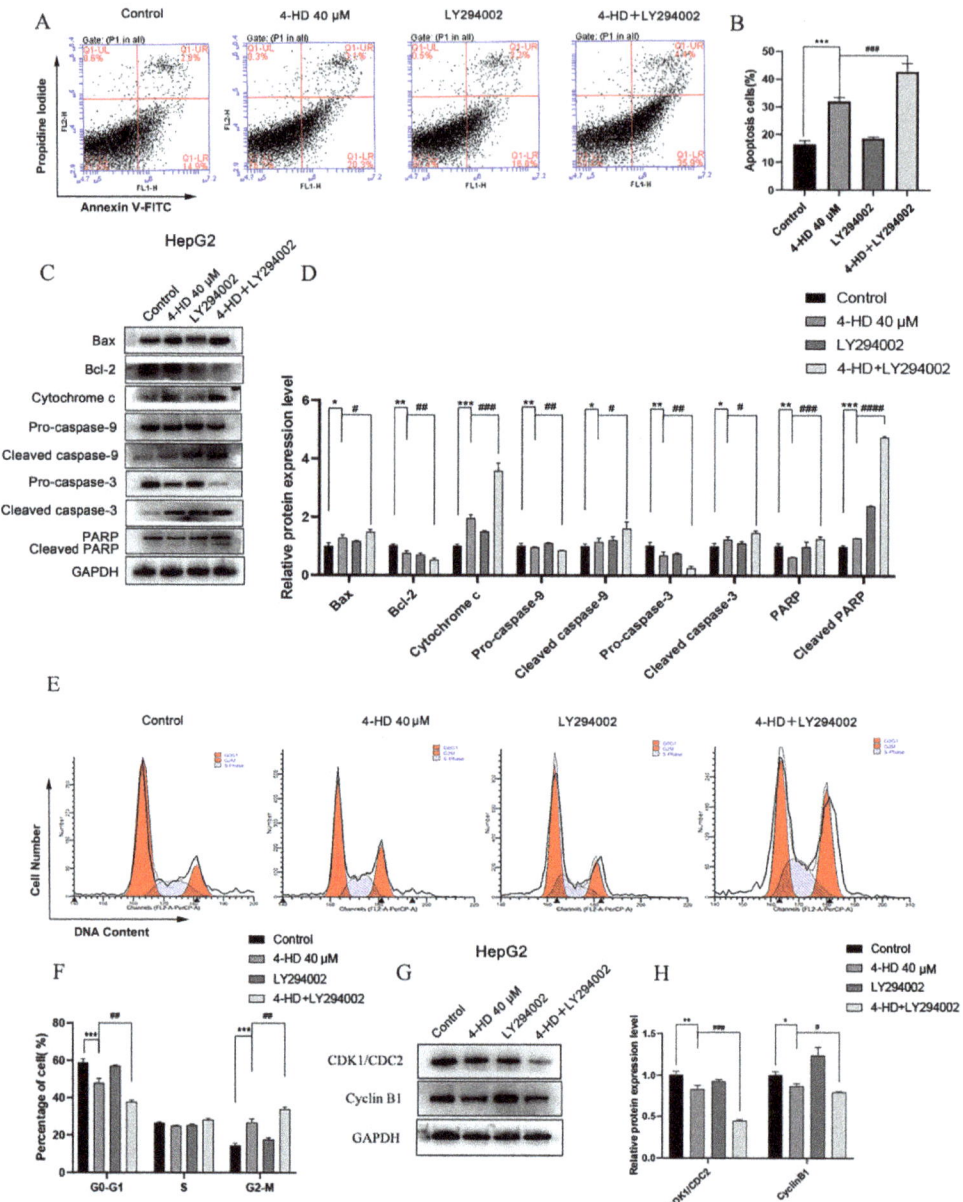

Figure 5. 4-HD induced apoptosis and cycle arrest of HCC cells by regulating the PI3K/AKT/mTOR signaling pathway. (**A**,**B**) Cell apoptosis proportion treated with 4-HD+LY294002 was detected by flow cytometry; (**C**,**D**) The effects of LY294002 on the expressions of Bax, Bcl-2, cytochrome c, caspase-9 and caspase-3 and PARP in HepG2 cells treated with 4-HD were evaluated by Western blotting; (**E**,**F**) Cell cycle distribution proportion treated with 4-HD+LY294002 was determined by flow cytometry; (**G**,**H**) Effects of LY294002 on the expressions of cyclin B1 and CDK1/CDC2 proteins in HepG2 cells treated with 4-HD were assessed by Western blotting. Relative expressions of the proteins were normalized to GAPDH. * $p < 0.05$, ** $p < 0.01$, *** $p < 0.001$ vs. control (0 μM); # $p < 0.05$, ## $p < 0.01$, ### $p < 0.001$, #### $p < 0.0001$ vs. 40 μM 4-HD.

4. Discussion

4-HD is one of the major natural chalcone isolated from *A. keiskei* with various functional properties, such as anti-tumor. To our knowledge, this is the first study evaluating the inhibitory effects of 4-HD on HCC cells. In two typical HCC cell lines, HepG2 and Huh7, we found 4-HD induced remarkable cell cycle arrest and apoptosis along with the inhibitory effect on the proliferation and metastasis. What's more, we proved that 4-HD may promote apoptosis and cell cycle arrest of the HCC cells by modulating the PI3K/AKT/mTOR signaling pathway.

Apoptosis is the main way of programmed cell death, which can restrain tumor cell growth [42]. Drug-induced cell apoptosis is mainly regulated by the mitochondrial mechanism through caspase activations [43]. Caspase and Bcl-2 family genes play an important regulatory role in the process of cell apoptosis [44]. Caspase-3 regulates the entire process of cell apoptosis [45]. Bcl-2 is a membrane protein that inhibits the release of cytochrome c (Cyto-C) by regulating the permeability of the mitochondrial membrane and restrains the activation of caspase-3 to exert anti-apoptotic effects [46]. Bax, as a pro-apoptotic protein, can be suppressed by forming a dimer with Bcl-2, while its activation destroys the integrity of the mitochondrial membrane. Subsequently, cyto-C is released from the mitochondrial membrane, which activates caspase-3 and induces mitochondrial apoptosis [47]. Our results showed that the expressions of pro-apoptotic proteins were up-regulated while those of anti-apoptotic proteins were down-regulated by 4-HD treatment. A previous study has shown that 4-HD can induce apoptotic death of HL60 cells through death receptor-mediated and mitochondrial pathways [48], which is consistent with our findings that 4-HD induces mitochondrial apoptotic cell death to exert anti-HCC cell proliferation effects.

The cell cycle is regulated by a protein complex composed of cyclins and cyclin-dependent kinases (CDK). It has been reported that the abnormal levels of CDK4, CDK6, and cyclin D1 in various human cancer cells are closely related to the abnormal proliferation of cancer cells [49,50]. Cell cycle arrest is considered as a potential target for cancer therapy in numerous malignant cancers [51]. Decrease in cyclin D inhibits growth and induces cell death in tumors such as esophageal, colon, and pancreatic cancers [52–54]. Many chemotherapeutic drugs exhibit anti-tumor effects by inducing cell cycle arrest [55]. The activation of CDK1/CDC2 and cyclin B1 plays a key role in the initial stages of mitosis. And the existing document indicates that the down-regulation of CDK1/CDC2 and cyclin B1 is related to the G2/M cycle arrest [56]. In this study, the expressions of cyclin B1 and CDK1/CDC2 proteins in HepG2 cells treated with 4-HD were down-regulated, while those of cyclin D1 and CDK4 were down-regulated in Huh7 cells, suggesting that 4-HD can trigger HepG2 cells arrest at G2/M phase and Huh7 cells arrest at G0/G1 phase.

PI3K/AKT/m-TOR cascade is a signal transduction pathway that regulates cancer cell growth, proliferation, cell energy, proliferation, senescence, and angiogenesis [57]. Inhibiting the different processes of the PI3K/AKT/m-TOR pathway is a common strategy for the treatment of human malignant tumors [58]. Many bioactive flavonoids, such as collagen and paclitaxel, have been reported to down-regulate the expressions of proteins such as p-PI3K, p-AKT, and p-mTOR by inhibiting PI3K/AKT/mTOR pathway in HCC [59,60]. Our study shows that the protein levels of PI3K, p-PI3K, p-AKT, and p-mTOR were down-regulated following 4-HD treatment, while the expressions of AKT and mTOR proteins were up-regulated, indicating that the suppression effect of 4-HD on the PI3K/AKT/mTOR signaling pathway. Moreover, we also co-treated the HCC cells with both 4-HD and LY294002, an inhibitor of PI3K. Both apoptosis and cycle arrest were exacerbated in the co-treatment groups. A previous study demonstrated that 4-HD could promote apoptosis and induces cycle arrest in melanoma by targeting PI3K [32], which complies with our findings. Collectively, all these results indicated that 4-HD induced apoptosis and cell cycle arrest of the HCC cells by modulating the PI3K/AKT/mTOR signaling pathway to inhibit the proliferation of HCC cells (Figure 6).

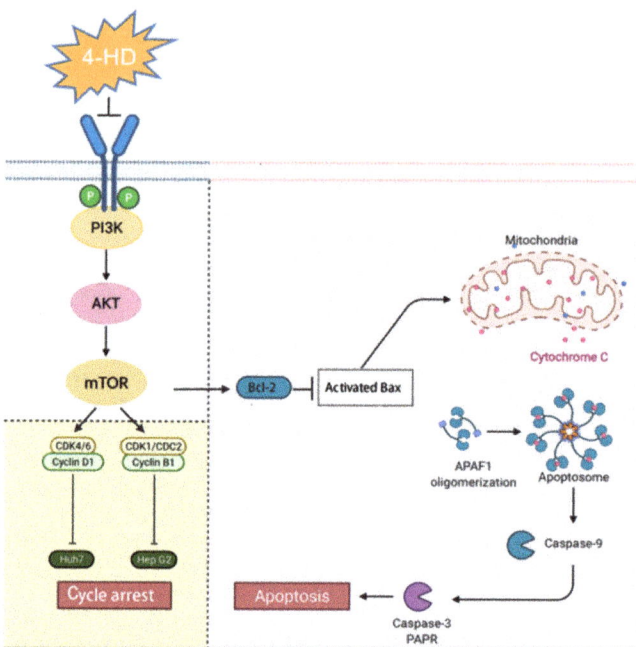

Figure 6. Proposed mechanism for 4-HD inducing apoptosis and cycle arrest in HCC cells through the PI3K/AKT/m-TOR signaling pathway. ⊥ indicates an inhibitory effect and → indicates a promoting effect.

5. Conclusions

In summary, this study revealed that 4-HD exhibited proliferation inhibitory effects in HepG2 and Huh7 cells in a dose-dependent manner. The potential mechanism may be related to the inhibition of the PI3K/AKT/m-TOR signaling pathway and the subsequent inducing of mitochondrial apoptosis and cell cycle arrest. This study provides a new strategy for the therapy of HCC and a theoretical basis for the exploiting of 4-HD as an anti-hepatoma natural functional ingredient.

Author Contributions: Conceptualization: Z.L.; data curation: X.G., Y.J. and F.L.; formal analysis: Y.J. and X.P.; funding acquisition: Z.L., M.W. and Q.L.; investigation: Y.J. and X.P.; methodology: Y.J., X.G. and Z.L.; project administration: Z.L.; resources: X.G., Q.X., F.L. and Z.L.; supervision: Z.L. and X.G.; validation: Z.L., X.G. and Q.X.; visualization: Y.J. and X.G.; writing—original draft preparation: Y.J. and X.G.; writing—review and editing: Z.L., X.G. and Q.X. All authors have read and agreed to the published version of the manuscript.

Funding: This research was funded by the Shandong Provincial Natural Science Foundation (ZR2019PH020) and Joint Institute of *Angelica keiskei* Health Industry Technology.

Data Availability Statement: All of the data are presented in the manuscript.

Conflicts of Interest: The authors declare no conflict of interest.

References

1. Ji, X.; Wei, X.; Qian, J.; Mo, X.; Kai, G.; An, F.; Lu, Y. 2′,4′-Dihydroxy-6′-methoxy-3′,5′-dimethylchalcone induced apoptosis and G1 cell cycle arrest through PI3K/AKT pathway in BEL-7402/5-FU cells. *Food Chem. Toxicol.* **2019**, *131*, 110533. [CrossRef]
2. Tang, A.; Hallouch, O.; Chernyak, V.; Kamaya, A.; Sirlin, C.B. Epidemiology of hepatocellular carcinoma: Target population for surveillance and diagnosis. *Abdom. Radiol.* **2018**, *43*, 13–25. [CrossRef]
3. Brunetti, O.; Gnoni, A.; Licchetta, A.; Longo, V.; Calabrese, A.; Argentiero, A.; Delcuratolo, S.; Solimando, A.G.; Casadei-Gardini, A.; Silvestris, N. Predictive and prognostic factors in HCC patients treated with sorafenib. *Medicina* **2019**, *55*, 707. [CrossRef]

4. Díez, D.V.; Zubiaur, A.G.; Montalvo, S.M. Reactive perforating collagenosis: A rare side effect associated with sorafenib. *Rev. Esp. Enferm. Dig.* **2020**, *112*, 960–961. [CrossRef]
5. Méndez-Blanco, C.; Fondevila, F.; Palomo, A.G.; González-Gallego, J.; Mauriz, J.L. Sorafenib resistance in hepatocarcinoma: Role of hypoxia-inducible factors. *Exp. Mol. Med.* **2018**, *50*, 1–9. [CrossRef]
6. Ma, L.; Liu, W.; Xu, A.; Ji, Q.; Ma, Y.; Tai, Y.; Wang, Y.; Shen, C.; Liu, Y.; Wang, T.; et al. Activator of thyroid and retinoid receptor increases sorafenib resistance in hepatocellular carcinoma by facilitating the Warburg effect. *Cancer Sci.* **2020**, *111*, 2028–2040. [CrossRef] [PubMed]
7. Sahin, I.D.; Christodoulou, M.S.; Guzelcan, E.A.; Koyas, A.; Karaca, C.; Passarella, D.; Cetin-Atalay, R. A small library of chalcones induce liver cancer cell death through Akt phosphorylation inhibition. *Sci. Rep.* **2020**, *10*, 11814. [CrossRef] [PubMed]
8. Hou, Y.-Q.; Yao, Y.; Bao, Y.-L.; Song, Z.-B.; Yang, C.; Gao, X.-L.; Zhang, W.-J.; Sun, L.-G.; Yu, C.-L.; Huang, Y.-X.; et al. Juglanthraquinone C induces intracellular ROS increase and apoptosis by activating the Akt/Foxo signal pathway in HCC cells. *Oxid. Med. Cell. Longev.* **2016**, *2016*, 4941623. [CrossRef]
9. He, Q.; Montalbano, J.; Corcoran, C.; Jin, W.; Huang, Y.; Sheikh, M.S. Effect of Bax deficiency on death receptor 5 and mitochondrial pathways during endoplasmic reticulum calcium pool depletion-induced apoptosis. *Oncogene* **2003**, *22*, 2674–2679. [CrossRef]
10. Reuter, S.; Eifes, S.; Dicato, M.; Aggarwal, B.B.; Diederich, M. Modulation of anti-apoptotic and survival pathways by curcumin as a strategy to induce apoptosis in cancer cells. *Biochem. Pharmacol.* **2008**, *76*, 1340–1351. [CrossRef]
11. Zhang, Y.-S.; Ma, Y.-L.; Thakur, K.; Hussain, S.S.; Wang, J.; Zhang, Q.; Zhang, J.-G.; Wei, Z.-J. Molecular mechanism and inhibitory targets of dioscin in HepG2 cells. *Food Chem. Toxicol.* **2018**, *120*, 143–154. [CrossRef] [PubMed]
12. Asghar, U.; Witkiewicz, A.K.; Turner, N.C.; Knudsen, E.S. The history and future of targeting cyclin-dependent kinases in cancer therapy. *Nat. Rev. Drug Discov.* **2015**, *14*, 130–146. [CrossRef]
13. Chen, J. The Cell-Cycle Arrest and apoptotic functions of p53 in tumor initiation and progression. *Cold Spring Harb. Perspect. Med.* **2016**, *6*, a026104. [CrossRef]
14. Barbieri, F.; Cagnoli, M.; Ragni, N.; Pedullà, F.; Foglia, G.; Alama, A. Expression of cyclin D1 correlates with malignancy in human ovarian tumours. *Br. J. Cancer* **1997**, *75*, 1263–1268. [CrossRef] [PubMed]
15. Shorning, B.Y.; Dass, M.S.; Smalley, M.J.; Pearson, H.B. The PI3K-AKT-mTOR Pathway and Prostate Cancer: At the Crossroads of AR, MAPK, and WNT Signaling. *Int. J. Mol. Sci.* **2020**, *21*, 4507. [CrossRef]
16. Zhao, Q.; Zhao, Y.; Hu, W.; Zhang, Y.; Wu, X.; Lu, J.; Li, M.; Li, W.; Wu, W.; Wang, J.; et al. m(6)A RNA modification modulates PI3K/Akt/mTOR signal pathway in gastrointestinal cancer. *Theranostics* **2020**, *10*, 9528–9543. [CrossRef]
17. Miricescu, D.; Totan, A.; Stanescu-Spinu, I.-I.; Badoiu, S.C.; Stefani, C.; Greabu, M. PI3K/AKT/mTOR signaling pathway in breast cancer: From molecular landscape to clinical aspects. *Int. J. Mol. Sci.* **2020**, *22*, 173. [CrossRef] [PubMed]
18. Wang, N.; Feng, T.; Liu, X.; Liu, Q. Curcumin inhibits migration and invasion of non-small cell lung cancer cells through up-regulation of miR-206 and suppression of PI3K/AKT/mTOR signaling pathway. *Acta Pharm.* **2020**, *70*, 399–409. [CrossRef]
19. Nepstad, I.; Hatfield, K.J.; Grønningsæter, I.S.; Reikvam, H. The PI3K-Akt-mTOR signaling pathway in human acute myeloid leukemia (AML) cells. *Int. J. Mol. Sci.* **2020**, *21*, 2907. [CrossRef]
20. Wang, W.; Dong, X.; Liu, Y.; Ni, B.; Sai, N.; You, L.; Sun, M.; Yao, Y.; Qu, C.; Yin, X.; et al. Itraconazole exerts anti-liver cancer potential through the Wnt, PI3K/AKT/mTOR, and ROS pathways. *Biomed. Pharmacother.* **2020**, *131*, 110661. [CrossRef]
21. Zhang, Y.-Y.; Zhang, F.; Zhang, Y.-S.; Thakur, K.; Zhang, J.-G.; Liu, Y.; Kan, H.; Wei, Z.-J. Mechanism of juglone-induced cell cycle arrest and apoptosis in Ishikawa human endometrial cancer cells. *J. Agric. Food Chem.* **2019**, *67*, 7378–7389. [CrossRef]
22. Li, L.; Liu, J.D.; Gao, G.D.; Zhang, K.; Song, Y.W.; Li, H.B. Puerarin 6″-O-xyloside suppressed HCC via regulating proliferation, stemness, and apoptosis with inhibited PI3K/AKT/mTOR. *Cancer Med.* **2020**, *9*, 6399–6410. [CrossRef]
23. Wu, T.; Dong, X.; Yu, D.; Shen, Z.; Yu, J.; Yan, S. Natural product pectolinarigenin inhibits proliferation, induces apoptosis, and causes G2/M phase arrest of HCC via PI3K/AKT/mTOR/ERK signaling pathway. *Onco. Targets Ther.* **2018**, *11*, 8633–8642. [CrossRef] [PubMed]
24. Li, Y.; Goto, T.; Ikutani, R.; Lin, S.; Takahashi, N.; Takahashi, H.; Jheng, H.-F.; Yu, R.; Taniguchi, M.; Baba, K.; et al. Xanthoangelol and 4-hydroxyderrcin suppress obesity-induced inflammatory responses. *Obesity* **2016**, *24*, 2351–2360. [CrossRef] [PubMed]
25. Pang, X.; Zhang, X.; Jiang, Y.; Su, Q.; Li, Q.; Li, Z. Autophagy: Mechanisms and therapeutic potential of flavonoids in cancer. *Biomolecules* **2021**, *11*, 135. [CrossRef] [PubMed]
26. Zhang, L.; Jiang, Y.; Pang, X.; Hua, P.; Gao, X.; Li, Q.; Li, Z. Simultaneous optimization of ultrasound-assisted extraction for flavonoids and antioxidant activity of *Angelica keiskei* using response surface methodology (RSM). *Molecules* **2019**, *24*, 3461. [CrossRef]
27. Zhang, W.; Gu, X.; Liu, X.; Wang, Z. Fabrication of Pickering emulsion based on particles combining pectin and zein: Effects of pectin methylation. *Carbohydr. Polym.* **2021**, *256*, 117515. [CrossRef]
28. Pang, X.; Gao, X.; Liu, F.; Jiang, Y.; Wang, M.; Li, Q.; Li, Z. Xanthoangelol modulates Caspase-1-dependent pyroptotic death among hepatocellular carcinoma cells with high expression of GSDMD. *J. Funct. Foods* **2021**, *84*, 104577. [CrossRef]
29. Inamori, Y.; Baba, K.; Tsujibo, H.; Taniguchi, M.; Nakata, K.; Kozawa, M. Antibacterial activity of two chalcones, xanthoangelol and 4-hydroxyderrcin, isolated from the root of *Angelica keiskei* Koidzumi. *Chem. Pharm. Bull.* **1991**, *39*, 1604–1605. [CrossRef]
30. Enoki, T.; Ohnogi, H.; Nagamine, K.; Kudo, Y.; Sugiyama, K.; Tanabe, M.; Kobayashi, E.; Sagawa, H.; Kato, I. Antidiabetic activities of chalcones isolated from a Japanese Herb, *Angelica keiskei*. *J. Agric. Food Chem.* **2007**, *55*, 6013–6017. [CrossRef]

31. Aulifa, D.L.; Adnyana, I.K.; Levita, J.; Sukrasno, S. 4-Hydroxyderricin isolated from the sap of *Angelica keiskei* Koidzumi: Evaluation of its inhibitory activity towards dipeptidyl peptidase-IV. *Sci. Pharm.* **2019**, *87*, 30. [CrossRef]
32. Ogawa, H.; Ohno, M.; Baba, K. Hypotensive and lipid regulatory actions of 4-hydroxyderricin, a chalcone from *Angelica keiskei*, in stroke-prone spontaneously hypertensive rats. *Clin. Exp. Pharmacol. Physiol.* **2005**, *32*, 19–23. [CrossRef] [PubMed]
33. Zhang, T.S.; Sawada, K.; Yamamoto, N.; Ashida, H. 4-Hydroxyderricin and xanthoangelol from Ashitaba (*Angelica keiskei*) suppress differentiation of preadiopocytes to adipocytes via AMPK and MAPK pathways. *Mol. Nutr. Food Res.* **2013**, *57*, 1729–1740. [CrossRef] [PubMed]
34. Li, Y.; Goto, T.; Yamakuni, K.; Takahashi, H.; Takahashi, N.; Jheng, H.-F.; Nomura, W.; Taniguchi, M.; Baba, K.; Murakami, S.; et al. 4-Hydroxyderricin, as a PPARγ agonist, promotes adipogenesis, adiponectin secretion, and glucose uptake in 3T3-L1 cells. *Lipids* **2016**, *51*, 787–795. [CrossRef]
35. Yoshioka, Y.; Samukawa, Y.; Yamashita, Y.; Ashida, H. 4-Hydroxyderricin and xanthoangelol isolated from *Angelica keiskei* prevent dexamethasone-induced muscle loss. *Food Funct.* **2020**, *11*, 5498–5512. [CrossRef]
36. Kweon, M.; Lee, H.; Park, C.; Choi, Y.H.; Ryu, J.-H. A chalcone from ashitaba (*Angelica keiskei*) stimulates myoblast differentiation and inhibits dexamethasone-induced muscle atrophy. *Nutrients* **2019**, *11*, 2419. [CrossRef] [PubMed]
37. Sumiyoshi, M.; Taniguchi, M.; Baba, K.; Kimura, Y. Antitumor and antimetastatic actions of xanthoangelol and 4-hydroxyderricin isolated from *Angelica keiskei* roots through the inhibited activation and differentiation of M2 macrophages. *Phytomedicine* **2015**, *22*, 759–767. [CrossRef]
38. Zhang, T.; Wang, Q.; Fredimoses, M.; Gao, G.; Wang, K.; Chen, H.; Wang, T.; Oi, N.; Zykova, T.A.; Reddy, K.; et al. The ashitaba (*Angelica keiskei*) chalcones 4-hydroxyderricin and xanthoangelol suppress melanomagenesis by targeting BRAF and PI3K. *Cancer Prev. Res.* **2018**, *11*, 607–620. [CrossRef]
39. Cai, J.; Luo, S.; Lv, X.; Deng, Y.; Huang, H.; Zhao, B.; Zhang, Q.; Li, G. Formulation of injectable glycyrrhizic acid-hydroxycamptothecin micelles as new generation of DNA topoisomerase I inhibitor for enhanced antitumor activity. *Int. J. Pharmaceut.* **2019**, *571*, 118693. [CrossRef]
40. Chen, Y.; Gao, X.; Wei, Y.; Liu, Q.; Jiang, Y.; Zhao, L.; Ulaah, S. Isolation, purification and the anti-hypertensive effect of a novel angiotensin I-converting enzyme (ACE) inhibitory peptide from *Ruditapes philippinarum* fermented with *Bacillus natto*. *Food Funct.* **2018**, *9*, 5230–5237. [CrossRef] [PubMed]
41. Gao, X.; Tian, Y.; Randell, E.; Zhou, H.; Sun, G. Unfavorable associations between serum trimethylamine N-oxide and L-carnitine levels with components of metabolic syndrome in the newfoundland population. *Front. Endocrinol.* **2019**, *10*, 168. [CrossRef]
42. Yang, J.; Yang, Y.; Wang, L.; Jin, Q.; Pan, M. Nobiletin selectively inhibits oral cancer cell growth by promoting apoptosis and DNA damage in vitro. *Oral Surg. Oral Med. O.* **2020**, *130*, 419–427. [CrossRef]
43. Reddy, D.; Kumavath, R.; Ghosh, P.; Barh, D. Lanatoside C induces G2/M cell cycle arrest and suppresses cancer cell growth by attenuating MAPK, Wnt, JAK-STAT, and PI3K/AKT/mTOR signaling pathways. *Biomolecules* **2019**, *9*, 792. [CrossRef]
44. Bertin-Ciftci, J.; Barré, B.; Le Pen, J.; Maillet, L.; Couriaud, C.; Juin, P.; Braun, F. pRb/E2F-1-mediated caspase-dependent induction of Noxa amplifies the apoptotic effects of the Bcl-2/Bcl-xL inhibitor ABT-737. *Cell Death Differ.* **2013**, *20*, 755–764. [CrossRef] [PubMed]
45. Wang, W.; Liu, Y.; Zhao, L. Tambulin targets histone deacetylase 1 inhibiting cell growth and inducing apoptosis in human lung squamous cell carcinoma. *Front. Pharmacol.* **2020**, *11*, 1188. [CrossRef] [PubMed]
46. Min, X.; Heng, H.; Yu, H.; Dan, M.; Jie, C.; Zeng, Y.; Ning, H.; Liu, Z.; Wang, Z.; Lin, W. Anticancer effects of 10-hydroxycamptothecin induce apoptosis of human osteosarcoma through activating caspase-3, p53 and cytochrome c pathways. *Oncol. Lett.* **2017**, *15*, 2459–2464. [CrossRef] [PubMed]
47. Zhang, M.; Zheng, J.; Nussinov, R.; Ma, B. Release of cytochrome C from Bax pores at the mitochondrial membrane. *Sci. Rep.* **2017**, *7*, 2635. [CrossRef] [PubMed]
48. Akihisa, T.; Kikuchi, T.; Nagai, H.; Ishii, K.; Tabata, K.; Suzuki, T. 4-Hydroxyderricin from *Angelica keiskei* roots induces caspase-dependent apoptotic cell death in HL60 human leukemia cells. *J. Oleo Sci.* **2011**, *60*, 71–77. [CrossRef]
49. Jia, X.; Han, C.; Chen, J. Effect of tea on preneoplastic lesions and cell cycle regulators in rat liver. *Cancer Epidem. Biomar.* **2003**, *11*, 1663–1667. [CrossRef]
50. Wu, G.-S.; Lu, J.-J.; Guo, J.-J.; Li, Y.-B.; Tan, W.; Dang, Y.-Y.; Zhong, Z.-F.; Xu, Z.-T.; Chen, X.-P.; Wang, Y.-T. Ganoderic acid DM, a natural triterpenoid, induces DNA damage, G1 cell cycle arrest and apoptosis in human breast cancer cells. *Fitoterapia* **2012**, *83*, 408–414. [CrossRef]
51. Mari, A.; Mani, G.; Nagabhishek, S.N.; Balaraman, G.; Subramanian, N.; Mirza, F.B.; Sundaram, J.; Thiruvengadam, D. Carvacrol promotes cell cycle arrest and apoptosis through PI3K/AKT signaling pathway in MCF-7 breast cancer cells. *Chin. J. Integr. Med.* **2020**, 1–8. [CrossRef]
52. Arber, N.; Doki, Y.; Han, E.K.; Sgambato, A.; Zhou, P.; Kim, N.H.; Delohery, T.; Klein, M.G.; Holt, P.; Weinstein, I.B. Antisense to cyclin D1 inhibits the growth and tumorigenicity of human colon cancer cells. *Cancer Res.* **1997**, *57*, 1569–1574. [PubMed]
53. Zhou, P.; Jiang, W.; Zhang, Y.J.; Kahn, S.M.; Schieren, I.; Santella, R.M.; Weinsteir, I.B. Antisense to cyclin D1 inhibits growth and reverses the transformed phenotype of human esophageal cancer cells. *Oncogene* **1995**, *11*, 571–580. [CrossRef] [PubMed]
54. Park, W.; Park, S.; Song, G.; Lim, W. Inhibitory effects of osthole on human breast cancer cell progression via induction of cell cycle arrest, mitochondrial dysfunction, and ER stress. *Nutrients* **2019**, *11*, 2777. [CrossRef] [PubMed]

55. Hao, W.-C.; Zhong, Q.-L.; Pang, W.-Q.; Dian, M.-J.; Li, J.; Han, L.-X.; Zhao, W.-T.; Zhang, X.-L.; Xiao, S.-J.; Xiao, D.; et al. MST4 inhibits human hepatocellular carcinoma cell proliferation and induces cell cycle arrest via suppression of PI3K/AKT pathway. *J. Cancer* **2020**, *11*, 5106–5117. [CrossRef]
56. Xu, J.; Zhang, M.; Lin, X.; Wang, Y.; He, X. A steroidal saponin isolated from Allium chinense simultaneously induces apoptosis and autophagy by modulating the PI3K/Akt/mTOR signaling pathway in human gastric adenocarcinoma. *Steroids* **2020**, *161*, 108672. [CrossRef]
57. Wang, R.; Lu, X.; Yu, R. Lycopene inhibits epithelial–mesenchymal transition and promotes apoptosis in oral cancer via PI3K/AKT/m-TOR signal pathway. *Drug Des. Devel. Ther.* **2020**, *14*, 2461–2471. [CrossRef] [PubMed]
58. Zhang, J.; Feng, M.; Guan, W. Naturally occurring aesculetin coumarin exerts antiproliferative effects in gastric cancer cells mediated via apoptotic cell death, cell cycle arrest and targeting PI3K/AKT/M-TOR signaling pathway. *Acta Biochim. Pol.* **2021**, *68*, 109–113. [CrossRef]
59. Yu, J.-Z.; Ying, Y.; Liu, Y.; Sun, C.-B.; Dai, C.; Zhao, S.; Tian, S.-Z.; Peng, J.; Han, N.-P.; Yuan, J.-L.; et al. Antifibrotic action of Yifei Sanjie formula enhanced autophagy via PI3K-AKT-mTOR signaling pathway in mouse model of pulmonary fibrosis. *Biomed. Pharmacother.* **2019**, *118*, 109293. [CrossRef]
60. Khan, K.; Quispe, C.; Javed, Z.; Iqbal, M.J.; Sadia, H.; Raza, S.; Irshad, A.; Salehi, B.; Reiner, Z.; Sharifi-Rad, J. Resveratrol, curcumin, paclitaxel and miRNAs mediated regulation of PI3K/Akt/mTOR pathway: Go four better to treat bladder cancer. *Cancer Cell Int.* **2020**, *20*, 560. [CrossRef]

Article

The Chemical, Structural, and Biological Properties of Crude Polysaccharides from Sweet Tea (*Lithocarpus litseifolius* (Hance) Chun) Based on Different Extraction Technologies

Huan Guo [1,2], Meng-Xi Fu [2,3], Yun-Xuan Zhao [3], Hang Li [1,2], Hua-Bin Li [4], Ding-Tao Wu [3,5,*] and Ren-You Gan [1,2,5,*]

1. National Agricultural Science & Technology Center, Chengdu 610213, China; ghscny@163.com (H.G.); tiantsai@sina.com (H.L.)
2. Research Center for Plants and Human Health, Institute of Urban Agriculture, Chinese Academy of Agricultural Sciences, Chengdu 610213, China; mxfu_1996@163.com
3. Institute of Food Processing and Safety, College of Food Science, Sichuan Agricultural University, Ya'an 625014, China; zhaoyunxuan0320@163.com
4. Guangdong Provincial Key Laboratory of Food, Nutrition and Health, Department of Nutrition, School of Public Health, Sun Yat-Sen University, Guangzhou 510080, China; lihuabin@mail.sysu.edu.cn
5. Sichuan Engineering & Technology Research Center of Coarse Cereal Industralization, Key Laboratory of Coarse Cereal Processing (Ministry of Agriculture and Rural Affairs), School of Food and Biological Engineering, Chengdu University, Chengdu 610106, China
* Correspondence: wudingtao@cdu.edu.cn (D.-T.W.); ganrenyou@yahoo.com (R.-Y.G.); Tel./Fax: +86-0835-2883219 (D.-T.W.); +86-28-80203191 (R.-Y.G.)

Abstract: Eight extraction technologies were used to extract sweet tea (*Lithocarpus litseifolius* (Hance) Chun) crude polysaccharides (STPs), and their chemical, structural, and biological properties were studied and compared. Results revealed that the compositions, structures, and biological properties of STPs varied dependent on different extraction technologies. Protein-bound polysaccharides and some hemicellulose could be extracted from sweet tea with diluted alkali solution. STPs extracted by deep-eutectic solvents and diluted alkali solution exhibited the most favorable biological properties. Moreover, according to the heat map, total phenolic content was most strongly correlated with biological properties, indicating that the presence of phenolic compounds in STPs might be the main contributor to their biological properties. To the best of our knowledge, this study reports the chemical, structural, and biological properties of STPs, and the results contribute to understanding the relationship between the chemical composition and biological properties of STPs.

Keywords: herbal tea; extraction methods; polysaccharides; structural properties; biological properties

1. Introduction

Lithocarpus litseifolius (Hance) Chun (*Fagaceae* family), commonly known as "sweet tea", is an underutilized plant distributed mainly in the mountainous area of South China, which has rich natural resources [1]. Sweet tea has both medicinal and edible functions, and has been used as a daily beverage and traditional herbal medicine to prevent and manage certain chronic diseases, especially diabetes [2]. In 2017, sweet tea was listed as a new food material by the National Health and Family Planning Commission of China. It contains plentiful polysaccharides, flavonoids, and polyphenolic ingredients, which have extensive biological properties, such as anti-diabetic, anti-hypertensive, neuroprotective, hepatoprotective, and anti-aging effects [3–5]. As some of the main bioactive components in sweet tea, sweet tea polysaccharides (STPs), however, have not been intensively investigated in respect of their structural and biological properties.

Different extraction technologies and extraction solvents can significantly affect the extraction yields, chemical compositions, and structural and biological properties of plant polysaccharides [6,7]. Traditional and simple hot water extraction is widely used to extract polysaccharides from plant cells. However, the shortcomings of the hot water extraction method are also obvious; that is, it is time-consuming and inefficient. Therefore, some new extraction technologies have gradually replaced traditional hot water extraction. In recent years, other extraction technologies, including microwave-assisted extraction, ultrasound-microwave-assisted extraction, pressurized water extraction, high-speed shearing homogenization extraction, deep eutectic solvent-assisted extraction, deep eutectic solvent-microwave-assisted extraction, and alkali-assisted extraction, have been efficiently used in extracting polysaccharides from plants. Numerous studies have demonstrated that the chemical, structural, and biological properties of natural polysaccharides can be influenced by extraction technologies [7–9]. However, to the best of our knowledge, the chemical, structural, and biological properties of sweet tea polysaccharides have not been explored, and whether these properties are affected by different extraction methods remains unknown.

In this study, we report the chemical, structural, and biological properties of STPs, as well as the impacts of different extraction technologies on them. Results from this study can contribute to understanding the relationship between the chemical composition and biological properties of STPs.

2. Materials and Methods

2.1. Materials and Chemicals

Sweet tea, the tender leaves of *L. litseifolius* after fixation, were obtained from Sichuan Mu Jiang Ye Ke Tea Co., Ltd. (Chengdu, China). The samples were dried to constant weight by an air-dryer (DHG-9246A, Jinghong Experimental Equipment Co., Ltd., Shanghai, China) at 50 °C, ground into fine powder with a mill (Guanze Biological Technology Co., Ltd., Shanghai, China), and passed through an 80-mesh sieve. Finally, the sweet tea powder was stored at −20 °C for further analysis.

The chemicals, including vitamin C (Vc), sodium azide, butylated hydroxytoluene (BHT), 2,2-diphenyl-1-picrylhydrazyl (DPPH), 3-ethylbenzthiazoline-6-sulphonic acid (ABTS), aminoguanidine (AG), acarbose, and α-glucosidase (10 U/mg), were purchased from Sigma-Aldrich ((St. Louis, MO, USA). Thermostable α-amylase (40,000 U/g) and pancreatic lipase (4000 U/g) were purchased from Solarbio (Beijing, China). All reagents and chemicals were of analytical grade.

2.2. Extraction of Crude Polysaccharides from Sweet Tea

2.2.1. Preparation of Raw Materials

The sweet tea (2.0 g) was accurately weighed and sonicated twice with 80% ethanol (v/v, 20 mL) by an ultrasonic cleaner (800 W, PL-S80T, Kangshijie Biotechnology Co., Ltd., Dongguan, China) for 30 min at room temperature (25 ± 1 °C) with the power of 80% to remove most of the alcohol-soluble compounds. After centrifugation, the obtained residues were separately subjected to the following extraction processes.

2.2.2. Hot Water Extraction (HWE)

HWE was carried out using our previously reported method [7]. Briefly, the residues were extracted twice with ultrapure water (1:20, w/v) at 95 °C for 2 h. After centrifugation, the obtained supernatant was collected to first remove starch by thermostable α-amylase (5 U/mL). Then, the water-extracted crude polysaccharides (STP-W) from sweet tea were obtained by alcohol-precipitation, dialysis, and lyophilization.

2.2.3. Microwave-Assisted Extraction (MAE)

MAE was performed based on our reported method [8]. In brief, the residues (2.0 g) were mixed with deionized water (1:30, w/v). Thereafter, the further extraction was executed twice by a microwave oven (MKJ-J1-3, Qingdao Makewave Microwave Applied Technology Co., Ltd., Shandong, China) at 500 W for 6.5 min. Then, the following treatment procedures were consistent with Section 2.2.2. Finally, the microwave-extracted crude polysaccharides were obtained and termed STP-M.

2.2.4. Ultrasound-Microwave Assisted Extraction (UMAE)

UMAE was carried out using our reported method [8]. Briefly, the residues (2.0 g) were extracted once with deionized water (1: 42, w/v) by ultrasonic homogenizer (650 W, JY92-IIN, Ningbo Scientz Biotechnology Co., Ltd., Ningbo, China) for 21 min at the ultrasonic amplitude of 68% and at room temperature (25 ± 1 °C). Then, the extraction solution was further extracted once by MAE, which was described in Section 2.2.3. Finally, the ultrasound-microwave-extracted crude polysaccharides were obtained and termed STP-U.

2.2.5. Pressurized Water Extraction (PWE)

PWE was carried out using our reported method [8]. Briefly, the residues (2.0 g) were extracted twice with deionized water (1: 30, w/v) by the laboratory-scale high pressure reactor (LEC-300, Shanghai Laibei Scientific Instruments Co., Ltd., Shanghai, China) for 40 min at 1.6 MPa and 55 °C. Then, the next treatment procedures were consistent with Section 2.2.2. Finally, the pressurized water-extracted crude polysaccharides were obtained and termed STP-P.

2.2.6. Alkali-Assisted Extraction (AAE)

AAE was performed based on the method previously reported by Yao et al. [10]. Briefly, the residues were extracted twice with 60 mL of 0.3 M NaOH solution containing 0.3% (w/w) $NaBH_4$ at room temperature (25 ± 1 °C) for 3 h. The extracting solution was neutralized by 1 mol/L HCl. Finally, the alkali-extracted crude polysaccharides (STP-A) were obtained.

2.2.7. High-Speed Shearing Homogenization Extraction (HSHE)

HSHE was performed by the optimized method with minor adjustments [11]. Briefly, the residues were mixed in ultrapure water at 80 °C for 5 min, and were further extracted by a high-speed shearing homogenizer (AD500S-H, ANGNI Instruments Co., Ltd., Shanghai, China) at 7500 rpm for 10 min. Finally, the high-speed shearing-extracted crude polysaccharides (STP-H) were obtained.

2.2.8. Deep Eutectic Solvent-Assisted Extraction (DAE)

DAE was conducted using the optimized method with minor adjustments [12]. Briefly, DES was prepared by mixing ethylene glycol (EG) with choline chloride (ChCl) (3:1, molar ratio). The DES was maintained at 80 °C and stirred until a clear solution was obtained. The DESs used as solvents were prepared by mingling DES with distilled water (7:3, v/v). Then, the residues were extracted twice by HWE, which was described in Section 2.2.2. Finally, the DES-extracted crude polysaccharides (STP-DW) were obtained.

2.2.9. Deep Eutectic Solvent Microwave-Assisted Extraction (DMAE)

DMAE was carried out using the optimized method with minor adjustments [12]. The preparation of DES was the same as the method mentioned in Section 2.2.8. The residues were extracted twice by MAE. Finally, the DES-microwave-extracted crude polysaccharides (STP-DM) were obtained.

2.3. Determination of Chemical Compositions of STPs

The proteins, total polysaccharides, uronic acids, and total phenolics in STPs were measured by Bradford's method, the phenol-sulfuric acid method, the m-hydroxydiphenyl method, and the Folin–Ciocalteu method based on our previous studies, respectively [7].

2.4. Determination of Structural Properties of STPs

2.4.1. Determination of Molecular Weights (M_w), Compositional Monosaccharides, and Apparent Viscosities

The weight-average M_w and M_w/M_n (polydispersities) of STPs were estimated by high-performance size exclusion chromatography coupled with multi-angle laser light scattering and a refractive index detector (HPSEC-RID, Wyatt Technology Co., Santa Barbara, CA, USA). A Shodex OHpak SB-806M HQ column was used at 30 °C. In addition, constituent monosaccharides of STPs were also measured by high-performance liquid chromatography (HPLC, Agilent Technologies, Santa Clara, CA, USA) analysis based on the previously reported method [8]. The apparent viscosities of STPs were also performed based on our previous reports [8].

2.4.2. Fourier Transform Infrared (FT-IR) Spectroscopy and Nuclear Magnetic Resonance (NMR) Analysis

The FT-IR spectroscopy analysis of STPs was conducted based on our previous reports [13]. The nmR analysis of STPs was performed based on the method previously reported by Nie et al. [14]. Briefly, 0.5 mL of D_2O containing 20.0 mg of sample was stored overnight before the nmR analysis. Furthermore, the 1D nmR spectra (^1H and ^{13}C) were measured by a Bruker Ascend 600 MHz spectrometer (Bruker, Rheinstetten, Germany) equipped with a z-gradient probe with frequencies of 600.13 MHz for protons and 150.90 MHz for carbon.

2.5. Evaluation of Biological Properties of STPs

2.5.1. In Vitro Antioxidant Assays

The in vitro antioxidant assays used in this study included reducing powers, DPPH, and ABTS assays, all of which were described in our previous work in detail [8]. BHT (mg/mL) was used as the positive control for the DPPH assay, and Vc (mg/mL) was used as the positive control for the reducing power and ABTS assays.

2.5.2. In Vitro α-Glucosidase Inhibitory Assay

The α-glucosidase inhibitory assays of STPs were analyzed at five concentrations according to the previously reported method [14]. Acarbose was used as the positive control.

2.5.3. In Vitro Antiglycation Assay

Inhibition of the formation of advanced glycation end-products (AGEs) by STPs was determined by the BSA-glucose model (BSA-Glc) as in the formerly reported method [15]. The quantification of AGEs was conducted by using fluorescence at wavelengths of 370 nm for excitation and 440 nm for emission. AG was used as the positive control.

2.6. Statistical Analysis

All the assays were conducted in triplicate, and data are presented as means ± standard deviation. Origin 9.0 software (OriginLab, Northampton, MA, USA) was applied for the statistical analysis, which was conducted by one-way analysis of variance (ANOVA) plus post hoc Duncan's test, with $p < 0.05$ defined as statistical significance. The heat map was drawn via Origin 9.0, and the correlation coefficient (r) was calculated.

3. Results and Discussion

3.1. Yields and Chemical Compositions of STPs

As shown in Table 1, STP-P had the highest extraction yield (3.98 ± 0.22%), and STP-W, STP-A, STP-H, and STP-DW had similar extraction yields, lower than STP-P. In addition, the STP-DM, STP-M, and STP-U had lower extraction yields than other STPs. These results suggest that the extraction methods and solvents had great effects on the extraction yields of STPs, in agreement with the result of a previous study [9]. In this study, the STP-P obtained by the PWE method showed the highest extraction yields, which might be due to the high pressure that increased the solubility of polysaccharides and subsequently led to infiltrating the solvent into the sample by reducing the surface tension and viscosity, thereby increasing the yields of polysaccharides [16].

Table 1. Chemical compositions of STPs extracted by different methods.

Samples	Extraction Yields (%)	Total Polysaccharides (%)	Protein Contents (%)	Degree of Esterification (%)	Total Uronic Acids (%)	TPC (mg GAE/g)
STP-W	3.65 ± 0.12 [b]	81.82 ± 0.93 [bc]	3.79 ± 0.18 [d]	27.97 ± 1.22 [e]	40.18 ± 1.09 [b]	17.44 ± 0.92 [e]
STP-M	2.27 ± 0.08 [d]	81.48 ± 1.36 [bc]	5.48 ± 0.24 [c]	34.08 ± 1.73 [c]	24.11 ± 0.78 [f]	40.22 ± 1.25 [c]
STP-U	1.96 ± 0.15 [e]	86.25 ± 2.12 [a]	6.43 ± 0.22 [b]	31.04 ± 1.38 [d]	34.90 ± 1.28 [d]	41.12 ± 2.09 [c]
STP-P	3.98 ± 0.22 [a]	80.93 ± 1.52 [c]	5.26 ± 0.19 [c]	29.09 ± 1.33 [de]	44.70 ± 0.84 [a]	28.33 ± 0.98 [d]
STP-H	3.51 ± 0.18 [b]	76.64 ± 0.77 [d]	5.83 ± 0.28 [c]	41.36 ± 1.65 [b]	29.32 ± 1.03 [e]	26.63 ± 1.17 [d]
STP-DW	3.44 ± 0.16 [b]	83.82 ± 1.89 [ab]	3.16 ± 0.16 [e]	46.43 ± 1.79 [a]	36.93 ± 1.67 [c]	47.52 ± 1.62 [b]
STP-DM	2.74 ± 0.21 [c]	81.29 ± 1.20 [bc]	5.48 ± 0.37 [c]	42.52 ± 1.88 [b]	36.66 ± 1.35 [c]	60.10 ± 1.70 [a]
STP-A	3.54 ± 0.16 [b]	77.56 ± 0.98 [d]	9.86 ± 0.61 [a]	-	9.42 ± 0.63 [g]	60.50 ± 1.36 [a]

STP-W, STP-M, STP-U, STP-P, STP-H, STP-DW, STP-DM, and STP-A indicate sweet tea polysaccharides extracted by hot water extraction, microwave-assisted extraction, ultrasound-microwave-assisted extraction, pressurized water extraction, high-speed shearing homogenization extraction, deep eutectic solvent-assisted extraction, deep eutectic solvent-microwave-assisted extraction and alkali-assisted extraction, respectively. Values represent mean ± standard deviation, and statistical analysis was carried out by ANOVA plus post hoc Duncan's test; statistical significance ($p < 0.05$) is indicated with different lowercase letters (a–g).

Furthermore, different extraction technologies can also affect the contents of total polysaccharides and proteins in samples. The contents of total polysaccharides in STPs ranged from 76.64 to 86.25%, suggesting that polysaccharides were the main ingredient in each sample. Besides, the contents of proteins in STPs ranged from 3.16 to 9.86%, and STP-A had a significantly ($p < 0.05$) higher protein content (9.86 ± 0.61%) compared to other STPs, which might be associated with the reason that alkaline conditions could destroy the hydrogen bridge and release the proteins into the alkaline solution [17–19]. Notably, the highest content of proteins and generally the lowest content of total polysaccharides in STP-A indicated that STP-A extracted by AAE comprised protein-bound polysaccharides, which was similar to previous research results [20,21].

The uronic acid contents of STPs ranged from 9.42 to 44.90%. STP-P (44.70 ± 0.84%) and STP-W (40.18 ± 1.09%) had the highest uronic acid contents, followed by STP-DW (36.93 ± 1.67%) and STP-DM (36.66 ± 1.35%), while STP-A (9.42 ± 0.63%) had the lowest uronic acid content, which might be due to the destruction of galacturonic acid and glucuronic acid by alkali, in accordance with previous studies [16,20,22]. Moreover, STP-DW (46.43 ± 1.79%) and STP-DM (42.52 ± 1.88%) had the highest degree of esterification, while STP-A had no degree of esterification. Although 80% ethanol was used to remove most of the small molecules, some phenolic compounds were still detectable and their contents were measured by the Folin–Ciocalteu method. Total phenolic content (TPC) in STPs ranged from 17.44 to 60.50 mg GAE/g. Compared with other STPs, relatively high TPC was found in STP-A (60.50 ± 1.36 mg GAE/g) and STP-DM (60.10 ± 1.70 mg GAE/g), with no statistical difference. Previous studies reported that deep-eutectic solvents could improve the solubility of plant bioactive ingredients and obtain different polysaccharides and flavonoids [23–25]. In addition, the alkaline solution could also increase the dissolution of phenolic compounds and flavonoids, which was similar to the result of the previous research [19].

3.2. Structural Properties of STPs

3.2.1. Molecular Weights of STPs

As shown in Figure 1, two fractions (fractions 1 and 2) were detected, and fraction 2 was determined as the major polysaccharide fraction in STPs. The weight-average M_w and M_w/M_n of two polysaccharide fractions in STPs are also presented in Table 1. The M_w of fractions 1 and 2 varied from 1.66×10^6 to 7.51×10^6 Da and 0.89×10^5 to 7.20×10^5 Da, respectively. The order of the M_w of fraction 1 in STPs was as follows: STP-H > STP-U> STP-A > STP-M > STP-W > STP-P > STP-DM > STP-DW. Different extraction technologies showed remarkable differences in molecular weights of STPs. The polysaccharide fraction 1 of STP-H (7.51 × 10^6 Da) had the highest M_w, followed by STP-U (5.08 × 10^6 Da) and STP-A (4.98 × 10^6 Da), while STP-DM (2.30 × 10^6 Da) and STP-DW (1.66 × 10^6 Da) had the lowest M_w. This phenomenon was similar to other studies, in that the molecular weight of polysaccharides extracted by HSHE was higher than that extracted by traditional hot water extraction [11]. The low M_w of fraction 1 in STP-DM and STP-DW might be related to the breakdown of glycoside bonds by DESs. In addition, the polydispersities of fractions 1 and 2 in STPs ranged from 1.20 to 1.94 and from 1.22 to 1.82, respectively, consistent with their HPSEC chromatograms.

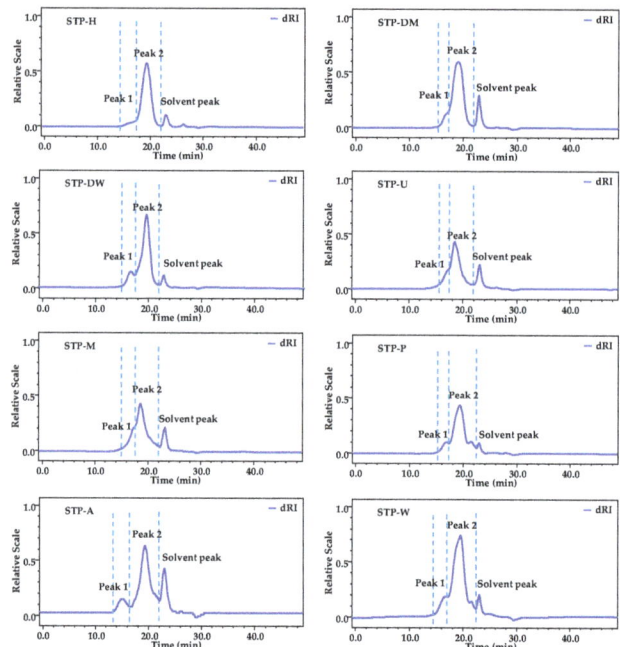

Figure 1. HPSEC chromatograms of STPs. STP-W, STP-M, STP-U, STP-P, STP-H, STP-DW, STP-DM, and STP-A indicate sweet tea polysaccharides extracted by hot water extraction, microwave-assisted extraction, ultrasound-microwave-assisted extraction, pressurized water extraction, high-speed shearing homogenization extraction, deep eutectic solvent-assisted extraction, deep eutectic solvent-microwave-assisted extraction, and alkali-assisted extraction, respectively.

3.2.2. Compositional Monosaccharides and Apparent Viscosities of STPs

The compositional monosaccharides and apparent viscosities of STPs were further investigated. Figure 2A shows the HPLC-UV profiles of STPs extracted by eight extraction technologies. In addition, the molar ratios of compositional monosaccharides are summarized in Table 2. The compositional monosaccharides of STP-A were Man, Rha, GlcA, GalA, Glc, Gal, Xyl, and Ara. Compared with STP-A, the Xyl in other STPs was almost not detected. In addition, except for Xyl, the ratio of Man in STP-A was also significantly higher than that in other STPs. The reason for this phenomenon was that hemicellulose in the cell wall could be extracted in alkaline solutions, which was similar to the result of the previous research [26]. The extraction in alkaline conditions causes cell wall swelling and hydrogen bond disruption between hemicellulose and cellulose. It can also disrupt ether bonds among hemicellulose and lignin, causing the release of hemicellulose. Moreover, the ratio of GalA in STP-A was significantly lower than that in other STPs, which was in accordance with the content of uronic acids. Furthermore, the ratio of Glc in STP-H was significantly increased, which might be due to the cellulose in the cell wall extracted by HSHE. With the combination of fierce collision, pressure differential relief, and high-intensity shearing force, the cell wall was destroyed, and polysaccharides and cellulose were extracted by HSHE [11]. These findings demonstrate that different extraction technologies, especially the alkali-assisted extraction method, had significant effects on the types and molar ratios of compositional monosaccharides in STPs. Sun et al. [26] and Yan et al. [27] also reported that extraction technologies can affect the compositional monosaccharides of polysaccharides.

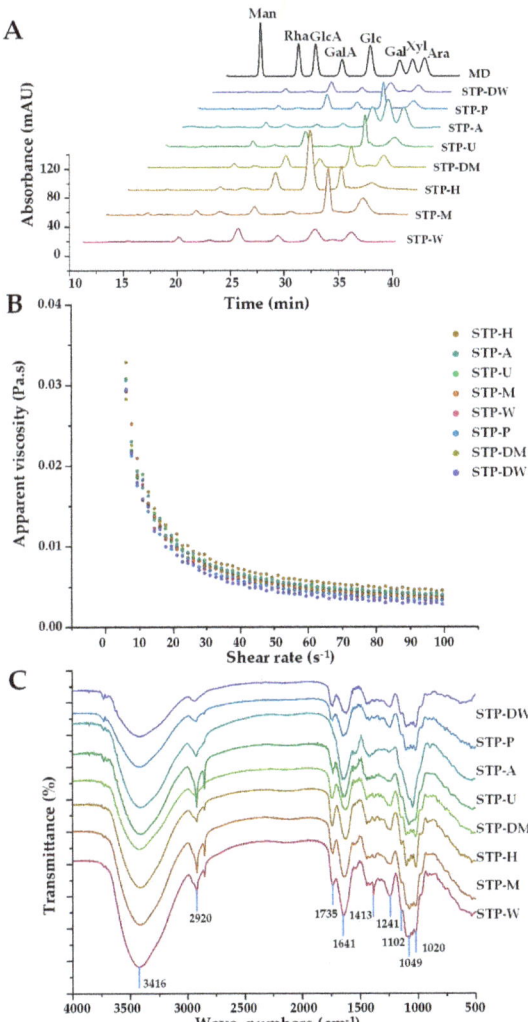

Figure 2. HPLC profiles (**A**), apparent viscosities (**B**), and FT-IR spectra (**C**) of STPs. STP-W, STP-M, STP-U, STP-P, STP-H, STP-DW, STP-DM, and STP-A indicate sweet tea polysaccharides extracted by hot water extraction, microwave-assisted extraction, ultrasound-microwave-assisted extraction, pressurized water extraction, high-speed shearing homogenization extraction, deep eutectic solvent-assisted extraction, deep eutectic solvent-microwave-assisted extraction, and alkali-assisted extraction, respectively. MD, mixed standard of monosaccharides; Man, mannose; Rha, rhamnose; GlcA, glucuronic acid; GalA, galacturonic acid; Glc, glucose; Gal, galactose; Xyl, xylose; Ara, arabinose.

Table 2. Molecular weight (M_w), polydispersity (M_w/M_n), and molar ratios of compositional monosaccharides of STP.

	STP-W	STP-M	STP-U	STP-P	STP-H	STP-DW	STP-DM	STP-A
M_w (Da)								
Fraction 1 ($\times 10^6$)	3.20 (±0.25%) [e]	3.57 (±0.29%) [d]	5.08 (±0.38%) [b]	2.92 (±0.20%) [f]	7.51 (±0.21%) [a]	1.66 (±0.28%) [h]	2.30 (±0.22%) [g]	4.98 (±0.26%) [c]
Fraction 2 ($\times 10^5$)	1.39 (±0.86%) [d]	7.20 (±0.36%) [a]	6.79 (±0.42%) [b]	1.16 (±0.61%) [f]	1.68 (±0.51%) [c]	0.89 (±1.25%) [g]	1.24 (±0.86%) [e]	1.23 (±2.82%) [e]
M_w/M_n								
Fraction 1	1.43 (±0.38%)	1.47 (±0.44%)	1.58 (±0.59%)	1.94 (±0.31%)	1.65 (±0.32%)	1.53 (±0.43%)	1.44 (±0.32%)	1.20 (±0.43%)
Fraction 2	1.50 (±1.43%)	1.12 (±0.48%)	1.13 (±0.60%)	1.41 (±0.93%)	1.51 (±0.80%)	1.82 (±2.13%)	1.56 (±1.56%)	1.66 (±5.07%)
Monosaccharides and Molar Ratios								
Mannose	0.13	0.06	0.03	0.03	0.05	0.09	0.03	0.71
Rhamnose	0.52	0.20	0.45	0.38	0.24	0.25	0.33	0.24
Glucuronic acid	0.17	0.16	0.15	0.15	0.21	0.25	0.17	0.20
Galacturonic acid	2.33	0.57	1.85	2.34	1.88	0.71	1.54	0.22
Glucose	0.54	0.17	0.13	0.62	3.49	0.12	0.62	0.16
Galactose	1.49	1.20	1.24	1.41	0.85	1.51	1.23	1.15
Xylose	0.08	0.00	0.00	0.09	0.00	0.00	0.18	1.54
Arabinose	1.00	1.00	1.00	1.00	1.00	1.00	1.00	1.00

STP-W, STP-M, STP-U, STP-P, STP-H, STP-DW, STP-DM, and STP-A indicate sweet tea polysaccharides extracted by hot water extraction, microwave-assisted extraction, ultrasound-microwave-assisted extraction, pressurized water extraction, high-speed shearing homogenization extraction, deep eutectic solvent-assisted extraction, deep eutectic solvent-microwave-assisted extraction, and alkali-assisted extraction, respectively. Values represent mean ± standard deviation, and statistical analysis was carried out by ANOVA plus post hoc Ducan's test; statistical significance ($p < 0.05$) is indicated with different lowercase letters (a–h).

Moreover, Figure 2B shows the effects of shear rate on the apparent viscosities of STPs (10 mg/mL) extracted by different extraction technologies. The order of the apparent viscosities of STPs was as follows: STP-H > STP-A > STP-U > STP-M > STP-W > STP-P > STP-DM > STP-DW. Results show that STP-H had a significantly higher viscosity compared to other STPs, similar to the results for pectin from pomelo [11]. Furthermore, the results illustrate that the apparent viscosities of STPs had a close association with their M_w, and different extraction technologies could change the apparent viscosities of STPs.

3.2.3. FT-IR Spectra of STPs

The FT-IR spectra (Figure 2C) were applied to compare the structural properties of STPs. Results show that the absorption peaks of eight STPs exhibited differences. Briefly, 2920 and 3416 cm^{-1} were the broad peaks caused by the stretching vibration of the hydroxyl group and the C-H asymmetric stretching vibration, respectively. In the FT-IR spectra of STP-W, STP-M, STP-U, STP-P, STP-H, STP-DW, and STP-DM, the signal at 1735 cm^{-1} was attributed to the stretching vibration of the esterified carboxylic groups (-COOR). As shown in Table 2, with the exception of STP-A, which had no degree of esterification, the DE of other STPs ranged from 27.97 to 46.43%, which was similar to the FT-IR spectra of STPs. Furthermore, the strong absorption peaks at approximately 1049, 1102, and 1167 cm^{-1} were assigned to the stretching vibrations of the C-O-C glycosidic band and the C-O-H side group vibration of a pyranose ring, suggesting that the eight extracted STPs contained pyranose sugar [28].

3.2.4. NMR Analysis

NMR analysis was applied to compare the structural properties of STPs. ^{1}H and ^{13}C analyses (1D nmR spectra) are shown in Figures 3 and 4, respectively. The signals at 5.25 and 1.25 ppm were tentatively deduced to be the H-1 and H-6 of 1,2-α-L-Rha, respectively [29]. The peaks at 5.09 and 4.13 ppm were tentatively deduced to be the H-1 and H-2 of 1,5-α-L-Ara, respectively [29]. The weak signal at 2.16 ppm was tentatively deduced to be the existence of acetyl groups [29], and the strong peak at 3.81 ppm was tentatively deduced to be the signal of methyl esters connecting to carboxyl groups of D-GalA [29]. In addition, the peak at 4.46 ppm was tentatively deduced to be the H-1 of 1,6-β-D-Gal, and the peaks at 4.23 and 3.72 ppm were tentatively deduced to be the H-4 and H-5 of 1,4-β-D-Gal [29,30], respectively. The peak at 3.95 ppm was tentatively deduced to be the H-3 of 1,4-α-D-Glc [29]. The peak at 4.18 ppm was tentatively deduced to be the H-3 of 1,4-α-D-GalA, and the peak at 4.01 ppm was tentatively deduced to be the H-4 of 1,4-β-D-GlcA [29,31]. It is worth noting that the difference between STP-A and other STPs is obvious. Compared to other STPs, the signal of methyl esters connecting to carboxyl groups of D-GalA at 3.81 ppm was not detected in STP-A. However, some new peaks were detected in STP-A between 3.19 and 3.78 ppm. The peak at 3.78 ppm was tentatively deduced to be the H-3 of 1,4-β-D-Man [32]. The signals at 3.55, 3.38, 3.29, and 3.19 ppm were tentatively deduced to be the H-4, H-3, H-5, and H-2 of 1,4-β-D-Xyl, respectively [33]. As shown in Table 2, each sample had at least six different compositional monosaccharides. However, only three to four anomeric peaks were present in the ^{1}H-NMR spectra. The reason for this phenomenon may be that the ^{1}H-NMR spectra (Figure 3) show peaks for very broad deuterated water (D_2O) located from 4.6 to 5.0 ppm, so some peaks corresponding to the anomeric region might be covered by D_2O. Previous studies have found that different extraction solvents including acid, hot water, and alkali extractions only have different effects on the degradation of polysaccharides without changing the major glycosidic linkages (the backbone) [20]. The reason for this phenomenon was that hemicellulose in the cell wall could be extracted by alkali-assisted extraction methods, which was consistent with the result for the compositional monosaccharides.

Likewise, some differences were observed between the ^{13}C nmR spectra of STP-A and other STPs. The ^{13}C nmR spectra of STP-A showed that the methyl of acetyl groups was at 19.43 ppm [14], and the signal at 83.87 ppm was tentatively deduced to be the C-4 of 1,5-α-L-Ara [34]. The signal at 64.14 ppm was tentatively deduced to be the C-6 of 1,4-β-D-Man [29]. The signals at 72.14 and 75.60 ppm were tentatively deduced to be the C-2 and C-4 of 1,4-β-D-Xyl [33], respectively. The signals at 65.90 and 99.41 ppm were tentatively deduced to be the C-6 of 1,2-α-D-Ara and C-1 of 1,2-α-L-Rha, respectively [35,36]. The signal at 68.11 ppm was tentatively deduced to be the C-2 of 1,4-α-D-GalA, while the signal at 76.56 ppm was tentatively deduced to be the C-3 of 1,4-α-D-Glc [31,37]. The C-1 signal for 1,6-β-D-Gal was at 104.51 ppm [38]. The C-3 signal for 1,4-β-D-GlcA was at 79.28 ppm [31]. The C-1 signals for 1,3-α-L-Ara and 1,5-α-L-Ara were at 112.34 and 110.34 ppm, and the signal at 86.93 ppm was tentatively deduced to be the C-2 of 1,5-α-L-Ara [39]. Similar to STP-A, the signals of other STPs were also detected at 64.25, 76.07, 79.40, 83.74, 86.77, 110.18, and 112.14 ppm. However, some new signals were detected in other STPs. For other STPs, the peak at 173.52 ppm was tentatively deduced to be the C-6 of un-esterified carbonyl groups of D-GalA [29]. The signal at 103.19 ppm was tentatively deduced to be the C-1 of 1,2-α-L-Rha [40]. The signals at 81.85 and 74.60 ppm were tentatively deduced to be the C-4 and C-3 of 1,4-α-D-GalA [41,42], respectively. The C-2 and C-4 signals for 1,4-β-D-GalA were detected at 70.00 and 77.34 ppm [35], respectively. The signals at 73.44 and 107.25 ppm were tentatively deduced to be the C-2 and C-1 of 1,4-β-D-Gal, respectively [36,38]. Furthermore, the signal at 55.85 ppm was the response to the presence of a methyl group esterified carboxyl group of GalA [39]. The peaks between 52 and 57 ppm were tentatively deduced to be the amino-substituted carbon signals of an amino sugar residue [43], suggesting the presence of proteins in the sample. Finally, results from the FT-IR spectra, the compositional monosaccharides, and the nmR spectra suggested that rhamnogalacturonan I (RG I), arabinogalactan (AG), and hemicellulose might exist in STP-A, and homogalacturonan (HG), AG, and RG I might exist in other STPs. However, 1D nmR spectra can only analyze the preliminary structure of tea polysaccharides, and the precise structures of STPs need further clarification (e.g., using methylation and 2D nmR) in the future. In short, the structure of polysaccharides extracted by alkali-assisted extraction methods from sweet tea was significantly different, suggesting that different extraction technologies can obtain different polysaccharides even in the same material.

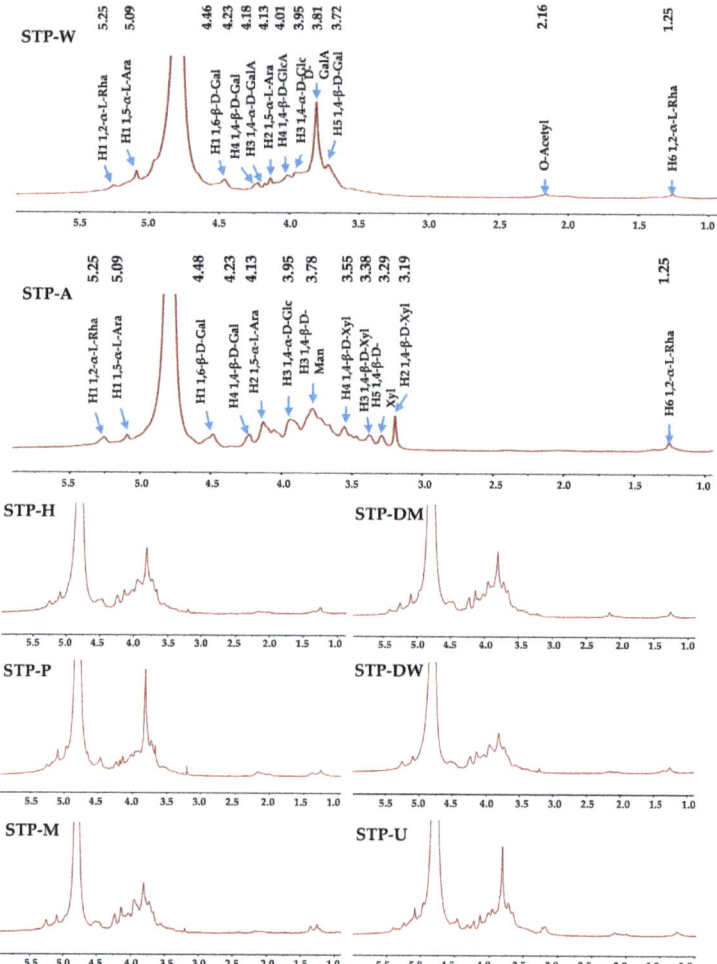

Figure 3. ^{1}H nmR spectra of STPs. STP-W, STP-M, STP-U, STP-P, STP-H, STP-DW, STP-DM, and STP-A indicate sweet tea polysaccharides extracted by hot water extraction, microwave-assisted extraction, ultrasound-microwave-assisted extraction, pressurized water extraction, high-speed shearing homogenization extraction, deep eutectic solvent-assisted extraction, deep eutectic solvent-microwave-assisted extraction, and alkali-assisted extraction, respectively.

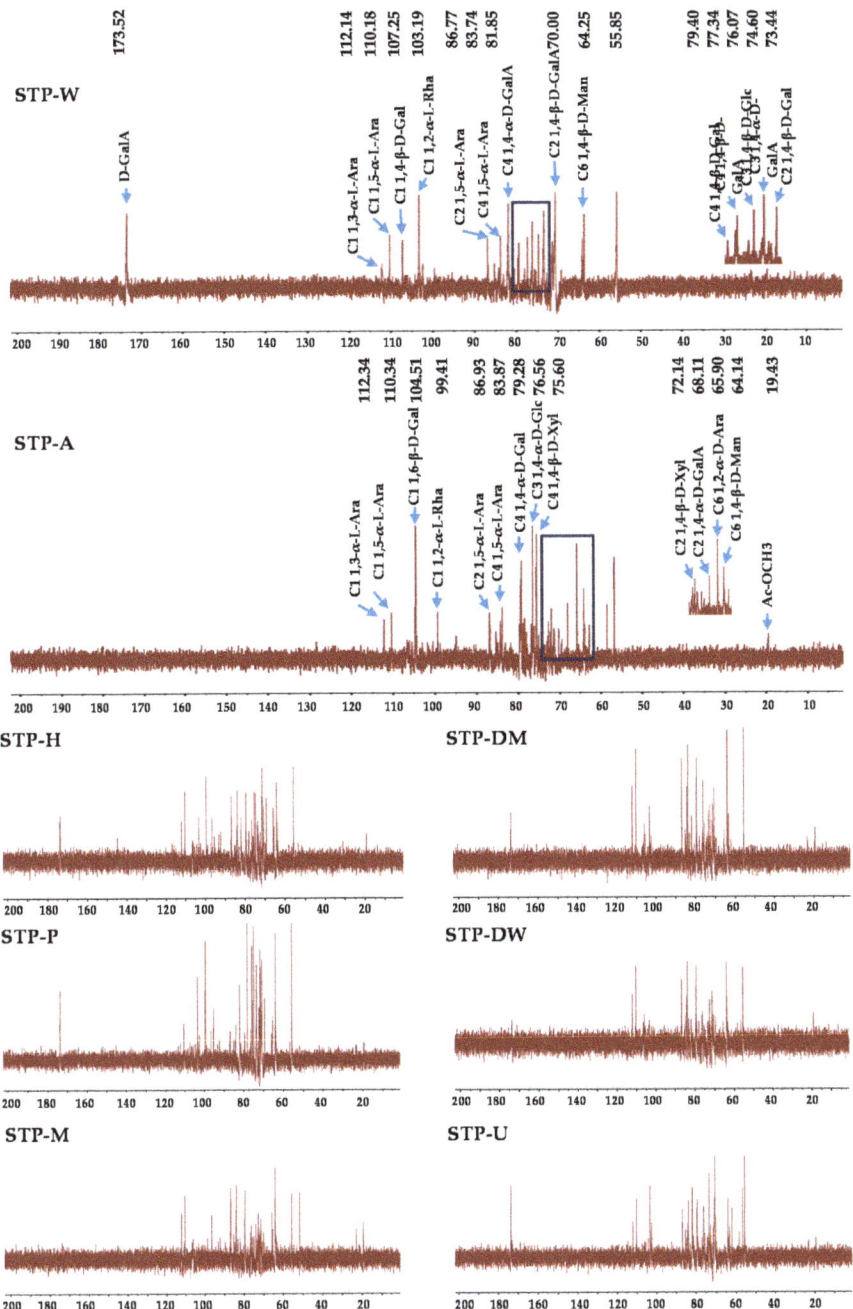

Figure 4. ^{13}C nmR spectra of STPs. STP-W, STP-M, STP-U, STP-P, STP-H, STP-DW, STP-DM, and STP-A indicate sweet tea polysaccharides extracted by hot water extraction, microwave-assisted extraction, ultrasound-microwave-assisted extraction, pressurized water extraction, high-speed shearing homogenization extraction, deep eutectic solvent-assisted extraction, deep eutectic solvent-microwave-assisted extraction, and alkali-assisted extraction, respectively.

3.3. Biological Properties of STPs

3.3.1. In Vitro Antioxidant Activities

Excessive reactive oxygen species (ROS) can lead to oxidative stress [44]. Elevated ROS levels can lead to the production of free radicals, which may have harmful effects on nucleic acids, proteins, and lipids [45]. Meanwhile, free-radical-induced oxidative stress is one of the important factors leading to various diseases, such as cancer, neurodegenerative disorders, and inflammatory diseases. Previous studies found that polysaccharides could protect from ROS-induced oxidative damage by scavenging free radicals [46]. Therefore, the antioxidant activities of different STPs were further studied. As shown in Figure 5A–C, the IC_{50} values of DPPH and ABTS radical scavenging activities of STPs ranged from 0.41 to 5.03 mg/mL and 0.60 to 3.26 mg/mL, respectively. The reducing powers of STPs (5.0 mg/mL) ranged from 60.97 to 120.77 μg Trolox/mg. STP-DW and STP-DM possessed the highest antioxidant activities among all STPs, while STP-W and STP-H showed the lowest antioxidant activities.

The antioxidant activities of polysaccharides depend on a variety of structural properties, such as compositional monosaccharides, chemical compositions, M_w, functional groups, and types of glycosidic linkages [9,21]. Compared with high-molecular-weight polysaccharides, STP-DM and STP-DW with relatively low molecular weights exhibited higher antioxidant activities, which might be due to their non-compact structure, exposing more free hydroxyl groups to react with free radicals [22]. In order to evaluate the correlation between chemical composition and biological properties of STPs, a heat map analysis was performed. As shown in Figure 5F, there was a positive correlation of the TPC with the DPPH ($r = 0.843$) and ABTS ($r = 0.763$) radical scavenging activities, suggesting that the presence of phenolic compounds in the STPs might be the main contributor to their antioxidant activities. Previous studies reported that the presence of some reducing compounds may cause the overestimation of TPC values due to the fact that they interfere with the determination of the Folin–Ciocalteu assay, such as protein [47]. According to the heat map, the correlation between protein content and antioxidant activities was not obvious ($r < 0.000$), indicating that proteins in the STPs might not significantly contribute to their antioxidant activities. As a result, STP-DM, STP-DW, and STP-A with the highest antioxidant activities might be closely related to the presence of phenolic compounds in the STPs. Indeed, many phenolic compounds, especially phenolic acids, exhibit good radical scavenging activities [48], which is similar to the result of the previous research [15]. Furthermore, the key factors for the antioxidant capacity of phenolic compounds might be related to the H-atom transfer, the metal chelation, and the electron transfer [49]. In the process of food grinding and processing, phenolic compounds can be spontaneously and quickly combined with cell wall polysaccharides in covalent and non-covalent manners when cell walls break down [50]. In the extraction processing, the combination of phenolic compounds and polysaccharides might be an important reason for the relatively high TPC in the STP-DM, STP-DW, and STP-A. In addition, Siu et al. [51] reported that phenolic and protein components instead of carbohydrates were mainly responsible for the antioxidant activities of mushroom polysaccharides. Consistent with that study, the antioxidant activities of STPs might be mainly due to STP-bound phenolic compounds. Overall, the antioxidant activities of STP-DM, STP-DW, and STP-A were significantly ($p < 0.05$) higher than those of other STPs, suggesting that DMAE, DAE, and AAE may be potential technologies for the extraction of STPs with relatively high antioxidant capacity.

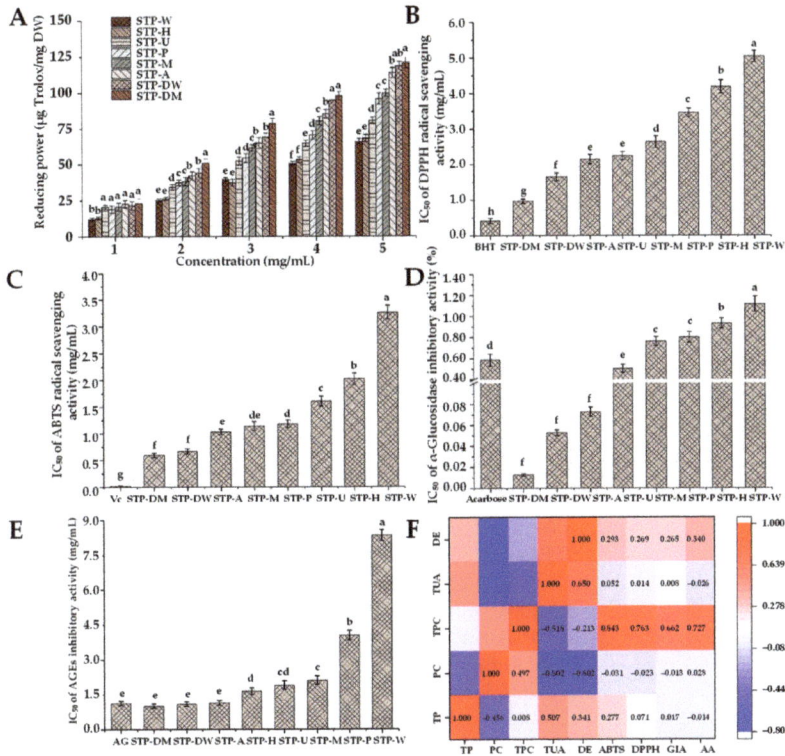

Figure 5. Reducing power (**A**), DPPH radical scavenging activity (**B**), ABTS radical scavenging activity (**C**), α-glucosidase inhibitory activity (**D**), and in vitro antiglycation activity (**E**) of STPs, and heat map analysis of the correlation between chemical composition and biological properties (**F**). STP-W, STP-M, STP-U, STP-P, STP-H, STP-DW, STP-DM, and STP-A indicate sweet tea polysaccharides extracted by hot water extraction, microwave-assisted extraction, ultrasound-microwave-assisted extraction, pressurized water extraction, high-speed shearing homogenization extraction, deep eutectic solvent-assisted extraction, deep eutectic solvent-microwave-assisted extraction, and alkali-assisted extraction, respectively. BHT, butylated hydroxytoluene; Vc, vitamin C; AG, aminoguanidine; TP, total polysaccharides; PC, protein contents; TPC, total phenolic content; TUA, total uronic acids; DE, degree of esterification; GIA, α-glucosidase inhibitory activity; AA, antiglycation activity. The error bars indicate standard deviation, and statistical analysis was carried out by ANOVA plus post hoc Duncan's test; statistical significance ($p < 0.05$) is indicated with different lowercase letters (a–h).

3.3.2. In Vitro α-Glucosidase Inhibitory Activity

Previous studies found that natural polysaccharides affect blood glucose levels by improving insulin resistance, promoting insulin secretion, and regulating the activity of related enzymes. Several findings have demonstrated that the degradation of carbohydrates can be effectively attenuated by α-glucosidase inhibitors [18]. Sweet tea has been widely used as a traditional Chinese herbal medicine to relieve hyperglycemia in the people of China. Previous studies indicated that the flavonoid extracts from sweet tea had significant in vitro hypoglycemic activity, and could significantly inhibit the activity of α-glycosidase [52,53]. However, there have been no reports about the α-glucosidase inhibitory activity of sweet tea polysaccharides. This study demonstrated that STPs had α-glucosidase inhibitory activity and different extraction technologies could differently influence the α-glucosidase inhibitory activity of STPs. As shown in Figure 5D, the IC_{50} of α-glucosidase inhibition by STPs was measured from 0.013 to 1.114 mg/mL. The order of the α-glucosidase inhibitory activity was as follows: STP-DM > STP-DW > STP-A > STP-U > STP-M > STP-P > STP-H > STP-W. Compared with the positive control acarbose (IC_{50} = 0.585 mg/mL), STP-DM, STP-DW, STP-A, and STP-U exhibited better inhibitory effects on α-glucosidase, while STP-M, STP-P, STP-H, and STP-W had relatively poor effects. The strong α-glucosidase inhibitory effects of STP-DM, STP-DW, STP-A, and STP-U might be related to their lower viscosity and molecular weight, and higher contents of uronic acids and TPC. Previous research reported that the more electric charges the polysaccharides have, the more easily they can form macromolecular complexes with α-glucosidase, leading to the blocking of enzyme activity [18]. Chen et al. [18] and Jia et al. [54] reported that polysaccharides with lower M_w had higher α-glucosidase inhibitory activities due to increased exposure to the active sites of the enzyme. In addition, Yuan et al. [55] reported that the α-glucosidase inhibitory effects might be related to high TPC, consistent with the results shown in Figure 5F.

3.3.3. In Vitro Antiglycation Activity

Reducing sugar could combine with the free amino groups in fats and proteins, leading to the formation of AGEs [56]. Elevated AGEs can lead to cell damage, and further promote the occurrence of various diseases, including cataracts, cancer, aging, neurodegenerative diseases, and cardiovascular diseases [57]. Therefore, the antiglycation activity of STPs was further investigated and compared. As shown in Figure 5E, the antiglycation activities of STPs as represented by IC_{50} were 1.01–8.33 mg/mL. Compared with the positive inhibitor AG (IC_{50} = 1.13 mg/mL), STP-DM, STP-DW, and STP-A exhibited similar inhibitory effects on AGEs. Furthermore, the antiglycation activities of STP-DM, STP-DW, and STP-A were significantly ($p < 0.05$) higher than those of other STPs, and STP-W also exhibited the lowest inhibitory effects. It was suggested that different extraction technologies can affect the antiglycation activity of STPs. Results also suggest that the DMAE, DAE, and AAE could be excellent technologies to obtain polysaccharides with relatively high antiglycation activity. As shown in Figure 5F, a positive correlation (r = 0.727) was found between the antiglycation activities of STPs and their TPC, suggesting that the presence of phenolic compounds in the STPs might also mainly contribute to their antiglycation activities.

4. Conclusions

In conclusion, this study systematically explored and compared the effects of eight extraction methods, including HWE, MAE, UMAE, PWE, HSHE, DAE, DMAE, and AAE, on the extraction yields, chemical compositions, structure properties, and biological properties of STPs. The results show that the pressurized hot water extraction method had the highest extraction yield among the selected methods. The chemical compositions, molecular weights, monosaccharide compositions, apparent viscosities, and biological properties of STPs were influenced by different extraction methods. Moreover, according to the heat map, TPC was most strongly correlated with the biological properties of STPs. It was speculated that the combination of phenolic compounds and polysaccharides during the extraction processing might be an important reason for the biological properties of STPs. Overall, this study for the first time interpreted polysaccharides extracted from sweet tea, and suitable extraction methods that can be applied to obtain STPs with high yields and biological properties, such as hypoglycemic and antiglycation activities, and which can be developed into functional foods to prevent and manage certain chronic diseases, such as diabetes. Results from this study can contribute to understanding the relationship between the chemical composition and biological properties of natural polysaccharides.

Author Contributions: Conceptualization, R.-Y.G. and D.-T.W.; data curation, R.-Y.G., D.-T.W., and H.G.; formal analysis, H.G., M.-X.F., Y.-X.Z., and H.L.; funding acquisition, R.-Y.G.; investigation, H.G., M.-X.F., Y.-X.Z., and H.L.; methodology, D.-T.W.; project administration, R.-Y.G.; resources, H.G. and M.-X.F.; software, H.G.; supervision, R.-Y.G. and D.-T.W.; validation, R.-Y.G., D.-T.W., and H.G.; writing—original draft, H.G.; writing—review and editing, R.-Y.G., D.-T.W., and H.-B.L. All authors have read and agreed to the published version of the manuscript.

Funding: This research was funded by the Local Financial Funds of the National Agricultural Science and Technology Center, Chengdu (No. NASC2020KR02), the Central Public-interest Scientific Institution Basal Research Fund (No. Y2020XK05), and the National Natural Science Foundation of China (grant number 31901690).

Institutional Review Board Statement: Not applicable.

Informed Consent Statement: Not applicable.

Conflicts of Interest: The authors declare no conflict of interest.

References

1. Zhou, C.J.; Huang, S.; Liu, J.Q.; Qiu, S.Q.; Xie, F.Y.; Song, H.P.; Li, Y.S.; Hou, S.Z.; Lai, X.P. Sweet tea leaves extract improves leptin resistance in diet-induced obese rats. *J. Ethnopharmacol.* **2013**, *145*, 386–392. [CrossRef] [PubMed]
2. Shang, A.; Liu, H.Y.; Luo, M.; Xia, Y.; Yang, X.; Li, H.Y.; Wu, D.T.; Sun, Q.; Geng, F.; Li, H.B.; et al. Sweet tea (*Lithocarpus polystachyus* rehd.) as a new natural source of bioactive dihydrochalcones with multiple health benefits. *Crit. Rev. Food Sci. Nutr.* **2020**, 1–18. [CrossRef] [PubMed]
3. Li, S.; Zeng, J.; Tan, J.; Zhang, J.; Wu, Q.; Wang, L.; Wu, X. Anti-oxidant and hepatoprotective effects of *Lithocarpus polystachyus* against carbon tetrachloride-induced injuries in rat. *Bangl. J. Pharmacol.* **2014**, *8*, 420–427.
4. Hou, S.Z.; Xu, S.J.; Jiang, D.X.; Chen, S.X.; Wang, L.L.; Huang, S.; Lai, X.P. Effect of the flavonoid fraction of *Lithocarpus polystachyus* Rehd. on spontaneously hypertensive and normotensive rats. *J. Ethnopharmacol.* **2012**, *143*, 441–447. [CrossRef]
5. Gao, J.; Xu, Y.; Zhang, J.; Shi, J.; Gong, Q. *Lithocarpus polystachyus* Rehd. leaves aqueous extract protects against hydrogen peroxideinduced SH-SY5Y cells injury through activation of Sirt3 signaling pathway. *Int. J. Mol. Med.* **2018**, *42*, 3485–3494.
6. Liu, Y.; Sun, Y.; Huang, G. Preparation and antioxidant activities of important traditional plant polysaccharides. *Int. J. Biol. Macromol.* **2018**, *111*, 780–786. [CrossRef] [PubMed]
7. Yuan, Q.; Lin, S.; Fu, Y.; Nie, X.R.; Liu, W.; Su, Y.; Han, Q.H.; Zhao, L.; Zhang, Q.; Lin, D.R.; et al. Effects of extraction methods on the physicochemical characteristics and biological activities of polysaccharides from okra (*Abelmoschus esculentus*). *Int. J. Biol. Macromol.* **2019**, *127*, 178–186. [CrossRef]
8. Guo, H.; Yuan, Q.; Fu, Y.; Liu, W.; Su, Y.H.; Liu, H.; Wu, C.Y.; Zhao, L.; Zhang, Q.; Lin, D.R.; et al. Extraction optimization and effects of extraction methods on the chemical structures and antioxidant activities of polysaccharides from snow Chrysanthemum (*Coreopsis tinctoria*). *Polym. (Basel)* **2019**, *11*, 215. [CrossRef] [PubMed]
9. He, J.L.; Guo, H.; Wei, S.Y.; Zhou, J.; Xiang, P.Y.; Liu, L.; Zhao, L.; Qin, W.; Gan, R.Y.; Wu, D.T. Effects of different extraction methods on the structural properties and bioactivities of polysaccharides extracted from Qingke (Tibetan hulless barley). *J. Cereal Sci.* **2020**, *92*, 102906. [CrossRef]
10. Yao, Y.; Zhu, Y.; Ren, G. Antioxidant and immunoregulatory activity of alkali-extractable polysaccharides from mung bean. *Int. J. Biol. Macromol.* **2016**, *84*, 289–294. [CrossRef]
11. Guo, X.; Zhao, W.; Liao, X.; Hu, X.; Wu, J.; Wang, X. Extraction of pectin from the peels of pomelo by high-speed shearing homogenization and its characteristics. *LWT-Food Sci. Technol.* **2017**, *79*, 640–646. [CrossRef]
12. Li, J.H.; Li, W.; Luo, S.; Ma, C.H.; Liu, S.X. Alternate ultrasound/microwave digestion for deep eutectic hydro-distillation extraction of essential oil and polysaccharide from *Schisandra chinensis* (Turcz.) Baill. *Molecules* **2019**, *24*, 1288. [CrossRef]
13. Guo, H.; Li, H.Y.; Liu, L.; Wu, C.Y.; Liu, H.; Zhao, L.; Zhang, Q.; Liu, Y.T.; Li, S.Q.; Qin, W.; et al. Effects of sulfated modification on the physicochemical properties and biological activities of β-glucans from Qingke (Tibetan hulless barley). *Int. J. Biol. Macromol.* **2019**, *141*, 41–50. [CrossRef] [PubMed]
14. Nie, X.R.; Fu, Y.; Wu, D.T.; Huang, T.T.; Jiang, Q.; Zhao, L.; Zhang, Q.; Lin, D.R.; Chen, H.; Qin, W. Ultrasonic-assisted extraction, structural characterization, chain conformation, and biological activities of a pectic-polysaccharide from okra (*Abelmoschus esculentus*). *Molecules* **2020**, *25*, 1155. [CrossRef] [PubMed]
15. Liu, W.; Li, F.; Wang, P.; Liu, X.; He, J.J.; Xian, M.L.; Zhao, L.; Qin, W.; Gan, R.Y.; Wu, D.T. Effects of drying methods on the physicochemical characteristics and bioactivities of polyphenolic-protein-polysaccharide conjugates from *Hovenia dulcis*. *Int. J. Biol. Macromol.* **2020**, *148*, 1211–1221. [CrossRef] [PubMed]
16. Cho, Y.J.; Getachew, A.T.; Saravana, P.S.; Chun, B.S. Optimization and characterization of polysaccharides extraction from Giant African snail (*Achatina fulica*) using pressurized hot water extraction (PHWE). *Bio. Carbohydr. Diet. Fibre* **2019**, *18*, 100179. [CrossRef]
17. Bai, L.; Zhu, P.; Wang, W.; Wang, M. The influence of extraction pH on the chemical compositions, macromolecular characteristics, and rheological properties of polysaccharide: The case of okra polysaccharide. *Food Hydrocolloid.* **2020**, *102*, 105586. [CrossRef]
18. Chen, X.; Chen, G.; Wang, Z.; Kan, J. A comparison of a polysaccharide extracted from ginger (*Zingiber officinale*) stems and leaves using different methods: Preparation, structure characteristics, and biological activities. *Int. J. Biol. Macromol.* **2020**, *151*, 635–649. [CrossRef]
19. Fang, C.C.; Chen, G.G.; Kan, J.Q. Comparison on characterization and biological activities of *Mentha haplocalyx* polysaccharides at different solvent extractions. *Int. J. Biol. Macromol.* **2020**, *154*, 916–928. [CrossRef]

20. Chen, C.; Wang, P.P.; Huang, Q.; You, L.J.; Liu, R.H.; Zhao, M.M.; Fu, X.; Luo, Z.G. A comparison study on polysaccharides extracted from *Fructus mori* using different methods: Structural characterization and glucose entrapment. *Food Funct.* **2019**, *10*, 3684–3695. [CrossRef]
21. Lin, X.; Ji, X.; Wang, M.; Yin, S.; Peng, Q. An alkali-extracted polysaccharide from *Zizyphus jujuba* cv. Muzao: Structural characterizations and antioxidant activities. *Int. J. Biol. Macromol.* **2019**, *136*, 607–615. [CrossRef]
22. Saravana, P.S.; Cho, Y.N.; Woo, H.C.; Chun, B.S. Green and efficient extraction of polysaccharides from brown seaweed by adding deep eutectic solvent in subcritical water hydrolysis. *J. Clean. Prod.* **2018**, *198*, 1474–1484. [CrossRef]
23. Luo, Q.; Zhang, J.R.; Li, H.B.; Wu, D.T.; Geng, F.; Corke, H.; Wei, X.L.; Gan, R.Y. Green Extraction of Antioxidant Polyphenols from Green Tea (Camellia sinensis). *Antioxid. (Basel)* **2020**, *9*, 785. [CrossRef]
24. Tang, B.; Zhang, H.; Row, K.H. Application of deep eutectic solvents in the extraction and separation of target compounds from various samples. *J. Sep. Sci.* **2015**, *38*, 1053–1064. [CrossRef]
25. Zhang, L.; Wang, M. Optimization of deep eutectic solvent-based ultrasound-assisted extraction of polysaccharides from *Dioscorea opposita* Thunb. *Int. J. Biol. Macromol.* **2017**, *95*, 675–681. [CrossRef]
26. Sun, Y.; Hou, S.; Song, S.; Zhang, B.; Ai, C.; Chen, X.; Liu, N. Impact of acidic, water and alkaline extraction on structural features, antioxidant activities of *Laminaria japonica* polysaccharides. *Int. J. Biol. Macromol.* **2018**, *112*, 985–995. [CrossRef] [PubMed]
27. Yan, J.K.; Yu, Y.B.; Wang, C.; Cai, W.D.; Wu, L.X.; Yang, Y.; Zhang, H.N. Production, physicochemical characteristics, and in vitro biological activities of polysaccharides obtained from fresh bitter gourd (*Momordica charantia* L.) via room temperature extraction techniques. *Food Chem.* **2021**, *337*, 127798. [CrossRef] [PubMed]
28. Yan, J.K.; Ding, Z.C.; Gao, X.; Wang, Y.Y.; Yang, Y.; Wu, D.; Zhang, H.N. Comparative study of physicochemical properties and bioactivity of *Hericium erinaceus* polysaccharides at different solvent extractions. *Carbohydr. Polym.* **2018**, *193*, 373–382. [CrossRef] [PubMed]
29. Li, F.; Feng, K.L.; Yang, J.C.; He, Y.S.; Guo, H.; Wang, S.P.; Gan, R.Y.; Wu, D.T. Polysaccharides from dandelion (*Taraxacum mongolicum*) leaves: Insights into innovative drying techniques on their structural characteristics and biological activities. *Int. J. Biol. Macromol.* **2021**, *167*, 995–1005. [CrossRef] [PubMed]
30. Yue, H.; Xu, Q.; Bian, G.; Guo, Q.; Fang, Z.; Wu, W. Structure characterization and immunomodulatory activity of a new neutral polysaccharide SMP-0b from *Solanum muricatum*. *Int. J. Biol. Macromol.* **2020**, *155*, 853–860. [CrossRef] [PubMed]
31. Ustyuzhanina, N.E.; Bilan, M.I.; Dmitrenok, A.S.; Nifantiev, N.E.; Usov, A.I. Two fucosylated chondroitin sulfates from the sea cucumber *Eupentacta fraudatrix*. *Carbohydr. Polym.* **2017**, *164*, 8–12. [CrossRef]
32. Shakhmatov, E.G.; Atukmaev, K.V.; Makarova, E.N. Structural characteristics of pectic polysaccharides and arabinogalactan proteins from *Heracleum sosnowskyi* Manden. *Carbohydr. Polym.* **2016**, *136*, 1358–1369. [CrossRef] [PubMed]
33. Yin, J.; Lin, H.; Li, J.; Wang, Y.; Cui, S.W.; Nie, S.; Xie, M. Structural characterization of a highly branched polysaccharide from the seeds of *Plantago asiatica* L. *Carbohydr. Polym.* **2012**, *87*, 2416–2424. [CrossRef]
34. Guo, C.; Zhang, S.; Wang, Y.; Li, M.; Ding, K. Isolation and structure characterization of a polysaccharide from Crataegus pinnatifida and its bioactivity on gut microbiota. *Int. J. Biol. Macromol.* **2020**, *154*, 82–91. [CrossRef] [PubMed]
35. Zhang, W.; Xiang, Q.; Zhao, J.; Mao, G.; Feng, W.; Chen, Y.; Li, Q.; Wu, X.; Yang, L.; Zhao, T. Purification, structural elucidation and physicochemical properties of a polysaccharide from *Abelmoschus esculentus* L (okra) flowers. *Int. J. Biol. Macromol.* **2020**, *155*, 740–750. [CrossRef] [PubMed]
36. Hu, C.; Li, H.X.; Zhang, M.T.; Liu, L.F. Structure characterization and anticoagulant activity of a novel polysaccharide from *Leonurus artemisia* (Laur.) S. Y. Hu F. *RSC Adv.* **2020**, *10*, 2254–2266. [CrossRef]
37. Austarheim, I.; Christensen, B.E.; Hegna, I.K.; Petersen, B.O.; Duus, J.O.; Bye, R.; Michaelsen, T.E.; Diallo, D.; Inngjerdingen, M.; Paulsen, B.S. Chemical and biological characterization of pectin-like polysaccharides from the bark of the Malian medicinal tree *Cola cordifolia*. *Carbohydr. Polym.* **2012**, *89*, 259–268. [CrossRef] [PubMed]
38. Han, K.; Jin, C.; Chen, H.; Wang, P.; Yu, M.; Ding, K. Structural characterization and anti-A549 lung cancer cells bioactivity of a polysaccharide from *Houttuynia cordata*. *Int. J. Biol. Macromol.* **2018**, *120*, 288–296. [CrossRef]
39. Shakhmatov, E.G.; Toukach, P.V.; Michailowa, C.; Makarova, E.N. Structural studies of arabinan-rich pectic polysaccharides from *Abies sibirica* L. biological activity of pectins of A. sibirica. *Carbohydr. Polym.* **2014**, *113*, 515–524. [CrossRef]
40. Liu, H.; Fan, H.; Zhang, J.; Zhang, S.; Zhao, W.; Liu, T.; Wang, D. Isolation, purification, structural characteristic and antioxidative property of polysaccharides from A. cepa L. var. agrogatum Don. *Food Sci. Hum. Well.* **2020**, *9*, 71–79. [CrossRef]
41. Arbatsky, N.P.; Kasimova, A.A.; Shashkov, A.S.; Shneider, M.M.; Popova, A.V.; Shagin, D.A.; Shelenkov, A.A.; Mikhailova, Y.V.; Yanushevich, Y.G.; Azizov, I.S.; et al. Structure of the K128 capsular polysaccharide produced by *Acinetobacter baumannii* KZ-1093 from Kazakhstan. *Carbohydr. Res.* **2019**, *485*, 107814. [CrossRef] [PubMed]
42. Liu, X.X.; Liu, H.M.; Yan, Y.Y.; Fan, L.Y.; Yang, J.N.; Wang, X.D.; Qin, G.Y. Structural characterization and antioxidant activity of polysaccharides extracted from jujube using subcritical water. *LWT-Food Sci. Technol.* **2020**, *117*, 108645. [CrossRef]
43. Agrawal, P.K. nmR-spectroscopy in the structural elucidation of oligosaccharides and glycosides. *Phytochemistry* **1992**, *30*, 3307–3330. [CrossRef]
44. Zheng, C.; Dong, Q.; Chen, H.; Cong, Q.; Ding, K. Structural characterization of a polysaccharide from *Chrysanthemum morifolium* flowers and its antioxidant activity. *Carbohydr. Polym.* **2015**, *130*, 113–121. [CrossRef]
45. Schieber, M.; Chandel, N.S. ROS function in redox signaling and oxidative stress. *Curr. Biol.* **2014**, *24*, R453–R462. [CrossRef]

46. Wang, Q.H.; Shu, Z.P.; Xu, B.Q.; Xing, N.; Jiao, W.J.; Yang, B.Y.; Kuang, H.X. Structural characterization and antioxidant activities of polysaccharides from *Citrus aurantium* L. *Int. J. Biol. Macromol.* **2014**, *67*, 112–123. [CrossRef]
47. Everette, J.D.; Bryant, Q.M.; Green, A.M.; Abbey, Y.A.; Wangila, G.W.; Walker, R.B. Thorough study of reactivity of various compound classes toward the Folin-Ciocalteu reagent. *J. Agr. Food Chem.* **2010**, *58*, 8139–8144. [CrossRef] [PubMed]
48. Cai, Y.Z.; Mei, S.; Jie, X.; Luo, Q.; Corke, H. Structure-radical scavenging activity relationships of phenolic compounds from traditional Chinese medicinal plants. *Life Sci.* **2006**, *78*, 2872–2888. [CrossRef]
49. Wu, D.T.; Liu, W.; Xian, M.L.; Du, G.; Liu, X.; He, J.J.; Wang, P.; Qin, W.; Zhao, L. Polyphenolic-protein-polysaccharide complexes from *Hovenia dulcis*: Insights into extraction methods on their physicochemical properties and *in vitro* bioactivities. *Foods* **2020**, *9*, 456. [CrossRef] [PubMed]
50. Bermudez-Oria, A.; Rodriguez-Gutierrez, G.; Fernandez-Prior, A.; Vioque, B.; Fernandez-Bolanos, J. Strawberry dietary fiber functionalized with phenolic antioxidants from olives. Interactions between polysaccharides and phenolic compounds. *Food Chem.* **2019**, *280*, 310–320. [CrossRef]
51. Siu, K.C.; Chen, X.; Wu, J.Y. Constituents actually responsible for the antioxidant activities of crude polysaccharides isolated from mushrooms. *J. Funct. Foods* **2014**, *11*, 548–556. [CrossRef]
52. Dong, H.Q.; Li, M.; Zhu, F.; Liu, F.L.; Huang, J.B. Inhibitory potential of trilobatin from *Lithocarpus polystachyus* Rehd against α-glucosidase and α-amylase linked to type 2 diabetes. *Food Chem.* **2012**, *130*, 261–266. [CrossRef]
53. Hou, S.Z.; Chen, S.X.; Huang, S.; Jiang, D.X.; Zhou, C.J.; Chen, C.Q.; Liang, Y.M.; Lai, X.P. The hypoglycemic activity of *Lithocarpus polystachyus* Rehd. leaves in the experimental hyperglycemic rats. *J. Ethnopharmacol.* **2011**, *138*, 142–149. [CrossRef] [PubMed]
54. Jia, Y.; Gao, X.; Xue, Z.; Wang, Y.; Lu, Y.; Zhang, M.; Panichayupakaranant, P.; Chen, H. Characterization, antioxidant activities, and inhibition on alpha-glucosidase activity of corn silk polysaccharides obtained by different extraction methods. *Int. J. Biol. Macromol.* **2020**, *163*, 1640–1648. [CrossRef]
55. Yuan, Q.; He, Y.; Xiang, P.Y.; Huang, Y.J.; Cao, Z.W.; Shen, S.W.; Zhao, L.; Zhang, Q.; Qin, W.; Wu, D.T. Influences of different drying methods on the structural characteristics and multiple bioactivities of polysaccharides from okra (*Abelmoschus esculentus*). *Int. J. Biol. Macromol.* **2020**, *147*, 1053–1063. [CrossRef]
56. Peng, X.; Ma, J.; Chen, F.; Wang, M. Naturally occurring inhibitors against the formation of advanced glycation end-products. *Food Funct.* **2011**, *2*, 289–301. [CrossRef]
57. Kulkarni, M.J.; Korwar, A.M.; Mary, S.; Bhonsle, H.S.; Giri, A.P. Glycated proteome: From reaction to intervention. *Proteom. Clin. Appl.* **2013**, *7*, 155–170. [CrossRef] [PubMed]

MDPI
St. Alban-Anlage 66
4052 Basel
Switzerland
Tel. +41 61 683 77 34
Fax +41 61 302 89 18
www.mdpi.com

Foods Editorial Office
E-mail: foods@mdpi.com
www.mdpi.com/journal/foods

www.ingramcontent.com/pod-product-compliance
Lightning Source LLC
LaVergne TN
LVHW070633100526
838202LV00012B/797